ELECTRICITY: HUMANITY'S LOW-CARBON FUTURE

Safeguarding Our Ecological Niche

About the Authors

Hans B. (Teddy) Püttgen is Distinguished Professor Emeritus at Georgia Institute of Technology, Georgia Tech (1981–2006) and Fellow of IEEE. He was Swiss Federal Institute of Technology, EPFL, Professor & Energy Center Director (2006–2013) and Visiting Professor & Senior Director of the Energy Research Institute at Nanyang Technological University, NTU (2013–2017).

Other past positions include: President IEEE Power and Energy Society, IEEE – PES, Member of the Swiss Federal Energy Research Commission (CORE), and Member of the Global Agenda Council for Alternative Energies – World Economic Forum.

Yves Bamberger is Fellow of the National Academy of Technologies of France. He was Head of Research & Development at Électricité de France (EDF), one of the largest utilities in the world operating on several continents. Before then, also at EDF, he was Information Systems Director then Deputy Director of the Generation and Transmission Division.

Other past positions include: Adjunct Professor at École Nationale des Ponts et Chaussées (1973–2000), Member of the Gartner Research Board (New York), Member of the Electric Power Research Institute Board (Palo Alto), and Member of the first Smart Grids European Technology Platform.

ELECTRICITY: HUMANITY'S LOW-CARBON FUTURE

Safeguarding Our Ecological Niche

Hans B. (Teddy) Püttgen

Georgia Institute of Technology, USA

Swiss Federal Institute of Technology, EPFL, Switzerland

Yves Bamberger

EDF – Electricité de France

National Academy of Technologies of France

World Scientific

NEW JERSEY · LONDON · SINGAPORE · BEIJING · SHANGHAI · HONG KONG · TAIPEI · CHENNAI · TOKYO

Published by

World Scientific Publishing Co. Pte. Ltd.

5 Toh Tuck Link, Singapore 596224

USA office: 27 Warren Street, Suite 401-402, Hackensack, NJ 07601

UK office: 57 Shelton Street, Covent Garden, London WC2H 9HE

Library of Congress Cataloging-in-Publication Data

Names: Püttgen, H. B. (Hans B.), 1948– author. | Bamberger, Yves, author.
Title: Electricity, humanity's low-carbon future : safeguarding our ecological niche /
    Hans B. (Teddy) Püttgen, Georgia Institute of Technology, USA,
    Swiss Federal Institute of Technology, EPFL, Switzerland,
    Yves Bamberger, EDF -- Electricité de France, National Academy of Technologies of France.
Description: New Jersey : World Scientific, [2021] | Includes bibliographical references and index.
Identifiers: LCCN 2021025914 | ISBN 9789811224355 (hardcover) |
    ISBN 9789811229305 (paperback) | ISBN 9789811227318 (ebook for institutions) |
    ISBN 9789811227301 (ebook for individuals)
Subjects: LCSH: Climate change mitigation. | Carbon dioxide mitigation. |
    Renewable energy sources. | Electric power--Government policy. | Energy policy.
Classification: LCC TD171.75 .P88 2021 | DDC 363.738/746--dc23
LC record available at https://lccn.loc.gov/2021025914

British Library Cataloguing-in-Publication Data
A catalogue record for this book is available from the British Library.

For the supplementary material, please visit
https://www.worldscientific.com/worldscibooks/10.1142/11939#t=suppl

Printed in Singapore

With Appreciation

Foremost, we wish to thank our families and friends for their unwavering encouragements and patience during the long adventure of writing this book.

We wish to thank the many colleagues and team members we have had the privilege and pleasure of working with in both industry and academia on three different continents. In many ways, this book is the result of what we have learned together during a wide variety of challenging and interesting projects spanning diverse scientific disciplines.

A special note of appreciation goes to Amanda Yun and Gregory Lee at World Scientific for their always professional and patient assistance and support during the editing, production and promotion of the book.

The book's many illustrations, charts and drawings are a tribute to Charles Raviaud's computer graphics talents and skills.

The names of the institutions, corporations and persons who have authorized us to incorporate their pictures and illustrations are mentioned in the relevant captions. Their contributions are very much appreciated.

We thank the Institut Pierre Lamoure for its financial support of the French language version publication at EPFL Press, Switzerland.

Financial support for the production and publication of this book was provided by Pictet Asset Management SA, Switzerland, and Solar Pacific Energy Corporation, Philippines; it is greatly appreciated.

Teddy and Yves

Preamble and Reader's Guide

Sustaining human life requires ambient temperatures and air quality to be maintained within certain boundaries that are, still, respected on a large portion of planet Earth. This "ecological niche" is also crucial for its fauna and flora; many of their species have disappeared following the depletion of their niches as consequences of events and/or evolutions beyond their control. Man could well be the first animal to have directly contributed to the extinction of its ecological niche through its own actions. This is precisely what we need to avoid!

Climate change is no longer deniable, and neither is the fact that greenhouse gases, due to human activities, continue to accentuate it nor that these activities have to be mitigated.

Since greenhouse gases typically remain in the atmosphere for decades, the challenge is to not only reduce our carbon footprint but to do so quickly.

The priority is thus to rapidly decarbonize our activities, especially in the energy sector that is responsible for the majority of greenhouse gas emissions.

The energy reaching the Earth's surface every year from solar radiation is equal to some 1 000 million TWh, i.e., almost 9 000 times more than our final energy consumption, which reached 115 600 TWh in 2018. Solar energy is thus quasi infinite at the human scale! In addition, other forms of renewable energy, such as from the oceans, have yet to be tapped.

We are not facing the risk of running out of energy.

The challenge lies in how we use it while safeguarding our ecological niche!

Using energy always has an impact on the environment. It therefore behooves us to use it as rationally as possible.

Man started to rely on natural renewable resources a long time ago to assist with daily activities, first burning wood, using river-flows or wind, and generally later coal followed by oil and gas. The advent of the industrial revolution in the 19th century saw the intensification of fossil energy uses to meet increasing industrial and transportation demands. Since the 20th century, fossil fuels have also been increasingly called upon to produce electricity.

In a growing number of countries and in hardly more than a century, the advent of electricity has changed the lives of women and men: better comfort at home, less drudgery at work, enhanced medical services, improved transportation systems, and broader industrial activities, all made possible by its flexibility of use. Electricity enabled the naissance of information and communication technologies, including the Internet.

The more than a billion women and men who do not yet have access to electricity await it eagerly; they know that it will bring a better life. However, worldwide, access to electricity unfortunately remains quite uneven.

In addition to the priority to rapidly decarbonize, energy-wise, it behooves Mankind to address and overcome two additional key challenges:

- The legitimate quest from a growing segment of Earth's population to have access to safe and affordable energy is a planetary challenge that must be resolved without irremediable damage to the environment. Access to a "clean and affordable energy" is one of the 17 Sustainable Development Goals (SDGs) set for 2030 by the United Nations [*sdgs.un.org*]. The on-going population increase makes this challenge even more poignant and complex.
- The overall reduction of our use of fossil resources is an inescapable necessity. Indeed, their natural renewal rate is fully incompatible with the rate of usage imposed by ourselves. Even as new reserves are discovered and as new technologies make their deployment economically accessible, it behooves us to leave some reserves for

future generations while restricting the present use of fossil resources to industrial activities, such as chemistry and materials, while no longer simply burning them.

The diversity of its end-use applications, along with the relative ease to deploy electrical infrastructures, compared to, for example, natural gas ones, makes electricity the preferred energy vector to valorize not only fossil resources — coal, oil and gas — but also, more importantly for the future, renewable resources — hydro, solar, wind, geothermal, marine and biomass — as well as nuclear energy.[1] A broader share of low emissions electricity in a number of energy end-uses not only contributes to the "clean and affordable energy" SDG but also toward the fulfillment of other UN SDGs, such as those related to water, health and education.

Toward an ever more electric world is Mankind's energy path!

Moving toward an electric world will thus enable our activities to be gradually decarbonized as our ecological footprint shrinks!

Often referred to as one of the most complex systems ever built by Mankind, today's electric power systems feature thousands of electric power plants, all delivering electricity to millions of loads by way of high voltage transmission and lower voltage distribution systems. Given their increasingly central role in future global energy systems, an overall understanding of how electric power systems are designed and operated will prepare citizens and decision-makers to voice their opinions as options for future developments are presented for their consideration.

This energy path toward a decarbonized future does not exonerate us from a more rational use of energy, broadly speaking, and of electricity, in particular. This implies encouraging, occasionally mandating, energy consumption reductions using new technologies, especially in regions with high-energy demands per capita. Decarbonizing implies the transfer of fossil fuel usages to other energy forms or vectors, primarily electricity,

---

[1] Nuclear power plants, while highly controversial, do not emit any greenhouse gases when in operation, similarly to renewable energies.

and, at the same time, the reduction of the "CO_2 content" of electricity. Fortunately, the long road that lies ahead of us is full of possibilities! Indeed, electricity only represents some 25% of all final energy consumption in Europe and has not reached 20% worldwide. In addition, lots remain to be done to decarbonize the production of electricity.

The designation "energy transition" is often used to describe the road to a more sustainable world, energy-wise. To better reflect both the depth and breadth of the transformations needed, we have elected to use the designation "energy mutation".

Since our dependency on energy in our daily lives has become ubiquitous, any energy mutation (i.e., the change of the energy systems we rely upon) is complex because it affects every one of us wherever we may be and whatever we may be doing. It also affects what we perceive to be our comfort zone. Private and public debates regarding individual rights versus the overall well-being of society take place at all decision levels and occasionally lead to conflicts. Energy mutations must take place within public policy and regulatory frameworks which are not only set for long periods but must also be clear and transparent to obtain buy-in from the population-at-large, along with private and public entities, thus leading to broad implementation.

Among the various paths toward emission reduction targets, it is important to strive toward the least costly ones, which requires global and systemic approaches. Investment and operational cost changes are inevitable, which induce the debate on the distribution of these costs among different sectors of the global economy. The dilemma remains between costs, on one hand, and local as well as global environmental impacts, on the other. This is especially true in emerging regions where broader and better access to electricity remains a priority. These debates and decisions are essential toward the resolution of the three societal challenges outlined above.

Planning Horizon

The investments involved, along with safety and reliability requirements, are such that technology and systems development and deployment cycles in the energy sector are often slower than in several other sectors, such as information and communication. Both the levels and durations of investments in major energy infrastructures are such that the related decisions can often only be reached in the context of long-range plans at a national level, or even broader, such as the European Union. The temptation is then to only conceive energy plans for 2050 or even 2100 while overseeing the urgency to decarbonize. Longer-term plans, while necessary, can also lead to an attitude of "by then new technologies will be available to solve our present day problems". While it is true that such new technologies might make further contributions by 2050, we do not know how useful these technologies will be. Furthermore, 2050 is too far out in time; it behooves us to act without further delay to use the technologies we already have access to.

By 2035, a determined deployment of technologies presently available, or already operational in pilot installations, can make significant contributions toward less carbon, improve energy access worldwide and reduce fossil fuel uses.

We have thus selected year 2035 as the horizon to assess the impact the best energy consumption, production and storage technologies already at hand can have at that horizon while not counting on the potential impact yet unknown or yet undeveloped technologies could have in 2035 or beyond.[2]

Our suggestions aim to address the challenges mentioned above in the best way while not impeding options beyond 2035.

Book Structure

While its primary focus is on electricity, the book addresses the entire energy spectrum: the primary energies, i.e., sources like coal, hydro or solar radiation, and their conversions into energy vectors, i.e., their suitable forms, like gasoline or electricity, and the energy end-uses and services we all rely upon.

---

[2] By 2035, three quarters of the time between 1990, the date of the Kyoto agreements, and 2050, a date often used in conjunction of emission reduction targets, will already have elapsed.

To illustrate the multitude of parameters and situations that influence our present-day energy uses and how they could evolve by 2035, a presentation of the lives of four families residing in different parts of the world — Africa, Asia, Europe and North America — launches the book. The description of Energia, a fictitious country "created" to further illustrate present-day situations and potential evolutions at a country level, follows. Energia is a modern and industrialized country of 50 million inhabitants who have decided to take the energy mutation head-on and who have therefore mandated successive governments to enact the necessary decisions.

The core of the book is introduced by the "Energy and Emissions — Where We Are" chapter, in which climate change is first discussed, including recent alarming IPCC observations. A brief introduction to the water energy nexus[3] precedes a presentation of key primary and final energy consumption statistics, which leads into the chapters describing presently deployable energy technologies. Contrary to the many books dealing with energy, we have deliberately opted to start with its uses before presenting electricity production and storage technologies, thereby deliberately putting the citizen-consumer-customer, thus the reader, first!

The description of the electric power system, which is crucial in the energy mutation context, follows naturally. The discussion related to the interaction between the electric system and other energy systems concludes the series of chapters that deal with presently available technologies.

Clearly, we had to make choices. We have limited the descriptions to the technologies we consider as being the most important and relevant for the present and predictable future. Having access to clearly defined and verifiable data and information is essential to reach well-founded decisions. We have therefore decided to rely on the data and information published on the websites of well-known and high quality entities, thus making it possible for the reader to update the information as time goes by, or to obtain information for a particular country of interest.

---

[3]The "energy for water" technologies are presented in the chapters dealing with energy consumption while the "water for energy" technologies are presented in the chapters dealing with electricity production.

Up until this point of the book, we have made every effort to remain factual in the description of the technologies presented. In the two chapters under "Where We Should Be Heading", we share our beliefs on what needs be done to set our course toward a more electrical, renewable and sustainable world that has a lesser ecological footprint by Mankind while also addressing the energy aspirations of the less fortunate. It is here that we share our belief that rapid and profound decarbonization requires that the Universal and Redistributed Carbon Fee we propose be broadly implemented worldwide. Energia's 2035 energy mutation highlights the contribution that existing technologies can bring when properly deployed toward the priority to rapidly decarbonize our way of living. The 2035 situation of four families, similar to those presented at the book's onset and living at the same locations, serves as the conclusion of the book.

The energy technologies and systems we have come to rely upon are the results of the creativity of a number of scientists, engineers and inventors, as well as the hard work of craftsmen, technicians and workers throughout history. We have therefore opted to provide the names of a few contributors who played particularly important roles, along with the periods they lived in.

Our aim in this book is to present the energy technologies and systems we routinely rely upon while largely taking them for granted, and to illustrate how they can be broadly deployed to lead us to a more decarbonized world while essentially sustaining the quality of life of the fortunate and providing better lives for the less fortunate.

We hope to have written a book for a reader interested and intrigued by the many energy and environmental challenges we face, and who also seeks information on the choices that she or he is to face as both a consumer and citizen. While the book does not specifically address our scientific colleagues, we hope they will also enjoy reading it! By way of the many illustrations, photographs, drawings and examples along with the many website references, we aim to encourage the reader to enhance her or his interest in the energy universe, which so largely conditions our daily lives and those of generations to come.

Teddy and **Yves**, Lutry and Paris

Reader's Guide

Main Text

The main text in each chapter is intended to be self-supported. Two types of inserts are interspersed to help the reader wishing to "dig in a bit more" or to refresh her or his knowledge of the topic at hand. The inserts with a light gray background provide a few illustrative examples of the technologies discussed while those in a black frame provide some supplemental technical or scientific background.

Companion Documents

A Companion Document is accessible at:

https://www.worldscientific.com/worldscibooks/10.1142/11939#t=suppl.

Identified and referenced in the main text, the Companion Document is not intended to be self-standing; it provides supplementary information related to specific topics:

- Supplementary examples to further illustrate some discussions.
- Slightly more in-depth presentations of a few scientific topics.
- Worksheets related to the families in 2015 and Energia in 2035, which are intended to be useful for readers who wish to revisit the computations using other input data.

Bibliography

Rather than including a traditional bibliography that would rapidly become outdated, we have elected to include extensive references available on publicly accessible Internet websites where the reader can access up-to-date information and data relevant to the countries of interest.

Contents

An Ever-more Ubiquitous Electricity

Where We Should Be Heading

To Conclude

Annex

Setting the Stage

The Lives of Four Families

An Energy Perspective

The housing situation, lifestyle and transportation means used by any family have a direct impact on its energy consumptions and resulting CO_2 emissions. To illustrate the large numbers of factors and personal decisions that influence their energy consumption as well as their CO_2 emissions, the circumstances of four imaginary families living in Africa, Asia, Europe and North America are described in the following pages.

> *We have tried to construct a few contrasting situations in terms of climates, countries and living circumstances. There is no judgmental intent on our part in terms of value nor any desire to symbolize any one country by way of each imaginary family described. Plausible situations are presented to showcase the diversity of energy situations as well as the large number of parameters which come into play when evaluating energy consumptions and the related emissions.*

The living context of each family is first presented without any numerical values. The energy consumption and CO_2 emissions tied to the dwelling, as well as those tied to the transportation needs, are presented in a graphical form or "cartoon". Only the energy consumptions and emissions directly tied to the activities of the families themselves are included but not those tied to their country's economy or the services they use. The data shown in the cartoon for each family is detailed under *"Families 2015 Worksheets"* in the Companion Document, to provide for each reader to repeat the computations using different assumptions more related to the lifestyles and/or equipment used, including her or his own.

The energy consumptions at home are evaluated first: (a) heating and air-conditioning, (b) hot water, (c) cooking, and (d) other consumptions that are specifically electric. Aside from the latter, the form of energy relied upon by the family for each of their consumptions is specified: electricity, biomass, natural gas, gasoline/diesel, coal, etc. Once the consumptions have been computed, one can use this information to compute the CO_2 emissions each family induces, also taking into account the local electricity generation mix (hydro, gas, coal, nuclear, solar, wind, etc.).

For transportation needs, personal trips, as well as daily commutes to work, are taken into account but not the trips taken for professional purposes. The fuel consumption, as well as the CO_2 emission, for private cars is evaluated. For public transport, the CO_2 emissions for each family member are computed based on the information found on public websites.

A few explanations for the different entries of the worksheets are provided below:

- To assess the heating requirements where the family lives, the Heating Degree Day (HDD) factor, described in the "Housing" chapter, is used. For each particular geographical location, it takes the number of days during which heating is required and the outside temperature during those days into account; it is expressed in degree·days. The higher the HDD, the more heating is required. The HDD cooling counterpart is referred to as the Cooling Degree Day (CDD); the higher it is, the more cooling is required. The average HDD and CDD values of three recent years are computed for the four locations where the families live and are based on the information found on [*degreedays.net*] where the relevant information can be found for several thousand meteorological stations around the world. The number of days of presence or absence from home is important when assessing the heating and cooling requirements; it is provided for each family.
- Once the location for a particular dwelling is known, one must also know its physical dimensions as well as its level of insulation to evaluate its heating and cooling requirements. This is done using a global factor, G; the related computational method is provided in the

"Housing" chapter. A low G factor, typically below 1, indicates that the dwelling is well insulated.

- The average energy consumption and power[1] data provided for house appliances are taken from manufacturers' or public administrations' websites. Using such sites, interested readers can find the information for their appliances of interest. Technological details can be found in the "Housing" chapter.

- The rated power, P_{rated}, for each appliance — stove, refrigerator, washer, vacuum cleaner, etc. — is generally indicated on the appliance itself; it is the maximum power drawn by the appliance under normal operating conditions. P_{rated} is also used to size the appliance protection fuse. The energy consumed by each appliance depends on its use, for example, the refrigerator compressor will work harder just after it is filled or if the door is often opened (one can then refer to the average power, $P_{average}$). The consumption of a cooktop depends on the number of plates used, the consumption of a light bulb depends on the number of hours it is turned on, or a television left on standby when not used will consume more than if it is fully turned off, etc. The assumptions used for each family are described in the worksheets.

Next, one can evaluate the CO_2 emissions resulting from the transportation needs of the entire family. From there, the average emissions of each family member can be computed.

- For cars, the technology-related information can be found in the "Transport and Travel" chapter. The fuel or electricity consumption can be found on constructors' or public administrations' websites. From the actual distances traveled, one can compute the fuel or electricity consumption and thus the CO_2 emissions.

- For public transport, the related authorities generally provide the CO_2 emissions per kilometer or passenger on their websites. These are

---

[1] A reminder of the definitions of force, energy, torque and power is provided under "*Elements of Classical Mechanics*" in the Companion Document. Throughout the book, the unit used for power is Watt (W) or its multiples (kW, MW and GW). The unit used for energy is Watt hour (Wh) or its multiples (kWh, MWh, GWh and TWh).

average values since the number of passengers varies. Overall, energy consumptions are less significant on a per passenger basis — the overall consumptions for trains, planes, etc., are provided in the "Transport and Travel" chapter. For air travel, the [*icao.int*] website provides the CO_2 emissions per passenger for a particular flight.

To compute the CO_2 emissions at home and for transport, the following values are used:

- The combustion of natural gas causes the emission of 200 g of CO_2 per kWh produced, or 200 gCO_2/kWh.
- For gasoline, diesel and kerosene, it is considered that, on average, the combustion of 1 liter produces 9 kWh, and each kWh produced causes the emission of 250 g of CO_2.
- For electricity, one takes the characteristics of the electricity production mix in the country or region into account. National websites or that of the International Energy Agency (IEA) [*iea.org*] provide the average annual CO_2 emission per kWh (or for a particular period, such as the heating season).

The cartoon for each family is completed by a few comments on its energy consumption at home, and its overall emissions, both at home and for transport leading to the average CO_2 emissions for each family member.

The situations of four similar families living at the same locations in 2035 is presented at the end of the book to illustrate the evolutions that are reachable using the already existing technologies described throughout the book.

Gilles Paire, Shutterstock

Diana Daley, Shutterstock

Wirestock Creators, Shutterstock

Dupont Family

Annie and Jean Dupont, a retired couple, live in Lille, an industrial city in the North of France with a population of 1.2 million. Annie was a primary school teacher and Jean taught at the high school level. They are sensitive to climate and environmental issues, which influence some of their decisions.

They take four weeks of vacation each year between May and September. As members of a group of retirees interested in the history of the Middle East, they travel there yearly during one week. The Duponts have three adult children, each married with children, who live in other cities in France. When they are not on vacation or traveling, Annie and Jean visit one of them and their families every other weekend, using the French high-speed train, TGV. As a result, they are only home some 290 days per year.

Annie and Jean live in a 135-m^2 house built in the 1960s, which they own. They have gradually improved the insulation of the house, first the attic and then the windows and doors, that resulted in a 30% decrease in the heating requirements. The house has a central circulating hot water heating system with an old gas heater. The Duponts do not heat the 35 m^2 of their children's rooms and cut the heating when they are not at home. Given the mild local climate and their concern for the environment, they have not installed any air-conditioning. As was usually installed during the 1960s, hot water is provided using two electric heaters with storage. Since the children no longer live in the house, the Duponts only use the hot water heater supplying the kitchen and their bathroom, where a towel dryer has been installed.

To reduce their energy consumption, Annie and Jean have an electric induction cooktop with an oven. They also have a steam cooker. Since they only shop once per week, they have an A+ rated refrigerator/freezer, which they turn off during their vacations. The kitchen has a dishwasher as well as other smaller appliances, such as a water kettle and a mixer. The Duponts have a washer and a dryer; they use sunshine on the terrace to dry the washes whenever possible.

Since recently, only LED lamps are used in the house, which has divided the Duponts' lighting consumption by six, compared to incandescent light bulbs. Annie and Jean have an LCD television in the living room, which they rarely watch, and a stereo. They have a second stereo in their bedroom. They each have a flat-screen personal computer (PC) that they turn off when not in use, as opposed to leaving it on standby. Each Dupont has a tablet and a mobile telephone but share a printer.

Since their house is close to a metro station, the Duponts use the local metro system to reach downtown Lille and the main train station. They each travel some 500 km/yr using the metro and tramway systems. They have a 2017 model of a rechargeable hybrid car, which they charge in their garage, that they use 6 000 km/yr, including 2 000 km/yr electrically.

To see their children and their families, Annie and Jean travel 1 600 km/month, or 20 000 km/yr, by TGV.

They fly economy class to the Middle East for their vacations, for example, to Cairo on a 6 400-km round trip.

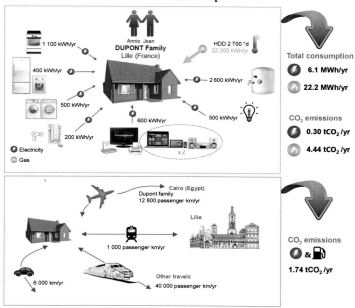

2015 energy consumption at home and CO$_2$ emissions

The yearly energy consumption of the Duponts at home is 28.3 MWh/yr. In conjunction with their house, daily commutes and travel, they induce 6.5 tCO$_2$ emissions each year, i.e., 3.25 tCO$_2$/yr for each one. This low emissions level, compared to similar families living elsewhere in industrialized countries, is due to their extensive use of electricity, at home and for their travel and transport, which has a low CO$_2$ content in France.

Jones Family

Seeking a high quality of life, including excellent schools and recreational facilities, the Jones live in Peachtree City, Georgia, even though it is located 50 km south from downtown Atlanta. Emma is an arts historian at the Atlanta High Museum and Chuck is a crew chief at the Delta Airlines aircraft maintenance facility at the Atlanta international airport, located 30 km north of Peachtree City. Their three children live at home: Billy works at Georgia Power Company, the local electric utility, at its downtown Atlanta headquarters, Ginette is a student at Georgia Institute of Technology, also at downtown Atlanta; they carpool from Peachtree City when Ginette goes to campus. John is still a student at the local high school. Emma and Chuck take three weeks of vacation, including one week when they travel to Europe with their three children. Billy has three weeks of vacation. Counting two more weeks of holidays per year, Emma, Chuck and Billy commute to work 47 weeks per year, i.e., 235 days per year. The house is fully inhabited all year round, even during the week when the entire family travels to Europe and when friends or family visit.

The Jones live in a 250-m^2 four-bedroom house. Since the house was built in 1994, during the local house construction boom, it is not well insulated. The Peachtree City climate is continental, requiring heating in the winter and air-conditioning during the summer. Pulsed air is used for both heating, using a gas furnace, and the electric air-conditioning. The house has a central gas hot water system.

In the kitchen, the Jones have a gas cooktop and oven along with a microwave that they use extensively to reheat ready-to-eat meals. Since they only go grocery shopping once per week, in addition to a large 540-liter refrigerator/freezer, they also have an additional freezer. The kitchen has a dishwasher and a few additional smaller electric appliances.

The washer and dryer are extensively used by the family. The Jones are in the process of gradually replacing incandescent light bulbs with LED ones.

In addition to a laptop, digital tablet and mobile phone for each member of the family, the Jones have a home theater system and one television set per bedroom and in the kitchen.

Peachtree City is not yet well served by public transport to Atlanta. The Jones have three cars of various sizes and ages. Emma drives 30 km to the

Atlanta Airport rapid rail station and then uses the local metro system, MARTA, to get to the Atlanta Arts Center, which is a further 20 km to the North; she returns home using the same route. Chuck drives a daily 60-km round trip from home to the Atlanta airport. Billy and Ginette's daily round trip car commute is 100 km. In total, the Jones drive 51 700 km just to commute to and from work. They drive another 15% to run errands and for leisure purposes.

Emma's family stems from Bavaria. The entire family travels once per year to Munich in economy class, a 15 400-km round trip from Atlanta using a modern passenger aircraft.

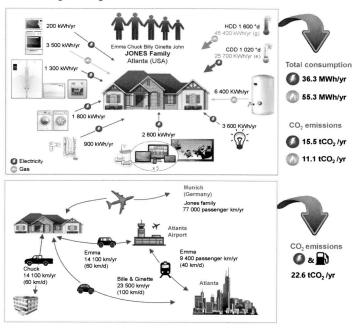

2015 energy consumption at home and CO_2 emissions

The yearly energy consumption of the Jones at home is 91.6 MWh/yr. In conjunction with their house, daily commutes and travel, they induce 49.2 tCO_2 emissions each year, i.e., 9.8 tCO_2/yr for each family member. These rather high values, compared to similar families living elsewhere in industrialized countries, is due to the size and age of their home and by the distances of their daily commutes to work. The comparatively high CO_2 emissions of the local electricity production mix also contributes.

Li Family

The Li family lives in Beijing, China, between the third and fourth Ring Roads on the 16th floor of an apartment building constructed in 2006. Li Ming, the father, was recently promoted to team leader in a factory, which is an hour away by the metro. Phan Zhang, the mother, works as a clerk at a local bank. Whenever possible, she commutes by bicycle; otherwise, the bus ride takes half an hour. Guo, the 13-year-old son, cycles 10 minutes to school. He is an excellent student and participates in chess championships. Hua, the three-year-old daughter, attends kindergarten; she walks there with her mother or a neighbor. The Li family lives close to a metro station, a public market and a shopping center; they walk or cycle for grocery shopping. They do not own a car.

Their 65-m^2 apartment has three bedrooms and a large living/dining room in addition to a kitchen and a bathroom. The Beijing climate requires heating during the winter. Their building is connected to the local district-heating network, which operates from mid-November to mid-March; the fuel is natural gas. During the coldest days, they use three supplemental electric heaters. The Beijing climate also makes cooling desirable during the summer. They have installed air-air air-conditioners in the children's bedrooms; they turn them on during hot days when the children return from school but turn them off during the nights. They do not have air-conditioners for their own bedroom or the living/dining room; however, their apartment has windows on two facades of the building, which improves natural airflow.

The apartment has a gas hot water heater; it serves the shower and sinks in the bathroom and the kitchen. The Lis have a washer that they use twice per week on average. Their clothes are dried naturally. Phan Zhang uses a hairdryer.

The Lis have a gas stove with four burners and an oven that they rarely use. They have an electric rice cooker that they use daily, and an electric kettle for tea. They have a small, old, and low-efficiency refrigerator-freezer that they brought from their previous apartment.

They have a television set that they watch every evening, but rarely more than two hours. The parents share one computer and Guo has his own. All three have a mobile phone; they do not have a landline since the installation of WiFi at home.

Li Ming commutes to work, i.e., a 20-km round trip 250 days per year, using the Beijing metro. The entire family uses the metro system during the weekends, a 10-km round trip.

The Lis are careful with expenses since they are assisting their aging parents, who live in Wuhan, 1 060 km away, and Kunming, 2 100 km away. They visit each set of parents once per year by high-speed train to Wuhan and by plane to Kunming.

Either Mr or Mrs Li travels with Guo to chess championships whenever possible, some 3 000 km by train yearly.

They travel once per year with the children to discover a province in China.

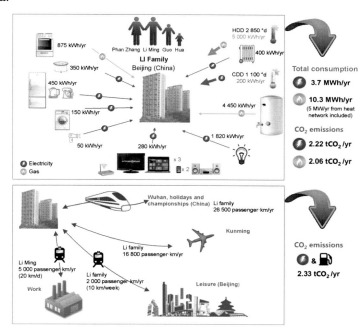

2015 energy consumption at home and CO₂ emissions

The yearly energy consumption of the Lis at home is 14.0 MWh/yr. In conjunction with their apartment, daily commutes and travel, they induce 6.6 tCO$_2$ emissions each year, i.e., 1.65 tCO$_2$/yr for each family member. The size of their dwelling and the fact that they do not have a car explain their relatively low levels of consumption and CO$_2$ emissions.

Menye Family

The Menye family lives in Djoum, a city of 6 000 inhabitants in the south of Cameroon. In addition to their five children, Frank and Rose also take care of two nieces who live in a bush village without schools. The two oldest children, Awa and Leke, walk 20 minutes to their secondary school while the three younger ones and their two cousins have a ten minute walk to their primary school. Rose's aunt also lives with them; she requires weekly medical attention not available in the village where she is from. Rose works in the fields, which is an hour's walk away. Frank has a neighborhood grocery store next to their home. He supplies the store with local agricultural and artisan products that he supplements by twice-a-week procurements in the local city, some five km away.

The Menyes live in a 100-m^2 house made of clay with a metal roof. The house has three bedrooms and a large room that they all share. The kitchen is outside, as are the latrines.

The climate in Djoum is equatorial with two rainy seasons. The local temperature ranges from 24°C to 35°C. Therefore, heating is not required. However, cooling would be welcome, but the Menyes cannot afford it.

The Menyes get their water from a well located 1 km away. The shower uses a bucket with water at ambient temperature without additional heat. The clothes are washed manually and dried outside. For cooking, they have installed an improved fireplace to reduce wood consumption. Frank and his eldest children collect the wood, or they purchase it at the local market.

The Menyes' home and grocery store are connected to the local electric grid. However, the grid availability is quite low with outages that can last up to three months. The quality of supply is also poor with under and over-voltages, which can damage their television set and the grocery's refrigerator since they do not have the means to purchase a voltage controller. When the grid supply is available, they use the grocery's refrigerator to store their food, light their home using incandescent bulbs and charge their two mobile phones. During the outages, they use five solar lights from which they can also charge their mobile phones.

Twice a week, Frank and Rose use tricycles to reach the fields and also to purchase goods and grocery items for the family from the nearby town. They use tricycles also when needed, especially for some of the children's activities.

Rose's aunt commutes weekly, a 10-km round trip, by moto taxi to the local hospital.

The Menyes do not take real vacations. The children help in the fields and the grocery store during their school vacations. The main vacations for the parents are to attend celebrations, family reunions and to visit their parents. To reduce costs, the entire family rarely travels together. Four times each year, five family members travel 1 500 km using bush taxis, on average.

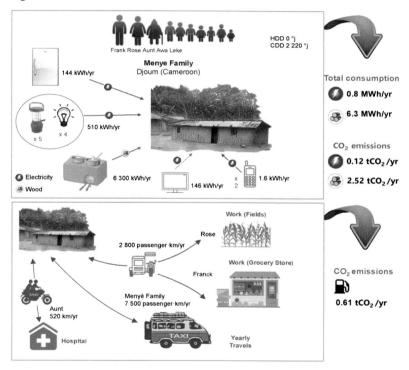

2015 energy consumption at home and CO_2 emissions

The energy consumption of the Menye family at home is 7.1 MWh/yr. In conjunction with their house and travel, they induce 3.3 tCO_2 emissions for three adults and seven children. The levels of consumption and CO_2 emissions are principally related to the use of wood for cooking.

Energia

A Modern Country with a Long History

Energia finds its origin in 1454 when the rulers of their respective principalities, Hans the Bear and Yves the Bald, decided to join forces to protect themselves against powerful neighbors and to better leverage their resources. A republic since 1800, as of 2015, its 50 million inhabitants enjoy a standard of living in line with OECD (Organization for Economic Cooperation and Development) countries. It is widely open to the world, economically and politically.

Located in the northern hemisphere between the 35th and 45th latitudes, Energia covers 330 000 km^2. The northern and southern extremities are separated by some 1 100 km.

The four main geographical regions are:

- Two coastal zones, one 500-km long — partial gentle slopes, partial 100-m cliffs — and facing mainly east with wide access to the ocean, and the second one facing south toward the sea along 300 km of primarily beaches. The cliffs along the ocean are listed as part of the UNESCO World Heritage.
- A vast plateau where over 75% of the population lives and most of the industrial and agricultural activities are concentrated.
- A mountainous region to the north and west with forests and some higher formations.
- An archipelago of some 30 islands to the east with 12 of them inhabited.

Several rivers flow from Energia's mountains toward the ocean and the sea; the largest one has its estuary on the ocean close to the country's main harbor.

Given the northern latitude of Energia, dwellings, businesses and stores require heating during the winters, even though they are mild due to the proximity of the ocean, except in the south, which benefits from a warmer microclimate. The demand for air-conditioning is increasing in the cities as it becomes economically accessible to a larger segment of the residents and also because the summer temperatures are increasing, not only due to intensifying activities and traffic but also as a consequence of climate change. Seawater desalination will become a necessity, especially in the south, driven by increased tourism, among other causes.

The capital of Energia, Flora, properly named after its gardens and public parks, which are famous worldwide, dates back to the creation of the country; its population is five million. Four other cities each have populations of one million. The five main cities have long histories with a number of buildings dating back several centuries. Half of the population lives in rural communities with less than 10 000 inhabitants. The average family size is 2.5 persons such that the total number of dwellings is 20 million with 60% being individual or row houses; the other 40% are apartments.

Energia's industrial sector is well established with strong worldwide exports of heavy equipment, precision mechanics, electronics as well as information technologies. To sustain its industrial activities, the country has a steel industry. Even though the agro-business is also prosperous, the country imports some food stock.

The total number of road vehicles, all categories included, is 32 million. The public rail transport system is well developed and mostly electrified.

Energy Resources

Aside from three coal mines, as confirmed by several exploration campaigns, Energia does not have any fossil or uranium resources that could be economically mined. Oil and natural gas are thus imported mainly for heating, transportation, and for the chemical industry.

Hydro resources along the rivers and in the mountains offer interesting development possibilities. Solar energy has strong potential — thermal on the plateau and photovoltaic in the south. Suitable wind energy sites have been identified along the coasts, in the mountains and also offshore in the archipelago; while attractive from a wind resource viewpoint, installing wind mills along the UNESCO-protected ocean cliffs would be met with strong public resistance. Previously identified geothermal resources remain to be leveraged. The tidal range is quite high along the ocean's coast.

Energia's Final Energy Consumption and CO_2 Emissions[1]

Energia's 2015 final energy consumption was 1 561 TWh, i.e., 31.2 MWh/cap, in line with OECD average values. The related emissions were 500 $MtCO_2$, i.e., 10 tCO_2/cap, also in line with OECD values.

Also, in 2015, 24.3% of the final energy consumption was covered using electricity, i.e., 380 TWh, or 7.6 MWh/cap, again in line with OECD data. The remaining 75.7% was provided from coal, oil and natural gas. The emissions due to the electricity production mix were 426 gCO_2/kWh, which was above the average emission levels for European Union (EU) countries at the time.

Ambitious Objectives for 2035

Energia signed the 2015 Paris Agreement on climate change. The outcome of the early 2015 elections was a large governmental coalition whose policies have set the country on a path leading to the complete elimination of fossil fuels by 2050. In this context and in line with the Paris Agreement, it has been decided to reduce the country's emissions from 500 to 170 $MtCO_2$ by 2035.

This overarching objective is to be reached not only by significantly decarbonizing the country's final energy consumption, including increased consumption of electricity, but also by decarbonizing the electricity production such that the emissions decrease to 111 gCO_2/kWh by 2035, which is at the core of the "Triple 1 society" public policy.

---

[1] Energy and emissions statistics are presented later in the book along with the related units.

"Triple 1 society" arguments

With the "Triple one society" moto, Energia's leadership seeks to align the country's electricity production mix emissions with those of the EU which, as of 2015, were at 315 gCO_2/kWh, i.e., much lower than Energia's at 426 gCO_2/kWh. To fully decarbonize its electricity production mix by 2050, in line with the Intergovernmental Panel on Climate Change (IPCC) recommendations, the EU would need to reach a target of 135 gCO_2/kWh by 2035.[2]

Using the slightly more ambitious target of 111 gCO_2/kWh, aside from the "catchy" moto, has the advantage of implementing a more drastic reduction from 426 to 111 gCO_2/kWh during the first 20 years as all coal and oil production is eased out, and while many attractive options to deploy new solar and wind resources are still available. The last reduction of 111 gCO_2/kWh to zero over the following 15 years will require the complete phasing out of all natural gas power plants even though suitable solar and wind sites will become increasingly difficult to find!

While deploying the best available technologies to reach these two overriding objectives is paramount, the decision-making processes involved must also seek to properly contain investments and operational costs without jeopardizing future development paths beyond 2035.

Energia's Energy and Emission Challenges

		2015 50 Million Inhabitants	2035 52.5 Million Inhabitants
Total Final Consumption	TWh	1 561	1 054
Incl. Electricity Consumption	TWh	380 (24.3%)	534 (50.7%)
Total Emissions	$MtCO_2$	500	170
Emissions, per Capita	tCO_2/cap	10.0	3.2
Emissions, Electricity Production	gCO_2/kWh	426	111

---

[2] Assuming a linear decrease between 2015 and 2050: $315 - [315 \cdot (20 / 35)]$.

Energy and Emissions

Where We Are

Since 1950, several evolutions have had significant impacts on our daily lives and energy challenges. Among them are:

- The impact of human activities, often designated as anthropogenic activities, on the environment (pollution, noise), climate change and the availability of natural resources, is increasingly taken into account to reach decisions at all levels. Broadly speaking, as the world population continues to grow, the need to curtail the use of fossil fuels and reduce emissions are ever more prominent in public debates worldwide and in public policy decision making.
- Simultaneously, discoveries of new deposits of fossil energy resources (coal, oil and gas), of new types of natural resources (shale gas and oil) along with the development of new extraction techniques of these resources, at reasonable costs, have provided for the quantitative coverage of all energy demands worldwide. This heavy reliance on fossil fuels could well continue for several decades unless deliberate and broadly coordinated policies result in energy mixes with rapidly decreasing carbon contents.
- Electrical technologies often have advantages in terms of efficiency, local pollution and noise. The emissions impacts may be positive or negative depending on the electricity production technologies involved. Specific uses of electricity such as information, communication and medical technologies, have increased and become crucial for society, thus creating new demands for electricity.

In addition, the worldwide population increase remains of key importance to energy consumption and thus regarding emissions levels.

Worldwide Population Growth

Worldwide population forecasts are regularly published by the United
Nations; they are accessible on [*population.un.org/wpp*].

The Earth's population reached one billion in 1804, then two billion
in 1927, i.e., 123 years later. As shown in Fig. 1, the worldwide population
reached seven billion in 2011, only 12 years after having reached six billion. By 2019, it reached 7.7 billion. The population size is expected to
increase at least until the end of this century. The forecast for 2035, i.e.,
the horizon for the book's energy mutation, is 8.8 billion. The trend is
toward a decrease in the rate by which the population will increase in the
future. The United Nations' mean forecast, in its June 2019 revision, is for
the worldwide population to reach 11.2 billion by 2100.[1]

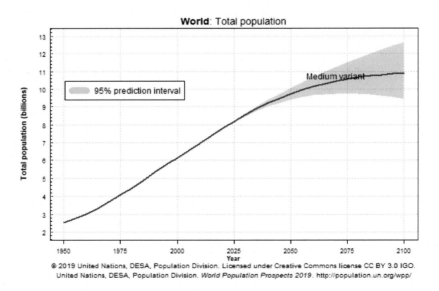

Figure 1.   UN World population projections / © *United Nations DESA*

---

[1]The 2050 forecast is 9.8 billion with a 95% probability that it will be between 9.4 and
10.2 billion; for 2100, the 95% probability is between 9.6 and 13.2 billion.

The population increase is not evenly distributed across the world; it will remain strong in Africa and Asia, particularly in India and Southeast Asia, moderate in North and Latin America while Europe's population has already stabilized and is predicted to decrease during the coming decades.

Climate Change

Greenhouse Gases

Greenhouse gases (GHGs) are gases in the atmosphere that absorb and emit radiation in the thermal infrared range. They are essential to sustain life on Earth. The issue is not to eliminate them but to ensure that they be kept within limits, resulting in temperatures that are compatible with Man's "ecological niche" and those of the flora and living species surrounding us.

The average overall solar radiation reaching the Earth's atmosphere, measured in W/m^2, is shown in Fig. 2. The left side of the figure illustrates

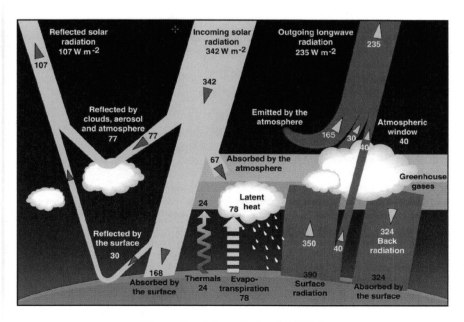

Figure 2. Solar radiation / © *IPCC*

the energy flow reaching the Earth's atmosphere: of the overall 342 W/m^2 average incoming solar radiation, 107 W/m^2 are directly reflected back by clouds (77 W/m^2) and from the surface (30 W/m^2). An additional 67 W/m^2 is absorbed by the atmosphere, such that only 168 W/m^2 are absorbed by the Earth's surface. As shown in the right side of the figure, the atmosphere sends 324 W/m^2 back to Earth, by way of the GHGs, which is absorbed by the Earth's surface. The total energy absorbed at the Earth's surface is thus: 168 + 324 = 492 W/m^2.

On the other hand, the energy flow away from Earth is composed of heat (24 W/m^2), evaporation (78 W/m^2) and radiation (390 W/m^2), for a total of 492 W/m^2.

The situation illustrated in the figure is thus an equilibrium between incoming and outgoing energy flows; the temperature on Earth would stay constant.

If the GHG's radiation back to the Earth's surface is too low, the average temperature would decrease; in fact, in their complete absence, the average temperature on Earth would be roughly –18°C. On the contrary, if the GHG radiation back to the Earth's surface is too high, the temperature on Earth increases, which is the situation we are presently witnessing.

The main greenhouse gases on Earth are: water vapor (H_2O), carbon dioxide (CO_2), methane (CH_4), nitrous oxide (NO_2) and ozone (O_3). Water evaporates from land, seas and lakes, and then falls back to Earth as rain or snow; it stays in the atmosphere only for a short time, typically 8–10 days. It is by far the main GHG; however, human activities play a minor role in the amount of water vapor created. Water natural evaporation is highly local and depends on the local humidity.[2] As the temperature increases, so does water evaporation; as a result, the effect of an increase of the Earth's average temperature due to CO_2 emissions is augmented by the water vapor effect.

The air we breathe is composed of: (a) nitrogen (78%), (b) oxygen (21%) and (c) others (1%), including argon (0.93%), CO_2 (0.04%) (400

---

[2] Under clear nightly skies and unhindered by clouds, heat escapes from the ground and the temperature drops faster than under cloud cover. In low precipitation regions, such as deserts, very little evaporation occurs during daytime; as a result, the temperature differences between day and night can be spectacular.

ppm),[3] and methane (0.000172%) (1.72 ppm). Nitrogen (N), oxygen (O) and argon (Ar) are not greenhouse gases.[4]

Combustion

Combustion is a chemical reaction between a fuel (solid, liquid or gaseous) and an oxidant (generally oxygen from air),[5] which is initiated by way of an outside energy such as a flame. The French chemist, philosopher and economist **Antoine Lavoisier** (1743–1794) described the role of oxygen during combustion. Combustions produce heat and are therefore described as exothermic.

Fossil fuels contain carbon (C), hydrogen (H), and in some instances, oxygen (O), water as well as other constituents in smaller quantities (sulfur, mercury, argon, etc.). For example:

- Anthracite, i.e., black coal, contains 90% of carbon as opposed to lignite with only 55%.
- Natural gas contains 80 to 90% of methane (CH_4), i.e., carbon and hydrogen.

The combustion of carbon can usually be described by the following equation: $C + O_2 \rightarrow CO_2$.[6]

It produces carbon dioxide (CO_2).

The "Units and Reference Values" annex provides the energy and CO_2 emissions resulting from the combustion of coal, oil, natural gas and wood.

The reaction described above is referred to as a complete combustion, which implies that enough oxygen is available. If this is not the case, the dangerous carbon monoxide (CO) gas is created. In addition, the presence

---

[3] Parts Per Million by volume.

[4] Molecules made out of two atoms of the same element, such as N_2 and O_2, as well as monatomic molecules, such as Ar, have no net change in their electrical charge distribution when they vibrate. As a result, they are largely unaffected by infrared radiation.

[5] As a result, combustion is sometimes referred to as oxidation.

[6] The combustion equation for methane is: $CH_4 + 2\,O_2 \rightarrow CO_2 + 2\,H_2O$. In addition to the CO_2, it also produces water, in the form of vapor.

of nitrogen and other impurities induces the creation of various nitrogen oxides and other oxides, some of which are dangerous, such as sulfur oxides. The combustion of wood, diesel and heating oil produces particular matters that can penetrate and remain in our lungs.

Air Pollution

Air pollution, referred to as particulate pollution, is generally given Particulate Matters (PMs), measured in millionth of grams per unit volume ($\mu g/m^3$). Two sets of PM data are generally collected:

- PM10, related to particles between 2.5 and 10 μm, which are primarily due to crushing and grinding operations along with the dust from vehicle movements on pavements, for example. Fossil fuel exploration and refining contribute to increasing the local PM10.
- PM2.5, related to particulates smaller than 2.5 μm, which are primarily due to various combustions such as in internal combustion engines (ICEs), power plants, residential wood burning and industry.

The actual PM values for any particular plant heavily depend on the type of fuel — coal or natural gas — used, especially regarding its sulfur content. The World Health Organization [*who.int*] maintains and regularly updates a large databank for both outdoor and indoor pollutants.

Greenhouse Gases Global Warming Potentials and CO_{2e}

The contribution to climate change of each greenhouse gas depends on its composition; their individual average lifetime in the atmosphere is also different. To compare their effects, the Global Warming Potential (GWP) is computed over 100 years for each one. With CO_2 as a reference, it is 25 times higher for CH_4 and 298 times higher for N_2O. Based on these observations, to simplify statistical tables, the CO_2 equivalent (CO_{2e}) notion is computed by the formula:

$$gCO_{2e} = gCO_2 + 25 \cdot gCH_4 + 298 \cdot gN_2O$$

GHG statistical data is generally given for CO_{2e}.

Land Use and Forest Implications

The actual use of land and forest plays an important role in the emission of greenhouse gases. Forests have a positive impact on climate change by absorbing CO_2 by way of photosynthesis.[7] Some of the CO_2 absorbed in forests is permanently stored in construction materials or furniture, for example, while the rest is returned to the atmosphere during decomposition and by combustion. Transforming a forest into agricultural land has a negative effect since the CO_2 absorption capacity of the crop is generally less than that of the forest it replaces. Urban development has a negative climate change impact. To quantify these effects, the Land Use Change and Forestry (LUCF) is computed in kg or ton of CO_{2e}. An in-depth discussion as to the actual computation of the LUCF is beyond the scope of this presentation. However, broadly speaking, it can be observed that in most emerging regions of the world the LUCF index is often positive, indicating that land use change and deforestation contribute to climate change whereas it is often negative in developed regions, indicating the positive impact of recent climate change policies.

Though the LUCF is decreasing at the worldwide level, further reductions remain a priority for future land and forestry public policies.

Greenhouse Gas Statistics

The computation of greenhouse gases is complex. Orders of magnitude are more important than actual, precise values. The evolution of GHG levels is also quite important; even though the computations incorporate some assumptions, the computational process generally remains unchanged from one period to the next such that comparisons remain relevant. The World Resources Institute (WRI) [*wri.org*][8] provides a wealth of greenhouse gas statistics; the following data was the latest update from this source at the time the book was written. In addition to the three main

---

[7] Photosynthesis is the process used by plants to grow. Using the light energy they receive, they combine CO_2 from the atmosphere with water drawn from their roots to create their organic matter.

[8] By way of its Climate Access Indicators Tool (CAIT).

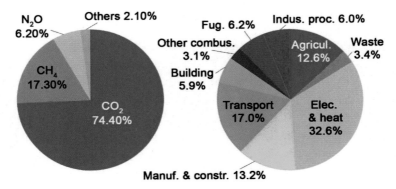

Figure 3.   GHG by types of gas and by sectors / *Data from World Resources Institute*

anthropogenic GHGs, the WRI statistics also include the impact of fluorinated gases, designed as F-gas.

Excluding the LUCF, the overall GHG emissions in 2016 were 46.2 $GtCO_{2e}$; they were 49.4 $GtCO_{2e}$ when including the LUCF.

As illustrated by the pie chart on the left hand side of Fig. 3, which shows the overall GHG distribution, eliminating combustions and the associated CO_2 emissions is key to overall GHG reductions. However, solely focusing on combustion reductions will not resolve overall greenhouse issues; indeed the related emissions only represent some 75%.

The right hand side of Fig. 3 shows the distribution among the GHG emission main causes.[9] Together, electricity and heat, transport, manufacturing and construction, industrial processes and other combustions represent 72%. Significant reduction opportunities exist in the first two.

Conference of the Parties (COP)

The Rio Earth Summit, held in 1992 in Rio de Janeiro, Brazil, set the formal stage, under the auspices of the United Nations Framework Convention on Climate Change (UNFCCC), to launch a worldwide effort to better understand and then mitigate climate changes.

---

[9] "Fug." stands for fugitive gases, due to intended or accidental leaks.

The UNFCCC came into force in March 1994. The 197 countries that ratified the convention are referred to as the Parties of the Convention; their yearly meetings are called Conference of the Parties (COP).

The Kyoto Protocol was adopted in December 1997 during the COP 3 in Kyoto, Japan. Its initial commitment period was from 2008 to 2012. The emissions target agreed upon was a worldwide GHG reduction of 5.2% from 1990 levels for 84 industrialized countries, essentially the OECD[10] members at the time, plus a few additional countries and the European Union; collectively referred to as Annex I countries. Several large developing countries, which were already significant contributors to GHG emissions, such as China, India and Brazil, were not included in Annex I; as a result, they were not bound to the agreed upon targets. Two major requirements had to be met for the Protocol to come into force: (a) 55 Parties to the Convention had to approve it and (b) the combined emissions from the Annex I countries having approved the Protocol had to represent at least 55% of the combined emissions of all Annex I countries in 1990. The Kyoto Protocol entered into force in February 2005.

Signed by 197 countries in December 2015 during the COP 21, the Paris Agreement sets forth a series of climate change mitigation commitments proposed by and specific to each country with the aim of "holding the increase in the global average temperature to well below 2°C above pre-industrial levels and pursuing efforts to limit the temperature increase to 1.5°C above pre-industrial levels, recognizing that this would significantly reduce the risks and impacts of climate change".[11] The distinction between Annex I and Non-Annex I countries was removed such that all signing Parties agreed to adhere to the Agreement's stipulations. The requirement set for the agreement to come into force was that more than 55 Parties to the Convention that collectively represented more than 55% of the total worldwide GHG emissions, had approved and formalized their instruments of ratification. This double goal was reached on October 5, 2016, such that the Paris Agreement entered into force on November 4, 2016.

The Paris Agreement stipulates that each signatory Party is to submit a report describing its GHG emissions mitigation actions and outcomes

---

[10] The acronym for the Organization for Economic Co-operation and Development.

[11] Article 2, paragraph 1a of the Paris Agreement, United Nations, December 2015.

every five years, along with the actions envisioned for the subsequent five years. The said reports are to be reviewed by independent commissions with the aim of proposing new coordinated actions deemed to be required to achieve the overall aim of the Paris Agreement. The first reports are to be submitted in 2025.

Intergovernmental Panel on Climate Change (IPCC)

The Intergovernmental Panel on Climate Change (IPCC) was created in 1988 by the United Nations Environment Program (UNEP) and the World Meteorological Organization (WMO); it has become the leading international body for the assessment of climate change. Also in 1988, the United Nations General Assembly endorsed the creation of IPCC. In 2016, the IPCC had 195 member countries; its General Secretariat is hosted by the WMO in Geneva, Switzerland.

The fifth and current IPCC Assessment reports were released in 2013 and 2014.[12] The IPCC website [*ipcc.ch*] can be consulted for more details regarding its activities and reports.

IPCC Emissions Targets

Figure 4, from the IPCC's latest release, shows the evolution of the total GHG emissions using the same classification as above. As shown in the figure, the total GHG emissions almost doubled in only 40 years from 1970 to 2010.

At the conclusion of the Paris COP 21, the IPCC was invited to provide a special report on the global warming impacts of a 1.5°C rise above pre-industrial levels and on the related global greenhouse gas emissions pathways. The IPCC published the Special Report in October 2018. In the report, the IPCC incorporated a number of scientific advances since the 2013–2014 Fifth Assessment Report. The report quantifies the increases of a number of impacts, such as extreme heat periods, sea level rise and species losses, should the temperature increase be limited to 2°C rather than 1.5°C.

---

[12]The Fourth Assessment Report was published in 2007; the sixth report is scheduled for 2022.

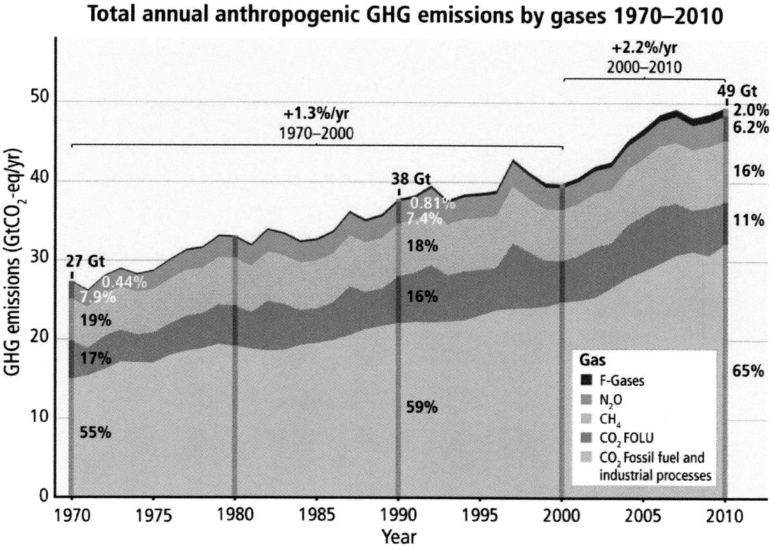

Total annual anthropogenic GHG emissions by gases 1970–2010

Figure 4. Greenhouse gas evolution / © *IPCC*

To have a medium chance, i.e., with a 50% likelihood, to limit the temperature increase to 1.5°C above pre-industrial levels, Mankind can still emit 770 GtCO₂ starting in 2018; for a 67% likelihood, the budget decreases to 570 GtCO₂.

From the WRI 2014 data above, i.e., emissions of some 46 GtCO₂/yr, versus an overall remaining budget of 570 or even 770 GtCO₂, it is clear that a "business as usual" emissions scenario would be irresponsible.

This is even more so since the worldwide energy consumption will, on the whole, continue to increase for at least several decades to come, as forecasted by all relevant organizations. Clearly, the increase in the world population plays a major role in these increases along with the indispensable quality of life improvements for the less fortunate populations on Earth.

Energy and Water Nexus

Energy is required to pump, purify and distribute the water we cannot live without — water is required for almost all processes to produce the energy we also cannot live without. This coupling, nexus,[13] has become critical and will become even more so to safeguard Man's ecological niche.

The 17 United Nations Sustainable Development Goals (SDGs) further emphasize the nexus:

- Goal # 6: Clean water and sanitation
- Goal # 7: Affordable and clean energy

We are not facing a shortage of water; there is lots of it in oceans and seas. However, since Man cannot survive on brackish or salt water, we will have to allocate, depending on the region, significantly more energy to purification, treatment and desalination processes to address the two UN SDGs mentioned above.

Further details are provided under *"Energy and Water Nexus"* in the Companion Document.

Energy Sources and Energy Vectors

The nine main energy sources, also designated as primary energy, accessible by Man are:

- Coal
- Oil
- Natural gas
- Nuclear, mainly uranium at the present time
- Biomass
- Hydro
- Wind
- Solar
    - Photovoltaic
    - Thermal
- Geothermal

---

[13] From the Latin verb *necto*, which means to tie, bind or unite.

These sources are often used after a transformation (also referred to as conversion) such as: refining of raw oil and purification of natural gas. One then refers to energy vectors. The four main energy vectors are:

- *Electricity*, which can be produced from any of the sources mentioned above and which provides for a broad range of end-use applications, some of which are only possible by way of electricity.
- *Liquid fuels*, primarily gasoline, diesel, heating oil and kerosene, almost exclusively obtained from oil refining even though biofuels are used in some regions. They remain the main energy vector for land, sea and air transport, aside from railways and subways.
- *Gas*, which is also an energy source. Raw natural gas, after purification, is almost exclusively used as the source for the gas energy vector even though biogas is used in some regions.
- *Heat and cold*. District heating networks have been in operation for decades in cities and industrial parks, primarily in industrialized countries, even though the deployment levels vary greatly from one country to the next. District cooling networks are being increasingly implemented.

Hydrogen is a developing energy vector; it is presently mainly used for industrial applications. It is emerging as an energy vector for transportation applications, mainly in cars, buses and trucks. As of 2019, the use of hydrogen for energy storage applications was at the experimental and demonstration stage.

Figure 5 illustrates the first set of conversions from energy sources to energy vectors, followed by the second set of conversions from energy vectors to the main energy consumption categories. The figure, along with the table below, illustrates the variety of possible technical solutions to address a particular energy end-use by way of a number of vectors as well as the number of sources that can be used to supply a particular energy vector. This flexibility is enhanced as new energy conversion and/or end-use technologies are implemented in the market. As further discussed in the two "Where We Should Be Heading" chapters, this increasing flexibility needs to be valorized taking local societal, economic and environmental situations into account along with strategic requirements.

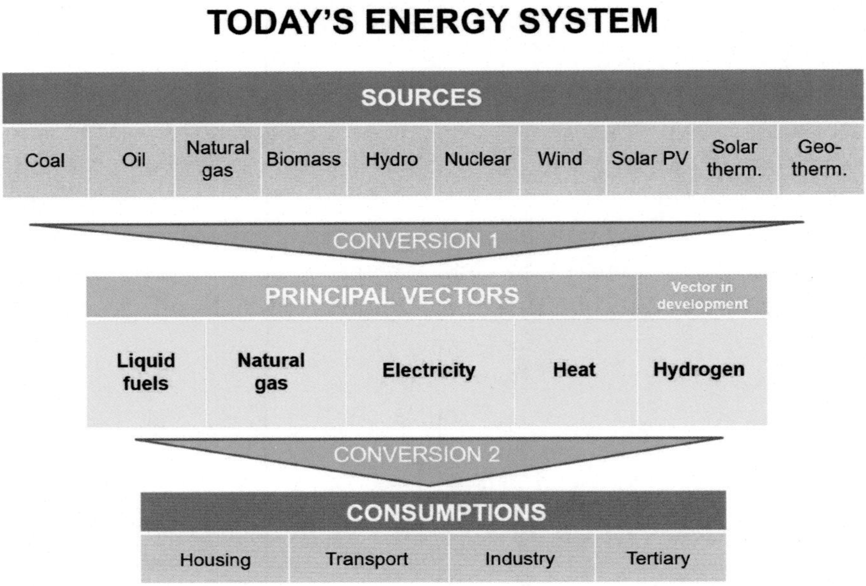

TODAY'S ENERGY SYSTEM

SOURCES									
Coal	Oil	Natural gas	Biomass	Hydro	Nuclear	Wind	Solar PV	Solar therm.	Geo-therm.

CONVERSION 1

PRINCIPAL VECTORS				Vector in development
Liquid fuels	Natural gas	Electricity	Heat	Hydrogen

CONVERSION 2

CONSUMPTIONS			
Housing	Transport	Industry	Tertiary

Figure 5.   Present energy system / © *French Academy of Technologies*

Source	Direct Use	Conversion to Energy Vectors				
		Liq. Fuel	Gas	Electricity	Heat/Cold	Hydrogen
Coal	×	×	×	×	×	
Oil		×		×	×	
Nat. Gas	×	×	×	×	×	×
Biomass	×	×	×	×	×	
Hydro	×			×		
Nuclear				×	×[a]	
Wind	×			×		
Solar PV				×		
Solar Th.	×			×	×	
Geother.				×	×	

[a]Nuclear power plants are presently not constructed to produce only heat

Final Energy Consumption and Primary Energy

When discussing energy consumption, we have deliberately opted to primarily focus on energy end-use and the related statistical data; indeed, it is the one actually consumed by the end-users. Consumers themselves can have a direct influence on the quantities of final energy they consume. For example, an end-user can decide to use efficient light bulbs, such as LEDs, and turn out the lights in vacant rooms. However, that same consumer has little influence over the amount of energy that was required to manufacture that light bulb, or over the energy required to transport it to the store where she or he bought it. As a further example, a consumer can decide to leave her or his car in the garage and use public transportation. However, that same consumer has no role in deciding whether the transportation is electric or not; if it is, the consumer does not know whether the electricity is produced by an old coal-fired central power plant or a state-of-art photovoltaic (PV) system, nor does she or he have any say in that decision.

Hence our deliberate decision to focus on energy end-uses, known as final energy consumption, when providing statistical energy information.

However, it is important to also keep in mind the energy required to extract, transport and convert the naturally available raw energy resources into energy products actually usable by individual residential, commercial and industrial consumers. For example, before we can put one liter of gasoline into our car, energy has to be consumed to explore the underground for potential oil pockets. The well then has to be drilled, the oil has to be pumped out of the ground or the sea and then transported to the refinery, where it will be refined into gasoline and then be transported to the fuel station, where it is pumped into the local fuel station's reservoir. Finally, the gasoline has to be pumped out of the local fuel station's reservoir into our car tank. A similar "transformation" path can be described for coal, for natural gas, etc.

This is where the notion of Total Primary Energy Supply (TPES) comes into play.

TPES Determination

At the worldwide level, the yearly TPES is the sum of the yearly primary energy supplies, i.e., of all sources. Stock variations must also be taken into account — positive if the stocks decrease, i.e., consumed during the year, or negative if they increase, i.e., not consumed (actually produced) during the year.

The determinations are made in one of the two following ways:

- For coal, oil and gas, the computations are based on the actual weight (for coal) or actual volumes (for oil and gas), extracted from underground or under the seafloor along with the calorific values of each one. Weight is also used for biomass, primarily wood. Since the actual weight/volume of coal or oil or natural gas consumed by power plants as well as by industrial, tertiary and residential sites is known, one can compute the related TPES.
- For the other energy sources, a coefficient is used to convert into primary energy the actual electricity or thermal energy produced at thermal and other electric power plants or at other installations. The coefficients listed in the following table are provided by the International Energy Agency (IEA) [*iea.org*][14] and have remained unchanged for decades.

Energy Source	Production — kWh	TPES — kWh
Nuclear	1	3
Geothermal — electricity	1	10
Geothermal — thermal	1	3
Hydro	1	1
Wind	1	1
Solar — PV	1	1
Solar — thermal	1	3

---

[14] For nuclear energy, the coefficient is based on the average efficiency of the thermal cycle of European nuclear plants, which was 33% at the time it was set. The coefficient for solar thermal is based on the average plant efficiencies also at the time it was set. For geothermal thermal power plants, the coefficient aims to take the heat losses from the source deep under the Earth's crust to the surface into account — 66% heat losses. Finally, for geothermal electric plants, the coefficient also takes the efficiency of the thermal steam producing cycle into account, i.e., 33%.

The TPES information provided later implicitly incorporates the coefficients listed above.

For a country[15] the TPES is computed similarly, including the incorporation of the stock changes, but with two differences:

- The import and export energies must be taken into account — worldwide they cancel out.
- As per international convention, the primary energy corresponding to the fuels delivered to ships and aircraft gearing up for international destinations, i.e., outside of the particular country, is subtracted from the annual TPES; they are referred to as international marine and aviation bunkers by the IEA.[16]

Formally, for a country:

TPES = indigenous primary energy supplies + imports
− exports − bunkers ± stock variations

Transformation Energy

Transformation energies are used to convert primary sources into energy vectors. They can be computed at worldwide or country levels, and are generally decomposed as follows:

- Energy consumed for the production of electricity and heat.
- Energy consumed by the other conversions — oil refining, gas purification, the transformation of coal and coke, etc.

From Primary Energy to Final Energy Consumption

The Total Final energy Consumption (TFC) worldwide or for a country is the TPES after the transformation energy is subtracted.

Formally: TFC = TPES − Transformation energy

---

[15] The designation "country" also includes "groups of countries".

[16] At the worldwide level, the bunkers of all countries are grouped together and included in the transportation total final energy consumption.

Determining the TFC for any one country is complex. Indeed, the vast majority of countries export and/or import manufactured products, services, energy and raw materials. In this context, for example, how can one compute the TFC related to the manufacturing of a car?

- If the car was entirely manufactured in the country where it is used, then the energy consumptions of the factories where it was made are included in the Industry Sector energy consumption of the particular country.
- However, if parts of the car, the engine and transmission, for example, were imported, then the country's Industry Sector energy consumption *under* represents the energy consumption related to the manufacturing of the car.
- On the other hand, if the local car manufacturing industry exports cars or subassemblies, then the country's Industry Sector energy consumption *over* represents the energy consumption related to the manufacturing of the cars sold in the country.

To summarize, at the level of a country, the final energy consumption is the sum of consumptions in each of the four sectors:

- Housing: The final energy consumed — fossil fuels, bio fuels, electricity and heat — includes the energy consumed in the dwellings and by their occupants, but not the energy required to prepare the energy vectors for consumption or for their transport to the dwellings.
- Transport and travel: The final energy consumed — fossil fuels, biofuels and electricity — includes the energy consumed by all transportation means (e.g., road, rail, maritime and air) within the country itself, including the transportation of people, goods as well as raw, refined and purified products. The energy consumed in the country's refineries to produce fuels from the sources is included in the transformation energy.
- Industry and agriculture: The final energy consumed — fossil fuels, renewable energies, electricity and heat — includes the energy consumed by all industries (e.g., automobile, chemical, food, metallurgy, paper, petrochemical, textile, etc.) as well as for agriculture activities, excluding the energy industry, the consumption of which is included in the transformation energy.

The energy consumed to transport the said goods and refined/purified products is included in the Transport and travel sector.
- Tertiary and services: Situation identical to the Housing sector.

Depending on the particular statistical data, under the "final energy consumption" designation for a particular end-use sector one may find three different values:

- The sector's final, direct, consumption, i.e., the consumption of coal, oil, gas, etc., but excluding electricity and heat.
- The sector's total final energy consumption, i.e., the same value as above to which the electricity and heat consumption of the particular sector is added.
- The TPES of the sector, i.e., the previous value to which the following energies are added:
  ○ The transformation energy related to the production of electricity and heat for the sector, including the related network losses.
  ○ The transformation energy for the other vectors consumed by the sector. It can also include the related losses.

Carbon Dioxide (CO_2) Emissions

The same categorization as outlined above for the energy consumptions can be applied to the related CO_2 emissions.

For a particular country, based on the observations above, the emissions related to energy imports and exports must also be taken into account while those related to international marine and aviation transport are not to be included.

At the worldwide level, the emissions from the transformation energies related to the production of electricity and heat represent some 12% of the overall CO_2 emissions;[17] roughly two-third are from refineries.

For any country, the proportion of emissions due to the transformations themselves, excluding the production of electricity and heat,

---

[17] As per Fig. 2 of the IPCC 2014 Summary Report for Political Leaders.

decreases if the use of fossil energies consumed during these transformations energies decreases.

Final Energy Consumption Mix and Electricity Mix

Final energy consumption mix

The final energy consumption mix is the decomposition of the total final energy consumption into the direct consumption of coal, oil, gas, non-electric renewables (i.e., biomass, waste), electricity and heat. It is expressed in percentage per vector.

Electricity mix

The electricity mix is the decomposition of the electricity production, i.e., of the direct electricity consumption augmented by the electricity transmission and distribution network losses, into sources: coal, oil, gas, nuclear, hydro, solar, wind and other renewables. It is expressed in percentage for each source. The losses typically are of the 10% order of magnitude.

Energy Statistics

Vast amounts of statistical data are required to study energy problems. Throughout the book, the data and information used are from several public access sites, especially from the IEA [*iea.org*], with the intent of making it convenient for the reader to either update information or find information more relevant to her or his circumstances.

Per capita information is important in addition to overall data; they facilitate comparisons while removing the size of the country. Indeed, comparing the energy data for China and Switzerland, for example, is not very useful while doing so at the per capita level could well be.

International Energy Agency (IEA)

Created in November 1974 in the wake of the 1973–1974 oil crisis, the IEA is part of the OECD; its headquarters are located in Paris, France. It has become the reference for a broad range of energy information related to energy resources and their main end-uses (residential, industry, transport, etc.). As of 2020, there were 30 IEA member countries.[18]

Statistical data often relied upon in the book come from the IEA Key World Energy Statistics document which is updated yearly.[19] Each publication is the outcome of a systematic compilation of worldwide consumption, production and emission information. As a result, the data included typically is for two years before the publication date; for example, at the time the book was completed, the latest edition was in 2020 and contained data for 2018.

Other statistical data with narrower subjects, such as for hydro energy, for example, may be more recent.

TPES Statistical Data

As shown in Fig. 6[20] and the following table between 1973 and 2018:

- The TPES more than doubled and the TPES per capita increased by 21%.
- The reliance on coal and natural gas increased. Indeed, their growths at 2.57 and 3.34, respectively, were higher than the overall TPES growth. On the other hand the reliance on oil decreased.[21]
- While their shares were still relatively small, the growths of both nuclear and "other", which includes solar and wind, were the highest.

---

[18] Member countries: Australia, Austria, Belgium, Canada, Czech Republic, Denmark, Estonia, Finland, France, Germany, Greece, Hungary, Ireland, Italy, Japan, Korea, Luxemburg, Mexico, New Zealand, Norway, Poland, Portugal, Slovak Republic, Spain, Sweden, Switzerland, the Netherlands, Turkey, United Kingdom and the United States.

[19] Typically early fall.

[20] The data provided by the IEA Key World Energy Statistics is given in toe, ton equivalent oil. One toe is the energy produced by the combustion of one ton of oil, based on a worldwide average quality of oil; it varies somewhat from one region to the next. The conversion factor is 1 toe = 11.63 MWh or 1 Mtoe = 11.63 TWh.

[21] Based on the IEA conversion coefficients between primary and final energy consumption, the entry for nuclear energy in the TPES data is equal to the electricity actually produced using nuclear power plants multiplied by three. The electricity produced in 2017 using nuclear power plants thus was: 7 962 / 3 = 2 654 TWh. The hydro energy multiplication factor is 1.

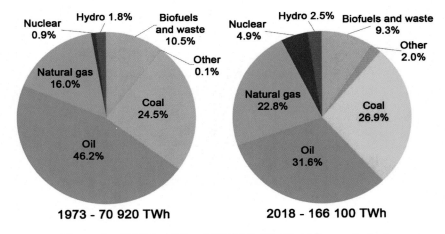

Figure 6.   TPES in 1973 and 2018 / © *IEA World Energy Outlook*

TPES					
	1973	70 920 TWh	Population	3.92 Billion	18.1 MWh/cap
	2018	166 100 TWh	Population	7.59 Billion	21.9 MWh/cap
Ratio		*2.34*		*1.94*	*1.21*

	1973		2018		
	%	TWh	%	TWh	Growth
Coal	24.5%	17 375	26.9%	44 681	2.57
Oil	46.2%	32 765	31.6%	52 488	1.60
Nat. Gas	16.0%	11 347	22.8%	37 871	3.34
Nuclear	0.9%	638	4.9%	8 139	12.75
Hydro	1.8%	1 277	2.5%	4 153	3.25
Bioenergy and Waste	10.5%	7 447	9.3%	15 447	2.07
Other	0.1%	71	2.0%	3 322	46.84
	100.0%	70 920	100.0%	166 100	

Final Energy Consumption Data

By vectors (referred to as "by sources" in the IEA data)

Since electricity is not an energy source but an energy vector, i.e., obtained after a conversion, it is not listed in the TPES data while nuclear and hydro energies are. On the contrary, electricity is listed in the total final energy consumption data while nuclear and hydro energies are not.

As seen in Fig. 7 and in the following table, between 1973 and 2018:

- The reliance on both coal and oil decreased; indeed, their growths were below that of the total final energy consumption. Only the reliance on natural gas increased among the fossil fuels.
- The reliance on electricity grew by more than fourfold. Electricity accounted for 19.3% in the final consumption in 2018. It was only 9.4% in 1973.
- The ratio between the final energy consumption and the TPES can be broadly viewed as being an indicator of the worldwide energy "efficiency". In 1973 it was: 54 196 / 70 920 = 76.4%. By 2018, it had *decreased* to 69.6%.[22]

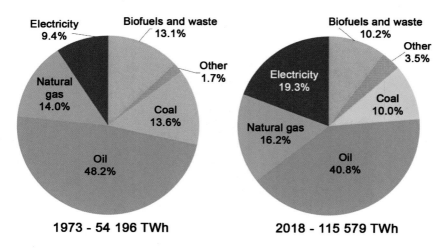

Figure 7.   Final energy consumption by energy vectors in 1973 and 2018 / © *IEA World Energy Outlook*

---

[22] It is encouraging to note that it has been increasing during the last few years!

The IEA conversion coefficients listed above influence this ratio; however, the impact is minor given the minor roles played by nuclear energy and geothermal energy worldwide.

- The fact that the growth of the TPES per capita was 21% between 1973 and 2018 and that it was only 10% for the final consumption confirms that, overall, humanity is not yet making progress toward a more rational use of the available energy resources.
- Between 1973 and 2018, the final energy consumption per capita only grew by 10% though the electricity consumption per capita more than doubled from 1.30 MWh/cap to 2.94 MWh/cap.

Total Final Consumption — By Vectors

	1973	54 196 TWh	Population	3.92 Billion	13.8 MWh/cap
	2018	115 579 TWh	Population	7.59 Billion	15.2 MWh/cap
Ratio		*2.13*		*1.94*	*1.10*

	1973		2018		
	%	**TWh**	**%**	**TWh**	**Growth**
Coal	13.6%	7 371	10.0%	11 558	1.57
Oil	48.2%	26 123	40.8%	47 156	1.81
Nat. Gas	14.0%	7 587	16.2%	18 724	2.47
Electricity	9.4%	5 094	19.3%	22 307	4.38
Biofuels and Waste	13.1%	7 100	10.2%	11 789	1.60
Other	1.7%	921	3.5%	4 045	4.39
	100.0%	54 196	100.0%	115 579	

By geographical regions

As seen in Fig. 8 and in the following table, between 1973 and 2018:

- The final energy consumption grew by 35% in the OECD countries between 1973 and 2018; it was multiplied by 3.33 outside the OECD region.

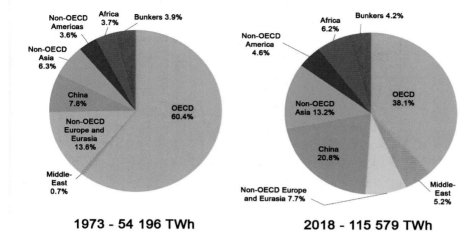

1973 - 54 196 TWh   **2018 - 115 579 TWh**

Figure 8. Final energy consumption by regions in 1973 and 2018 / © *IEA World Energy Outlook*

Total Final Consumption — By Regions

	1973	54 196 TWh	Population	3.92 Billion	13.8 MWh/cap
	2018	115 579 TWh	Population	7.59 Billion	15.2 MWh/cap
Ratio		*2.13*		*1.94*	*1.10*

	1973		2018		
	%	TWh	%	TWh	Growth
OECD	60.4%	32 735	38.1%	44 036	1.35
Africa	3.7%	2 005	6.2%	7 166	3.57
Middle East	0.7%	379	5.2%	6 010	15.84
Non-OECD Americans	3.6%	1 951	4.6%	5 317	2.73
Non-OECD Asia	6.3%	3 414	13.2%	15 256	4.47
Non-OECD Europe and Eurasia	13.6%	7 371	7.7%	8 900	1.21
China	7.8%	4 227	20.8%	24 040	5.69
Bunkers	3.9%	2 114	4.2%	4 854	2.30
	100.0%	54 196	100.0%	115 579	

- While the overall final energy consumption per capita hardly increased (13.8 to 15.2 MWh/cap), a more in-depth analysis of the average final energy consumptions reveals the diversity of the energy challenges facing Humanity, for example:
  - In OECD countries, although it slightly decreased from 35.3 to 33.9 MWh/cap, it remained far above the worldwide average.
  - In China, it grew from 4.7 to 16.8 MWh/cap, slightly exceeding the worldwide average.
  - The increase was greater in the Middle East where the per capita consumption grew from 3.4 to 10.2 MWh/cap.
  - In Africa, it only grew from 5.1 to 5.53 MWh/cap, remaining very far below the worldwide average, even though, overall, Africa's final energy consumption was multiplied by 3.52, again highlighting the role of population evolutions in energy data; it grew from 395 million in 1973 to 1 276 million in 2018.

By activity sector

As of 2018, the total final energy consumption distribution among activity sectors was:

- Housing: 21.2%
- Transport and travel: 29.1%
- Industry and agriculture: 30.8%
- Tertiary and services: 8.1%
- Non specified and non-energy use: 10.8%

Further statistical information as to the final energy consumption in each of the four end-use sectors is provided in the dedicated chapters.

Key Energy Indicators

A number of key indicators are relied upon to characterize the energy consumptions and emissions of countries. Depending on the situation, they may be given on absolute levels, on a per capita basis or related to the Gross Domestic Product (GDP).

Further details are provided under *"Key Energy Indicators"* in the Companion Document.

Energy Consumptions and Emissions

Housing

At Home

We spend an important part of our lives in our homes; it is where we consume the most energy, not just for our comfort (heating and air-conditioning) or our health (forced ventilation), but also for daily routines such as turning the lights on, taking a shower, having breakfast, and turning on the radio and our mobile telephones. Our energy consumption, whether we live in a house or in an apartment, depends on three factors: the dwelling itself, the appliances we use, and our behavior. It is, therefore, inherently complex, diverse and tied to our personal lives.

Only our energy consumptions and emissions they cause are discussed in this chapter. The local energy productions, mainly solar, are discussed in the related "How Electricity Is Produced" chapters, whereas the dwelling seen as an energy system is discussed in the "Where We Should Be Heading" chapters.

Statistics

Residential final energy consumption is a major component of the total final energy consumption[1]: 24 530 TWh or 21.2% worldwide in 2018.[2]

---

[1] Worldwide statistics are from [*iea.org*] while the ones concerning Europe are from [*ec. europa.eu/eurostat*]. To obtain energy consumption and emissions data for dwellings, one must often add two sets of data related to the consumption in the dwelling itself (oil, gas, wood, etc.), on one hand, and related to the electricity and district heating, as may be the case, consumed in it, on the other.

[2] In 2016, it was 25.4% in the European Union (EU).

The 2018 decomposition was:

Source/Vector	%
Coal	3.6
Oil and oil products	10.3
Natural gas	22.8
Wind, solar PV, other non-hydro renewables	1.7
Bioenergy and waste	31.9
Electricity	24.5
Network heat	5.2

The 2018 residential electricity consumption was 6 008 TWh, which represented 26.9% of the worldwide total electricity consumption.

In 2017, worldwide, the CO_2 emissions due to dwellings, either by direct combustion of fossil fuels, such as coal, oil or natural gas, or induced by electricity and heat consumption, were 5 390 $MtCO_2$, i.e., 16.4% of the overall emissions.[3]

In countries with moderate or cold climates, heating and hot water generally represent over 70% of a dwelling's energy consumption. In Europe, in 2016, heating represented 64.7% of the energy consumed in dwellings. Venting and air-conditioning consumptions are increasing in warm countries driven by climate change and the desire for enhanced comfort; the installation of facade-mounted air-conditioners is growing in emerging countries as illustrated by Fig. 1. In poor and warm countries, cooking is the highest energy consumption. End-uses specific to electricity, such as information and communication technologies, are increasing worldwide.

The basic need for drinking water at home generally represents only a small portion of the overall energy consumption, except if sea water desalination is required. Still, in 2018, some two billion inhabitants on Earth did not have suitable sanitation or access to drinking water at home; ref. "Energy and Emissions — Where We Are" chapter.

---

[3] IEA CO_2 emissions from fuel combustion, Highlights, 2019.

Figure 1. Facades with window air-conditioning systems in Hong Kong / © *Lee Yiu Tung, Shutterstock*

Design, Comfort and Energy Consumption

The comfort of a dwelling, the pleasures of living in it (or its disadvantages) and its cost (to an owner or renter) primarily depend on its location, design and type of construction. The home's thermal comfort (need for heating and/or cooling), brightness (need for lighting), and acoustical comfort directly depend on these considerations.

The reduction of our environmental footprint requires considering not only new building construction techniques but also the renovation of older buildings. Indeed, in Europe, for example, barring proactive replacement policies, a number of existing buildings will still be operational by 2050. Individual houses and apartments in buildings should be considered separately; indeed, some technologies are inherently different, such as for heating and cooling, while others are inherently similar, such as for lighting. Even though collective housing is developing with urbanization and city densification, the number of individual houses remains high. In 2018, 70% of all Europeans lived in individual houses [*ec.europa.eu/eurostat*]; the proportion was similar in the United States (US).

Insulation quality is essential for space heating and cooling. Since a long time, houses have been built taking the local climate and situation into account, such as orientation, thermal inertia, roof inclination, or positioning of windows and doors. As illustrated by Fig. 2, local materials for the roof and the walls played important roles. During intense city construction periods, for example in Europe after the Second World War, these fundamental notions were sometimes overlooked.

(a)   (b)

Figure 2.   (a) Swiss wood houses under snow / © *Swissdrone, Shutterstock,* (b) White Greek house / © *Alexandros Michailidis, Shutterstock*

Materials improvements and creation of new materials, as well as new construction and industrial processes, along with the deployment of new information and communication technologies, enable the construction of better performing buildings. Technological advancements, along with regulatory evolutions, have led to a division by two or more of the heating requirements in a number of industrialized countries since the beginning of the 20th century.

Heating

The perception of well-being is subjective. It depends on the ambient temperature and its uniformity within a room as well as on the humidity, airspeed, radiation and environment. Ideally, the temperature should be homogeneous between 18 and 22°C; in addition, to avoid the "cold wall" impression, the wall temperatures should be within a few degrees of that of the surrounding air. The hygrometry[4] should be within 40 and 60%. Cold air at the feet level and hot air at the head level can be unpleasant as may be the case with some electric heating systems. Floor heating systems can be perceived as uncomfortable; this can also be the case due to excessive hot air from forced air systems. One can feel comfortable in a cold room in front of an open fire that radiates heat or in the Sun when the temperature is –5°C. However, as soon as there is shade (i.e., when an obstacle blocks radiation), the cold sensation returns. This illustrates the comfort of infrared radiation[5] and its limits.

Heating systems rely upon: (a) radiation, i.e., heating by reception of radiation, (b) conduction, i.e., heating by contact between two solid bodies at different temperatures, and (c) convection, i.e., heating by way of contact with a fluid, generally air. In each case, the actual size of the exchange surface is very important; this is why heating elements feature their sometime strange designs.

Selecting a heating system is a long-term decision; it is tied to the construction of the dwelling and is therefore difficult to modify later. Once installed, a central heating system, with a single heat source and a hot water system distributing the heat to the various rooms, is generally left unchanged even during extensive renovations. It might be possible to replace the heat source, for example, an oil furnace with a natural gas furnace or a heat pump, or with a hybrid solar-electric system. A floor heating system using embedded pipes or a forced-air system where the heating fluid is air circulating in dedicated ducts, is difficult to replace. Similarly, once heat sources have been installed in each room, replacing them with a central heating system would be difficult and costly.

---

[4] Hygrometry is the ratio between the amount of water vapor actually contained in the air at a certain temperature and the maximum water vapor it could contain at that same temperature. It is generally given as a percentage.

[5] Infrared radiation, which is not visible, is well absorbed by water, which constitutes 70% of our bodies, hence the well-being sensation in its presence.

Several criteria can be used to characterize heating systems:

- The amount of greenhouse gases emitted yearly.
- The associated costs, installation, consumption and maintenance.
- The energy source or vector used:
  - Heat production: Biomass (wood, for example), fossil (coal, oil and gas) and electricity.
  - Heat transfer: From the ground, water or outside air.
  Dual energy systems are also implemented.
- The system's efficiency, i.e., the ratio of the energy provided to the system and the energy delivered to the house. Warm water pipes of a central heating system induce losses, especially when installed outside of the heated space.
- The system's control possibilities — for each room, remotely, manual or automatic (using monitors, for example) — are increasingly important considerations in the context of a more rational energy end-use. For example, the replacement of traditional thermostats by electronic ones in a number of domestic equipment, such as furnaces, electric heating and refrigerators, have enabled more precise temperature controls leading to energy and often emission reductions of 5 to 10%.
- The rate at which a room's temperature can be increased in cold conditions; this directly depends on the system's installed capacity and the thermal inertia of the walls.

The presentation focuses on three heating systems:

- Furnaces, which burn fossil fuels, biomass or waste. They are most commonly used even though CO_2 is emitted since combustion is involved.
- Electric.
- Heat pumps, which are increasingly deployed for both heating and cooling purposes.

Electric heating and heat pumps induce remote emissions, which depend on the CO_2 content of the electricity production mix.

Before the three systems are discussed, a method is presented to determine the energy consumption required to heat a dwelling to a comfortable temperature and the associated required installed capacity of a furnace and a heat pump as examples.

Determination of a Heating System's Yearly Energy Consumption and Installed Capacity

The three-stage process consists in assessing the impact of:

- The geographical location of the dwelling — where to heat.
- The characteristics of the dwelling to be heated — what to heat.
- The type of heating system to be implemented — how to heat.

Where to heat — the heating degree days

The impact of the local climate on the quantity of heat necessary is quantified by way of the Heating Degree Day (HDD) index.

Without heating, a dwelling's temperature decreases toward the outdoor temperature, more or less rapidly depending on the insulation. When the outdoors temperature descends below a certain level, referred to as the reference temperature, heating the dwelling is required to maintain a comfortable temperature. The reference temperature is thus the outdoors temperature below which the heating system must be turned on to maintain a comfortable inside temperature. A day is said to be cold if its average temperature is below the reference temperature.

The HDD characterizes the entire cold period, taking into account its duration, i.e., the number of cold days, and intensity, i.e., the difference between the reference temperature and the average temperature for the cold days. It is generally expressed in °C·day for an entire year; it changes from year to year.[6]

A HDD value of 1 800°C·day with a reference temperature of 18°C can correspond to:

- 100 days at 0°C, i.e., 18°C below the reference temperature, or

---

[6] In European countries and in the US, where temperatures have been recorded for decades, the HDD is decreasing, which illustrates global warming.

- 180 days at 8°C, i.e., 10°C below the reference temperature, or
- 120 days at 8°C, i.e., 10°C below the reference temperature and 30 days at −2°C, i.e., 20°C below the reference temperature, or
- More complex combinations, in reality.

Further information regarding the HDD computations can be found under "*HDD*" in the Companion Document.

Increasing the reference temperature increases the HDD and the energy required to maintain a comfortable inside temperature.

The HDD varies from 0°C·day around the equator to over 8 000°C·day close to the poles. As illustration, the average HDD values from 2013 to 2015 and using 18°C as the reference temperature are given below for a few cities.

Atlanta (US): 1 600, Bergen (Norway): 3 700, Buenos-Aires (Argentina): 770, Djoum (Cameroon): 0, Lille (France): 2 700, Beijing (China): 2 850.

The HDD values for the cities mentioned above, as well as for a large number of cities worldwide, can be found on the [*degreedays.net*] site. They provide a convenient way to compare heating requirements between locations and also to assess the impact of selecting various reference temperatures.

What to heat — the characteristics of the dwelling to be heated

Assessing the amount of heat the heating system must provide implies assessing the losses between inside and outside of the dwelling during cold days.

The losses depend on:

- The characteristics of the dwelling to be heated: Its size and shape along with its insulation, i.e., the losses through the roof and the floors, through its walls, windows and doors. Intuitively and as a first approximation, the losses can be considered as proportional to the volume of the dwelling and to a coefficient, designated as G, describing its insulation level. As further detailed in the following insert, G is measured in $W/m^3 \cdot {}^\circ C$. Renovating a dwelling to improve its insulations implies reducing its G value.

- The indoor-outdoor temperature differences and their durations as indicated by the local HDD.

The G coefficient — the dwelling's insulation level

The G coefficient characterizes the dwelling itself as well as its situation with respect to surrounding dwellings such as walls between houses, for example. A 135-m^2 house with a ceiling height of 2.7 m, i.e., a volume, V, of 364 m^3 and an insulation level such that $G = 1.2$ $W/m^3 \cdot {}^\circ C$ has heat losses of 437 W per degree temperature difference with the outdoors.

The G coefficient varies between 2.5 $W/m^3 \cdot {}^\circ C$ for a particularly poorly insulated house to 0.3 $W/m^3 \cdot {}^\circ C$ for a particularly well-insulated one. It is generally lower, between 0.1 to 0.6 $W/m^3 \cdot {}^\circ C$, for apartments between floors since the ceiling and floor losses will be lower.

How to heat — the characteristics of the heating system

The heating system's installed capacity needs to be determined such that the desired indoor temperature can be maintained even during very cold days.

This requires selecting:

- The desired indoor temperature, for example, 20°C.
- The coldest temperature for which the temperature selected above is to be maintained. One could, for example, select the average of the lowest temperatures reached during the five coldest days of the year during the past five years.

If the heat is distributed via hot water or air ducts, the related network losses must also be taken into account.

In conclusion

While simplified, the description above illustrates the complexity involved when determining the energy consumed and the system's installed capacity.

The selections of the reference temperature and the comfortable temperature to be maintained inside the dwelling play important roles, along with the lowest outdoor temperature for which it should be maintained. It is important to select a reasonable system installed capacity. If it is under dimensioned, it might occasionally be overloaded, or, alternatively, supplemental heating may be required too often. It must also be able to increase the dwelling's temperature reasonably fast when turned on. On the other hand, over-dimensioning implies additional installation costs; the system might then also often be running at low output, i.e., at lower efficiencies.

The expressions for the energy consumed by a furnace or a heat pump and their installed capacities are provided in a dedicated insert further on in the chapter. As an illustration, the computations for the three steps described above that are related to the Duponts' house can be found under "*Families 2015 Worksheets*" in the Companion Document.

Furnaces

The combustion of fossil fuels, biomass or waste occurs in a furnace. The heat from the combustion is transferred to a fluid, generally water for home furnaces. Many types of furnaces exist for houses, apartments, buildings and industry. Individual apartment furnaces are generally wall-mounted whereas furnaces for houses can be installed on the floor.

In view of their widespread use, home gas furnaces are presented below. The left hand side of Fig. 3 illustrates the operation and main components of standard gas furnaces:

- The burner, the flames of which are directed toward a pipe containing cold water returned from the heating elements or the heated floor.

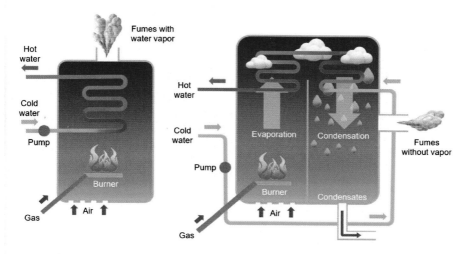

Figure 3. Gas furnaces, non-condensing and condensing / © *HBP and YB*

- The pump, which provides for the circulation of the water through the heating circuit.[7]
- The air intake and the fumes' evacuation.

The furnace efficiency is limited by heat lost in the fumes and water vapor resulting from the combustion. Condensing furnaces seek to remedy some of these losses. The right hand side of Fig. 3 illustrates the basic operation of condensing furnaces: the pipe containing cold water returning to the furnace first goes through the fumes to recuperate as much lost heat as possible before being reheated by the burner. The condensation can increase the furnace efficiency to roughly 90%,[8] thereby increasing the efficiency by 10% compared to standard furnaces while reducing the

---

[7] Since a circulation pump is required but also in view of the automatic relighting of the furnace, the heating system cannot operate without electricity. This was not the case for older systems where the circulation was gravity based (the hot water rising to the top) and the furnace lighting was manual.

[8] This is the ratio between the heat energy produced by the furnace and the heat provided by the source. In some countries, the efficiency information provided by manufacturers is the ratio between the heat produced with condensation and that without condensation; this can therefore result in a misleading efficiency above 100%!

CO_2 emissions by an equal proportion. This heat recuperation, which requires specific operating conditions, condenses the water, hence the designation, which then requires the evacuation of the acid water condensates.

Similarly to any equipment involving combustion, furnaces require proper ventilation to allow for the supply of oxygen as well as for the evacuation of fumes and condensates, as may be required. The mitigation against the risk of carbon monoxide (CO) requires that particular attention be paid to proper maintenance.

Electric Heating Systems Based on the Joule Effect

Electric heating elements were invented over 100 years ago; they rely on the heat generated by an electric current circulating through a resistance. This phenomenon is referred to as the Joule effect.[9] Electric heating elements can be the main source of heat and are then generally wall-mounted or can provide supplemental heat when they are often mobile.

The construction of electric heating elements, sometimes referred to as convectors, is straight forward and inexpensive: a metallic enclosure with an air entry wire rack at the bottom, resistances inside, often with fins to augment the contact with air, and an air exit wire rack at the top. As the temperature of the resistances increases, by the Joule effect, the air warms up; as its density decreases, it rises to the exit wire rack while colder air enters the bottom one. While the convector radiates to some extent, the heat is primarily transferred to the surrounding air.

Convectors rapidly heat the air but not directly the occupants, and neither the walls nor the objects in the room. The quick heating and low investment cost are advantages. The more significant stratification of the air, when compared to central heating elements, is a disadvantage. In addition, if the dwelling insulation is poor and if extensive heating is required during prolonged periods, the electricity consumption can become quite high. On the other hand, if the house is properly insulated and/or is only

---

[9] Named after the British physicist **James Prescott Joule** (1818–1889).

occupied sporadically and thus requiring lower amounts of heat, electric heating elements can be a cost-attractive solution.

Some thermal inertia can be added by the inclusion of refractory plates in the convector itself so that some heat can be stored and given back after the electric supply has been shut off — these are energy storage heating elements.

Recently available materials and new manufacturing technologies, for example thin-film deposition, have enabled the development of improved electric heating systems, which can transform an increasing amount of the heat obtained using the Joule effect into radiation, thus reducing the stratification. These elements are referred to as radiating panels; modern panels can valorize more than 50% of the electric energy into radiation.

Gases, Liquids and Solids — Phase Changes

Phase shifts are the core concept underlying heat pumps, which are further described next. They are also crucial in the heat transfer context of thermal power plants — fossil or nuclear — which are discussed in the relevant "How Electricity Is Produced" chapters.

The three usual phases of a body are: gas, liquid and solid. We are familiar with water — vapor, water and snow/ice. In simplified terms: molecules in a gas are distant from each other and mobile, while molecules in a liquid are closer and less mobile; in a solid, the molecules are almost stationary at their equilibrium state. Depending on its temperature and pressure, a body is in one of the three phases.

When heating a liquid, its temperature increases so does the kinetic energy of its molecules; they become increasingly mobile until they separate from each other such that the liquid becomes a gas. This happens at the vaporization temperature, i.e., 100°C for water when we boil it. During vaporization, the heat provided is used for the vaporization itself; the liquid proportion of the liquid-gas mixture decreases as the gas proportion increases while the mixture's temperature remains constant. When the water has been fully vaporized, the temperature increases again provided heating is maintained. The phenomenon is the same for the other phase shifts during overall temperature increases or decreases: the temperature remains constant during the phase shifts themselves.

Furthermore, as the pressure of a liquid increases, more heat is required to free the molecules. As a result, a liquid's vaporization temperature increases with pressure. Conversely, with higher gas temperatures, the molecules have a higher tendency to separate from each other; as a result, higher pressures are required to compress a gas into a liquid.

The heat required during phase shifts is generally far higher than that required to increase a material's temperature by a few degrees. Intuitively, we know that vaporizing water requires a significant amount of heat: evaporating perspiration or a wet garment, which cools when we wear it, results in a cooling sensation.

Further clarifications regarding phase shifts can be found under "*Phase Shifts*" in the Companion Document.

Heat Pumps

We all commonly use one type of heat pumps: the refrigerator, which "pumps" heat from its inside to cool it, toward the exterior, generally toward the kitchen where the temperature therefore slightly increases.

In nature, heat goes from a warm body toward a cold one. On the contrary, heat pumps transfer heat from where it is cold toward where it is warm:

- During winter, they extract heat from the outside (cold), transferring it inside (warm). This is the heating heat pump application, generally referred to as the heat pump.
- During summer, they extract heat from the inside (cold), transferring it outside (warm). This is the air-conditioning heat pump application.

Heat pumps contain a heat transfer liquid, or coolant,[10] which goes through two phase shifts before coming back to its initial state and starting

---

[10]The heat transfer fluid is also occasionally referred to as refrigerant.

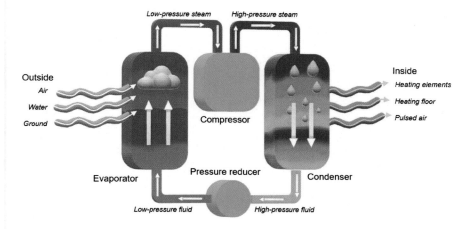

Figure 4.   Heat pump operating principle / © *HBP and YB*

the cycle again. Figure 4 illustrates the four steps of a heating heat pump cycle:

- During Step 1, the fluid, which is initially cold and liquid, receives heat from the outside.[11] Its temperature increases, and it vaporizes.
- During Step 2, the fluid is compressed in the compressor. Its pressure increases; its temperature further increases.[12] This is the only step that requires energy, typically in the form of electricity for the compressor.
- During Step 3, the fluid transfers its heat inside to the heating network; it becomes liquid again.
- During Step 4, the fluid's pressure is decreased in the expander, such that its pressure and temperature return to their initial values.

The key to a proper heat pump operation is the selection of the coolant, which must go through gas-liquid and liquid-gas phase shifts at the

---

[11] Even at 0°C, the air outside contains lots of heat since it is at 273 K (ref. "Units and Reference Values" annex)!

[12] When air is rapidly compressed, its temperature increases, as illustrated by bicycle pumps, for example.

proper temperature-pressure combinations. The following insert provides a few additional clarifications with regard to heat pump operations.

Heat pumps demystified

The heat transfer fluid is selected such that it vaporizes at the outside temperature, where it is cold, and such that it liquefies at the inside temperature, where it is warm. Since the vaporization temperature increases with pressure, the fluid must thus be at low pressure when it is outside and high pressure when it is inside. The fluid must thus be compressed when it transits from the outdoors to indoors.

As illustrated by Fig. 4, the heat pump fluid goes through a cycle in four steps:

Step 1: The heat transfer fluid enters the evaporator, which is located outdoors, as a low pressure and low temperature liquid. The fluid and its pressure are selected such that it vaporizes at the outside temperature. The vaporization heat is thus extracted from the outside. The fluid exits the evaporator as a gas and at a higher temperature than when it entered it.

Step 2: The compressor, or pump, hence the name "heat pump", compresses the gas, which further increases its temperature. Once compressed, the fluid enters the inside at a higher pressure and temperature than when it entered the compressor. The temperature is selected to be higher than that of the heating system inside.

Step 3: Indoors, the heat transfer fluid transfers its heat either to water or air, depending on the dwelling's heating system. Having been carefully selected to do so, the fluid becomes liquid again as it cools down and it condenses under these temperature and pressure conditions. Hence, the designation condenser for this type of heat exchanger.

Step 4: Finally, the fluid flows through a pressure reducer, designated as an expander, to reduce its pressure to reach a pressure and temperature combination such that it vaporizes outside.

The key advantage of a heat pump is that the heat energy it provides inside is greater than the energy consumed by the compressor. This ratio, referred to as the Coefficient of Performance (COP), is regularly higher than 3. Stating that the COP is equal to 3, for example, means that for 1 kWh of electricity consumed by the compressor one obtains 3 kWh of heat, of which 2 kWh is obtained from the outside. The 2 kWh from the outside is generally considered as renewable; however, in reality, a very slight decrease of the outside temperature results. Losses due to the fluid's viscosity and mechanical friction are unavoidable. However, the COP is always above 1 under normal operational conditions. Further clarifications can be found under *"Thermodynamic Cycles"* in the Companion Document.

The COP depends on both the inside and outside temperatures; it is higher when the temperature difference is lower. In addition, when the outside temperature decreases, it becomes more difficult to recuperate heat and the COP decreases, which is clearly contrary to the desired outcome!

A wide variety of heat pumps exist; they are categorized by:

- The fluid/medium from which the heat is extracted: air, water (aquifers) or ground, using underground horizontal or vertical pipes;[13] ref. Fig. 5.
- The fluid/medium to which the heat is provided: air (pulsed air system) or water (heated floor requiring water at some 35°C or a central heating system using heating elements requiring water between 45 and 70°C), depending on the age of the system and its capacity.

Heat pumps are thus referred to as air-air, or air-water, etc.; the first word designating the source, and the second one, the arrival fluid. Each type of heat pump has its advantages and disadvantages. The aquifer water temperature and that of the ground are generally higher than the outdoor air temperature during winters, and more importantly, are generally more stable, which are advantages. However, one does not always have convenient access to aquifers or the ground, especially in an apartment.

The performances of heat pumps have improved considerably during the last 50 years: higher COPs, lower costs, less noise and more compact packaging. New heat transfer fluids and compressor speeds electronically

---

[13] One generally does not drill beyond 100 m. Indeed, the purpose is not to capture the Earth's heat very deep underground but rather to provide a supply at the temperature of the ground, which is rather stable, i.e., close to the yearly average; warmer than the outside air in the winter and cooler during the summer.

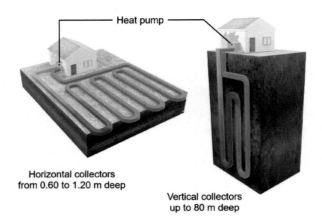

Horizontal collectors
from 0.60 to 1.20 m deep

Vertical collectors
up to 80 m deep

Figure 5.   Two examples of geothermal heat pumps / © *HBP and YB*

adjusted to the temperature differences, also contribute to performance enhancements.

Practical deployment considerations

Heat pumps can have ratings from a few kW to over 10 kW for individual houses. They can also have ratings of up to several 100 kW for apartments or commercial buildings, as well as factories. Recently, specially designed air-water heat pumps have been introduced for the renovation of existing central heating systems; they fulfill the need for 60°C water temperatures, as often required for heating system renovations.

One should differentiate between the COP indicated by manufacturers, which corresponds to standard test conditions, often 4 or 5 (or even higher), from the yearly average COP after installation. The average yearly COP depends on the installation location and varies from year to year. As for furnaces, a heat pump's installed capacity must be selected such that the inside comfort temperature can be maintained during the coldest days, i.e., when the COP is the lowest.

While heat pumps do not themselves emit any CO_2, the heat transfer fluid can be a greenhouse gas, including CO_2, and, as a result, has an environmental impact in case of leaks. International efforts have resulted in increased use of less harmful refrigerants.

Selected Heating System Energy and Emissions Comparisons

The installation of a heat pump rather than a furnace in a new dwelling or replacing an oil or gas furnace by a heat pump during a renovation, increases the electricity consumption but decreases the final energy consumptions by ratios that depend on the heat pump's average COP.

Figure 6 provides comparative information based on the carbon intensity of the electricity production mix. If the electricity production mix has a low carbon content, specifically below 200 gCO_2/kWh during the heating periods, then electric heating systems have lower emissions than gas furnace systems.[14] The same is true for heat pump systems with an average COP of 3 over a year's operation, provided the electricity production mix emissions are below 600 gCO_2/kWh, which is the case in the majority of countries. Biomass solutions are also attractive, provided transportation to

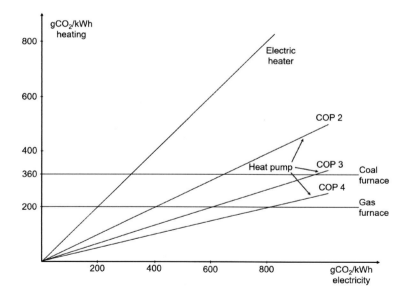

Figure 6. Heating systems CO_2 emissions comparison / © *HBP and YB*

---

[14]To simplify the computations, the gas heating system is assumed to be 100% efficient.

the consumption location is not overly CO_2 intensive; its combustion always produces CO_2.

Details regarding the computation of the energy consumed and the required installed capacities for a furnace and a heat pump are provided in the following insert. Actual computations for the Duponts' house are provided under "*Families 2015 Worksheet*" in the Companion Document.

Expressions for the energy consumed by a furnace or a heat pump and their installed capacities

The observations from above can be transcribed as follows for the energy, Q, consumed by a furnace:

$$Q = G \cdot V \cdot HDD \cdot 24 \cdot 1/\eta_f \cdot 1/\eta_s$$

where:

Q is the energy consumed, in Wh.

G is the coefficient characterizing the dwelling's insulation, in $W/m^3 \cdot °C$.

V is the volume of the dwelling to be heated, in m^3.

Multiplying the HDD by 24 is required to convert it from °C·day to °C·hour.

η_f is the furnace's efficiency, i.e., a number below 1. The consumption increases as η_f decreases.

η_s is the heating system efficiency, i.e., the impact of the heat losses. It is also below 1.

The furnace installed capacity required to overcome the difference between the coldest selected temperature, T_c, and the desired indoor temperature, T_i, can be expressed as:

$$P = G \cdot V \cdot (T_i - T_c) \cdot 1/\eta_f \cdot 1/\eta_s$$

where:

P is the furnace installed capacity, in W.

T_i is the desired indoor temperature, in °C.

T_c is the coldest exterior temperature retained, in °C.

For a heat pump, the furnace efficiency, η_f, is replaced by the heat pump's COP.

• In the expression related to the consumption, the average COP over an entire year should be used, typically 3 or more.

(Continued)

(Continued)

- In the expression related to the installed capacity, the COP corresponding to the exterior and interior temperatures retained for the computation should be used, typically 2 or more; it is lower than the one retained for the consumption computation due to the lower temperature involved.

The furnace's efficiency, η_f, is of the order of 90%, such that both the consumption and installed capacity of a heat pump are always lower than for a furnace; the first one by a factor of three (or more) and the second one by a factor of two (or more).

Residential Building and District Heating Systems

Residential building heating systems also feature furnaces or heat pumps, appropriately dimensioned. These larger furnaces and heat pumps can take advantage of higher performance technologies such as multi-stage heat exchangers. District heating systems for an urban neighborhood or a city can use furnaces burning waste, or valorizing waste heat from an industrial plant or heat from electricity-producing co-generation systems, for example. Improved duct insulations allow for the transport of heat over longer distances with lower losses, typically less than 1% per km.

Air-Conditioning and Ventilation

Air-Conditioning

Worldwide, evolving lifestyles combined with climate change and decreasing costs induce increased air-conditioning installations.

The approach used to assess heating needs is also used to assess the cooling energy consumption. Having selected a reference temperature, the days when the average outdoors temperature is above it are said to be warm. The Cooling Degree Days (CDD) index is then used, which characterizes the entire warm period during a year, taking into account the number of warm days and their intensity, i.e., the difference between the reference temperature and the average temperature during warm days. As for the HDD, it is also expressed in °C·days. Below, the CDDs and HDDs are provided for a few cities; the reference temperature is 18°C in both cases.

City	Atlanta	Beijing	Bergen	Buenos Aires	Djoum	Lille
CDD	1 020	1 100	20	960	2 220	190
HDD	1 600	2 850	3 700	770	0	2 700

These values illustrate the need for heating in some locations, cooling in others, and elsewhere both.

The aim then is to "pump" heat out of the dwelling's interior toward the outside, i.e., the reverse of heating. The most commonly used system is the air-air heat pump, also referred to as an air-conditioner, which uses a properly selected heat transfer fluid, referred to as the coolant in this case.

A split system is commonly used; it features two subsystems, one located outside, ref. Fig. 1, and the other located inside the dwelling — both are connected by conduits for the coolant to circulate. Other systems, called multi-splits, feature a single subsystem outside which is connected to interior subsystems in each air-conditioned space.

The system's performance is characterized by the Energy Efficiency Rating (EER), which is equivalent to the COP for heating systems. As for heating systems, the installed capacity depends on the desired temperature difference between the interior and exterior.

Reversible heat pumps are also being developed, which can both heat and cool dwellings. The system is then designed to provide for the reversal of the fluid flow cycle. When heating, the heat transfer fluid evaporates outside before reaching the compressor, and when cooling it evaporates inside before reaching the compressor. Generally, the efficiency is higher for the main utilization, heating or cooling, as it may be the case.

Ventilation

Ensuring proper and regular air replacement in a dwelling is important for health reasons. Thus, the importance of proper ventilation.

Windows and doors are frequently not fully airtight; along with some air intakes and occasionally a chimney, they ensure the renewal of fresh air, elimination of humidity and the evacuation of fumes. This natural ventilation method depends on weather conditions and might induce significant heat losses.

High-performance insulation requires proper sealing, which in turn, requires forced ventilation to evacuate stale air and draw fresh air in. In the most basic systems, stale air is evacuated by way of air vents and fans; fresh air arrives through vents above the windows. When it is cold outside, heat losses occur when warm stale air is evacuated and the air entry vents are cold. To remedy this situation, using improved systems, the stale and fresh air exit and enter the space by way of separate ducts, and a heat exchanger allows for the partial recuperation of the stale air heat. However, the improved systems require fans that are two to three times more powerful than those in more basic systems. The installed capacity of the fans is of the order of a few tens of watts for a 100-m^2 dwelling. If the heating and cooling systems use pulsed air, they also provide for the ventilation requirements.

Hot Water Systems

In countries where running water is available, a person uses between 100 and 250 liters of water per day, one-third of which is for hot water. The energy required to treat, pump and circulate the water can be estimated at 2 kWh/m^3; should desalination also be required, the energy can be estimated at 6 kWh/m^3; ref. "Industry and Agriculture" chapter. Either consumption is lower than the energy required for hot water production.

For a family of four persons, each taking a daily shower, roughly 400 liters of hot water at 40°C is required daily. A few orders of magnitude regarding the 40°C hot water consumptions and flow rates are provided:

- Washing of hands: 2–5 (7 l/min)
- Dish washing: 40 l (9 l/min)
- Shower: 60 l (3 l/min)
- Bath: 150 l (20 l/min)

Selecting a hot water heating system requires addressing the following questions:

- *Instant heating or heating including storage?*
  Instantaneous hot water heaters produce hot water as the faucet is opened. Aside from the water temperature increase (very fast for modern heaters), sufficient power must be available to ensure the required hot water flow at the desired temperature. Hot water systems integrating some storage capacity can heat the water over longer durations; as a result, the installed capacity requirements are lower. Furthermore, if it is electric, heating may be restricted to periods when the electricity required is inexpensive. On the other hand, hot water storage tanks require space, and even though the overall efficiency is better than for instantaneous heaters, losses are associated with the storage tanks depending on their insulation and where they are installed.[15] Additional losses occur in the pipes transporting the hot water.[16]

- *Natural gas or electric or solar?*
  Natural gas systems rely on combustion; the principle is the same as for furnaces. Electrical systems can either use the Joule effect, i.e., a resistance heating the water, or heat pumps. One can readily achieve a yearly average COP of 2.5; the electricity consumption is therefore lower than that of a hot water heater using the Joule effect. However, the latter is easier and less expensive to install.
  Solar hot water heaters, as shown in Fig. 7, use solar energy collectors/panels to heat a heat transfer fluid, which heats the water by circulating in a heat exchanging pipe inside the hot water storage tank. The solar panel is transparent on its external face and black on the other, to absorb as much heat as possible. Supplemental heating may be required

---

[15] For 200-liter hot water storage systems manufactured after 2000 and installed where the temperature is 20°C, the losses are below 15% per 24 hours even if the water is not used. In reality, the losses are lower and depend on the daily hot water usages.

[16] The losses are due to the transport of the hot water and the fact that after the faucet is closed, warm water stays in the pipe and slowly cools down. These pipes are often poorly insulated or not at all. When the dwelling has several hot water distribution points, an attractive solution may be to use several heaters.

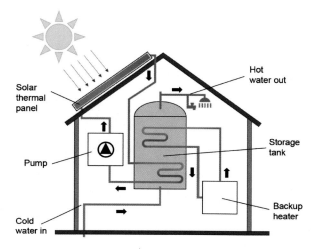

Figure 7. Indirect solar water heater with heating backup / © *HBP and YB*

during low solar energy availability; it can be provided either by the furnace or using an electric resistance. If the climate is sunny, natural convection can provide for the circulation of the primary fluid, but the hot water storage tank must be placed above the panel. Some systems can heat the water directly, as shown in Fig. 8. This robust, emissions-free and economic solution[17] is increasingly used worldwide.

- *Separate hot water system or integrated with the central heating system?*
  When using a single furnace for the central heating and hot water, the most common solution is a natural gas or oil furnace, which heats the water of both circuits. To ensure that the hot water heats up sufficiently fast, the furnace installed capacity is often dictated by the hot water requirement.[18] One can also use a hot water storage reservoir to provide

---

[17] Efficiencies of 60 to 70% can be reached, i.e., higher than what could be reached using electricity produced from a photovoltaic panel of the same size as the thermal panel.

[18] As an example, one can compare the power requirements for heating and hot water in the Duponts' house. The predominant role of the hot water power requirement increases as the insulation of the house improves.

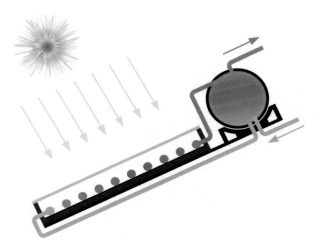

Figure 8.   Direct solar water heater / © *Robert Paul Laschon, Shutterstock*

for faster hot water availability. For such installations, the problem is the efficiency during periods when space heating is not required, which might fall below 80% since many "stop and go"s will be required.

- *One central hot water heater or distributed?*
  In buildings, similarly to heating systems, individual hot water heaters or a central system with a common hot water storage reservoir can be used. Using solar heat, a number of thermal panels can be installed on the roof supplying a central hot water storage reservoir connected, in turn, to a distribution network. Alternatively, the solar thermal panels can be directly connected to the distribution network, which in turn, supplies individual storage reservoirs.

To conclude the discussion regarding hot water systems, it is important to note that:

- From a sanitation point of view, the water must be stored at a sufficiently high temperature to avoid the development of dangerous bacteria while also avoiding burning users.

- From an emissions point of view, the best solution is solar thermal, followed by solar thermal combined with low-carbon electricity. Neglecting the system losses, in either case, natural gas hot water heaters emit more CO_2 than electric solutions, provided the emissions of the electricity production mix are below the 200 gCO_2/kWh due to the combustion of natural gas; for heat pumps, this value is to be multiplied by their COP.

Other End-uses

Cooking

The main cooking performance criteria are the flexibility of the temperature settings, the rate of temperature increase and decay along with the installed capacity (to limit the cooking duration). Safety during and after the cooking itself is also important. These criteria, along with the cooking method (with or without a lid, etc.) influence the amount of energy required. A cooktop is typically rated between 700 and 3 000 W and a cooking range with four elements/burners can reach 10 kW.

Natural gas cooking ranges and cooktops generally fulfill the requirements listed above but induce CO_2 emissions, except if biogas is used.

Electric cast iron heating elements heated using the Joule effect by way of resistances still dominate. The cast iron's thermal inertia does not provide for fast temperature increases and decays. In addition, the elements remain hot after they have been shot off thereby inducing safety hazards. The more recent induction cooktops fulfill the criteria mentioned above.[19] They consume roughly half as much energy as gas or electric cast iron cooktops.

---

[19] Induction cooktops have wound coils covered by a transparent ceramic glass. The currents circulating in the coils induce eddy currents in the pan, which acts as the magnetic core; this requires that the pan be made of a magnetic material. In turn, the eddy currents in the pan heat up its bottom by the Joule effect. The thermal inertia is quite low such that this is presently the fastest heating method; it fulfills the usability criteria mentioned above quite well. In addition, the heating element does not remain hot after the cooking. The

---

CO₂ emissions: Natural gas versus induction

Since the combustion of natural gas yielding 1 kWh of heat emits 200 gCO_2, (ref. "Units and Reference Values" annex), an induction cooktop, consuming 50% of a gas cooktop, produces less CO_2 than a natural gas one as long as the electricity production mix is such that 1 kWh corresponds to less than:

$$200 \text{ gCO}_2/\text{kWh} \cdot (100/50) = 400 \text{ gCO}_2/\text{kWh}.$$

Other food cooking methods exist. For example, microwave[20] ovens heat or cook food by way of electromagnetic waves. One can also use water steam, produced using natural gas or electricity, taking advantage of the fact that the heat exchange between food and steam is better than using air. Finally, some appliances are specially designed for particular applications such as rice cookers and fryers.

When discussing these technologies, it is important to remember that two billion people only have access to biomass, in particular wood, to cook their meals. Some still use open fireplaces, which are dangerous and inefficient. Improved cooking appliances are gradually deployed featuring better heat containments to allow for better combustion, thus reducing energy consumption, and more importantly, increasing safety.

Lighting

The overarching objective is to obtain lighting of sufficient intensity, with a luminescent spectrum as close as possible to that of daylight and at an acceptable cost while consuming as little electricity as possible. Other objectives, such as long lifetimes and fast turn-on (for high transit locations), are also important.

---

lower energy consumption is mainly due to the fact that only the pan itself is heated irrespective of the heating element's diameter.

[20] Its wavelength is between that of infrared and radio waves. Microwaves excite and heat the food's water, fat and sugar molecules at a frequency of 2.5 GHz. The rest of the food is heated by way of conduction from the part that is heated by the microwaves.

Several technologies have been developed over time, including the first (which is still most widely used) developed technology and another one, which is increasingly displacing it:

- *Incandescent light bulbs* in which the Joule effect is used to heat a tungsten element at up to 2 500°C, which in turn, is placed in an inert gas (argon or krypton) to prevent it from burning. At this temperature, the element emits a continuous light spectrum close to that of daylight; it also emits significant infrared radiation such that the light bulb temperature reaches 150°C and 95% of the energy is transformed into heat, which could be useful in winter while not so much during summers!
- *Light Emitting Diode (LED)* lamps rely on the reverse solar photovoltaic phenomenon by transforming electricity into light radiation. Their performance is steadily improving: they consume six times less energy compared to an incandescent bulb and lasts up to ten times longer while having comparable turn-on delays. Materials are available for all colors, such that combinations achieve light spectra comparable to natural light. One can also affect the color by changing the LED enclosure.

It remains, however, true that hundreds of millions of people, especially in Africa and India, without access to electricity still use paraffin, candles or kerosene lamps for lighting. The lower consumption of LEDs, coupled with battery and photovoltaic solar panels, provides for the broader distribution of solar kits to less-favored populations.[21] The recharging of mobile telephones is generally also possible using such kits.

Appliances, Multimedia and Communications Devices

The number of electrical applications keeps growing; three emblematic ones are further briefly discussed:

- Refrigerators overwhelmingly are air-air heat pumps that extract heat from their inside toward the room in which they are located. Their

---

[21] With the support of governments and/or NGOs, if necessary.

functionalities have increased: From the simple refrigerator, it first became possible to keep frozen food and then to deep-freeze foods. Their noise levels have also been reduced. At the same time, the energy consumption has been reduced to less than half between 1990 and 2010 for the same size and functionalities; this is important given their 24-hour and 7/7 operational use.

The search for better insulation[22] and compressor performance improvements is ongoing. Modern combined refrigerator-freezers with volumes of 300 liters/100 liters consume some 160 kWh/yr in a 20°C room, i.e., similar to a 20 W light bulb that is lit all year round.

- Televisions have also gone through drastic evolutions since the cathode ray black-and-white sets of the 1950s that had a limited number of pixels. Today, flat-screen displays provide image and color qualities that have nothing in common with those from the 1990s. However, the 100 W of the cathode ray tube has given way to 100 to 250 W liquid crystal 102 cm (40 in) displays or to 250 to 350 W plasma displays of the same size.[23]

- Computers, mobile telephones and their technology ecosystems have been in constant evolution during the last 25 years, both from a performance point of view and regarding their worldwide deployment. Their typical energy consumption is generally provided in most countries; it heavily depends on the type of equipment and its usage. A personal computer (PC) typically requires 300 W when used for a video game but only 60 W when used in an office environment. Presently, a PC that utilizes its loudspeakers for eight hours each day consumes some 500 kWh/yr; typically it is between 100 and 200 kWh/yr for a laptop.

Most televisions and computers incorporate LEDs that indicate their operational status. They consume energy even when in standby mode.

---

[22] The thermal inertia thus made possible is important in conjunction with food safety in case of voluntary or accidental power supply interruptions, from a few minutes to several hours.

[23] Consumption comparisons can vary depending on the types of usage emulated during the tests.

Taken together, these standby consumptions can add up to several tens of watts for a four-person household. Additional consumptions are due to the batteries of mobile telephones and laptops.

Additional information regarding the energy consumption of these devices is provided under *"Families 2015 Worksheets"* in the Companion Document. Similarly, additional information concerning the energy consumption of the "digital system" — Internet, computer and data centers, and their terminals — can be found in the "Industry and Agriculture" chapter.

Transport and Travel

Daily Commutes, Travels and Transport

Up until the beginning of the 20th century, the daily activities and chores of most women and men generally took place close to home; transport was not an integral part of life. More recently, public transport and private cars have made daily and often long commutes between one's home and workplace possible, or even a necessity, especially when both partners in a household are employed. While explorers, adventurers, craftsmen and soldiers have been traveling to faraway places for centuries, only since the end of the Second World War has long-distance travel become accessible to the public-at-large. Participation in international meetings, conferences, exhibitions and forums induce travels for professional reasons even though electronic meetings are increasingly implemented as the capabilities of the underlying technologies advance. Tourism to faraway locations, sustained by high-speed trains, mammoth cruise ships and affordable air travel, is a fast-growing industry worldwide. Enhanced land, sea and air goods transportation infrastructures and technologies have enabled the broad development of international trade and economic cooperation. Multimodal (land, air and sea) transportation hubs are driving the economic development of entire regions, not only addressing the needs of local industries and populations but also for transfers of goods between transportation modes.

Modern transport propulsion systems primarily use combustion engines burning carbonated fuels — fossil, biofuels or synthetic fuels — or electric propulsion systems, or various combinations of both, referred to as hybrid systems. Combustion engines can be categorized as:

- *External combustion*, where combustion is used to heat a working fluid, in general water or steam. The steam engine is the most common example of an external combustion engine.
- *Internal combustion*, where the combustion occurs within the working fluid itself. Modern cars, trucks, ships as well as most propeller aircraft use internal combustion engines (ICEs).

Or as,

- *Intermittent combustion*, where the combustion only occurs at carefully spaced instances of the propulsion cycle. Gasoline and diesel engines are intermittent combustion engines.
- *Continuous combustion*, where the combustion is on-going all of the time. Modern transport aircraft use continuous internal combustion turbines, commonly referred to as "jet engines".

Combustion occurs between a fuel and an oxidizer. For most transport and travel applications, the oxygen contained in air is used as the oxidizer such that, broadly speaking, only the fuel has to be transported on board the vehicle itself.[1]

While the available sun-exposed areas on vehicles are generally not sufficient to propel them using only solar energy, some auxiliary services, such as navigational systems and/or lighting, may be supplied using solar energy. While sail passenger and cargo ships are no longer used, the use of wind energy aboard ships is again envisioned.

In this chapter, only civilian transport systems are discussed while focusing on the energy and environmental impacts of the various technologies and applications. Technologies specific to military and/or space applications are not addressed.

---

[1] On the contrary, most underwater and outer space applications require that both the fuel and its oxidizer be carried on board.

Statistics

Transport activities are related either to:

- Passenger transport, measured in passenger·km; i.e., the number of passengers multiplied by the distance transported in kilometers; one passenger·km corresponds to one passenger transported for one kilometer.

or

- Freight transport, measured in ton·km; i.e., the number of tons multiplied by the distance transported in kilometers; one ton·km corresponds to one ton of goods transported one kilometer.

The 2015 statistics, summarized in the following two tables, were jointly developed by the International Union of Railways [*uic.org*] and the International Energy Agency (IEA) [*iea.org*].[2]

	passenger·km (%)	**ton·km (%)**
Road	79.6	20.2
Rail	6.7	6.9
Navigation	0	72.1
Air	13.7	0.8

	Energy (%)	**CO_2 (%)**
Road	75.8	72.7
Rail	1.9	4.2
Navigation	9.5	10.2
Air	10.7	10.8
Other, primarily pipelines	2.1	2.1

---

[2] Railway Handbook, 2017; the most recent edition when the book was written.

In 2018, the worldwide total final energy consumption for transport worldwide was 33 621 TWh, i.e., 29.1% of the worldwide final energy consumption [*iea.org*].[3] The 2018 decomposition was:

Source/Vector	%
Coal	—
Oil and oil products	91.6
Natural gas	4.1
Wind, solar PV, other non-hydro renewables	—
Bioenergy and waste	3.1
Electricity	1.2
Network heat	—

The 2018 electricity consumption of the transport sector was 390 TWh, which represented 1.7% of the worldwide electricity consumption.

In 2017, according to the IEA,[4] the transport sector was responsible for emissions of 8 040 $MtCO_2$, or 24.5% of worldwide emissions. The emissions due to road traffic was 5 958 $MtCO_2$, i.e., 74.1% of all transport emissions.

Commonly Used Transportation System Performance Indicators

The overall performance of transportation systems can be assessed in terms of:

- Speed, or velocity, measured in m/s, or more commonly in km/h or miles/h.
- Acceleration, i.e., how quickly the cruising speed can be reached, measured in m/s^2.

---

[3] Key world energy statistics, 2020.
[4] CO_2 emissions from fuel combustions, Highlights, 2019.

- Fuel consumption, typically in liters/100 km or miles/gallon, primarily in the United States (US).
- Emission, typically measured in gCO_2/km. As will be seen later, other pollutants are also monitored.

For road and train transportation, i.e., vehicles on wheels, torque, as illustrated in Fig. 1, is applied directly to the axles of one or more wheels. If the torque is superior to the vehicle's rolling resistance against the ground and drag through air, the vehicle accelerates; otherwise, it slows down, sometimes helped by brakes.

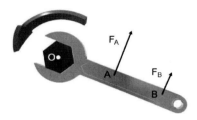

Figure 1.   Wrench illustrating the principle of torque / © *HBP and YB*

For airplanes, as illustrated in Fig. 2, thrust, which is a force, is applied to move them forward and also to keep them airborne. The horizontal thrust provided by the engines overcomes drag while the wing lift overcomes gravity.

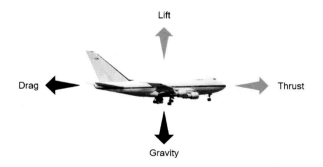

Figure 2.   Aircraft thrust, drag, lift and drag / © *HBP and YB*

In 1687, the British philosopher, mathematician, physicist and astronomer **Sir Isaac Newton** published the three laws of motion, which are at the foundation of classical mechanics and thus of the various transportation systems to be discussed next. Further details on the three laws can be found under *"Elements of Classical Mechanics"* in the Companion Document.

Four-cycle Internal Combustion Engine (ICE)

Presently, the most commonly used road vehicle engine is the four-cycle, also referred to as the "four-stroke" engine, that consumes gasoline or diesel fuel.

As illustrated in Fig. 3, the main components of a four-cycle engine are:

- Pistons that travel linearly inside the cylinders and close the combustion chamber. The piston seals limit the leakage of gases from the combustion chambers to the crankshaft space.

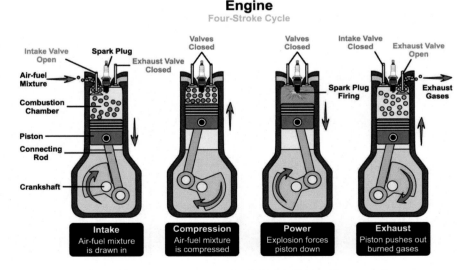

Figure 3.   Four strokes gasoline ICE / © *Udaix, Shutterstock*

- Connecting rods that mechanically connect the pistons to the crankshaft.
- The crankshaft that converts the linear piston motion into the rotational motion which is transferred to the wheels to propel the vehicle.
- Intake and outlet valves.

For gasoline engines, spark plugs ignite the combustion mixture. As further explained below, for diesel engines, the fuel mixture spontaneously ignites such that spark plugs are not necessary.

The four stages of a gasoline internal combustion engine

Nikolaus Otto (1832–1891), a German inventor, developed the first operational four-cycle internal combustion engine. The Otto cycle, illustrated in Fig. 3 for a gasoline engine, goes through the following stages:

- **Intake**: The intake valve is open to inject a mixture of fuel and air into the combustion chamber; direct fuel injectors ensure precise fuel mixtures and injection timing. Driven by inertia, the piston moves away from the valves and the combustion chamber volume increases, filling it with the fuel mixture.
- **Compression**: When the piston reaches its lowest point, the intake valve closes and the piston moves upward, again driven by inertia. The pressure in the combustion chamber increases along with the temperature of the fuel mixture.
- **Power**: When the piston again reaches its highest point, with both valves still closed and the pressure and temperature inside the compression chamber having reached their maximum levels, the spark plug ignites the fuel mixture. The combustion gas' volume rapidly increases driving the piston downward.
- **Exhaust**: At the end of the power phase, when the piston again reaches its lowest point, the exhaust valve opens, allowing the combustion gases to leave the combustion chamber as the piston moves upward, driven by inertia.

Aside from transferring the linear piston motion into a rotational motion, the crankshaft provides for the proper time spacing of the four phases in multi-cylinder engines; it also provides the inertia required to have each piston complete the three "inactive" phases between each power phase.

The theoretical efficiency of a gasoline engine with a 10:1 compression ratio,[5] which is typical for modern ICE gasoline engines, is 60%. However, in reality it is only in the 30% range given the numerous losses, including the non-perfect seals between the moving pistons and the cylinders, as well as the mechanical frictions in the engine itself.

Internal combustion engines are not self-starting. An external torque must be applied to the crankshaft to start the piston movement as fuel is supplied. At the advent of ICEs, this was most often done manually by "cranking" the engine before the ICE "kicks in". Modern engines use an electric motor to provide the starting torque, hence the need for an electric battery in any ICE vehicle.

Diesel ICE

In 1892, **Rudolph Diesel**, a German engineer, was granted a patent for the ICE, which was eventually named after him. The Diesel cycle is similar to the Otto cycle except that heated air is admitted into the combustion chamber during the intake phase and further heated by the high compression. As the air compression reaches its highest value, fuel is injected under high pressure into the combustion chamber, thus setting up a spontaneous ignition. A special fuel, often called diesel fuel, is required to ensure the spontaneous combustion at the higher pressure. The compression ratio of diesel engines is generally in the 14:1 to 16:1 range. Since it operates at higher temperatures, its efficiency is generally better than for gasoline engines: 45% as opposed to 30% for a gasoline ICE.

Road Vehicles

Road vehicles include personal cars and buses as well as delivery vehicles and trucks. Along with various forms of public transport, the automobile is undoubtedly one of the key transformational technologies of the 20th century.

---

[5] Ratio between the combustion chamber volumes when the piston is either all the way down or all the way up.

Worldwide Vehicle Statistics

The website of the Organisation Internationale des Constructeurs d'Automobiles (OICA) [*oica.net*] indicates that in 2015, over 947 million private cars and 335 million commercial vehicles were in circulation worldwide. In total, over 68 million private cars and 22 million commercial vehicles were manufactured in 2015, with China being the largest producer with 24 million total vehicles, followed by the United States (US) with 12 million.

Personal Cars

As of 2018, the ICE remains the main engine of choice for personal cars. For many years, the key customer car selection criteria were style and performance. More recently, along with safety features that are now taken for granted based on increasingly stringent regulations, fuel consumption and emission levels play increasingly important roles.

Official fuel consumption information, which is increasingly used as a sales argument, is provided using what is referred to as "administrative" test circuits designed to mimic highway or city driving, or a combination of both. These circuits are generally different from one country to another, making comparisons difficult. The regulations associated with the official qualification tests allow the manufacturers to specially prepare the vehicles to be tested. As a result, the actual fuel consumption and emissions achieved by production cars when driven in actual traffic are generally higher than the official test results. However, since the tests are consistently conducted, the results are valuable to compare performances between vehicles from different manufacturers in a specific country.[6] Examples of consumer information labels that must be displayed on all new cars in the US and Europe are shown in Fig. 4.

---

[6]Fuel consumption is generally given in liters consumed per 100 km; elsewhere, particularly in the US, it is given in miles driven (1 mile = 1.6 km) with 1 gallon of fuel (1 US gallon = 3.78 liters). For example:

10 liters/100 km = 23.5 US miles/US gallons

30 US miles/US gallon = 7.8 liters/100 km.

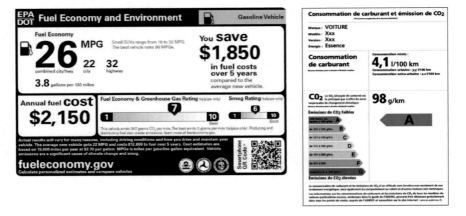

Figure 4.   Customer automobile information / © *Public domain*

Emissions information is generally given in gCO_2/km or $gCO_2/$ mile. In the US, the Corporate Average Fuel Economy (CAFE) standards are promulgated by the Federal Government and enforced by the Environmental Protection Agency (EPA) [*epa.gov*]. In Europe, the European Commission has promulgated a series of increasingly severe emission regulations. Starting with the Euro III standard, promulgated in 2000, separate limits were imposed for gasoline and diesel engines; separate limits were also enforced for hydrocarbons and nitrogen oxides (NO_x). The following table, for the European standards, illustrate the progress made in less than 15 years; similar progress is also true for the EPA standards.

	Euro III (2000)		Euro VI (2014)	
	Gasoline	**Diesel**	**Gasoline**	**Diesel**
CO (g/km)	2.3	0.64	1.0	0.5
NO_x (g/km)	0.15	0.50	0.006	0.08
PM (g/km)	Not imposed	0.05	0.005	0.005

Propulsion systems comparisons for a particular car model

An increasing number of manufacturers offer several propulsion systems for a particular car model. It is thus interesting to compare the performances provided by the manufacturer based on the Worldwide harmonized Light vehicle Test Procedures (WLTP) carried out under laboratory conditions. The data provided below is from a manufacturer's technical specifications for a medium-sized, five-passenger and five-door car with the standard transmission for each model.

Engine Displacement and Power	Consumption	Emissions
Gasoline 1.5 liter — 96 kW	4.5 l/100 km	101 gCO_2/km
Diesel 1.6 liter — 85 kW	4.1 l/100 km	106 gCO_2/km
Natural gas 1.5 liter — 96 kW	3.5 kg/100 km	95 gCO_2/km
Hybrid 1.4 liter gasoline ICE and 75 kW electric	1.9 l/100 km and 12.2 kWh/100 km	43 gCO_2/km 0 for the car itself
Electric 100 kW — variable speed	15.5 kWh/100 km	0 for the car itself

The emissions due to the electric vehicle depend on the emissions of the local electricity production mix. In general, when underway, the emissions of an electric vehicle will be less than from a gasoline ICE vehicle, as long as the emissions from the electricity production mix is less than some 700 gCO_2/kWh, which is the case in a number of countries.

Personal car performance enhancements

Figure 5 illustrates the overall energy consumption of an ICE when driven in an urban environment or on a highway — the situation when the car is equipped with an electric or hybrid propulsion system will be further discussed. A first conclusion from the figure is that only 13%, respectively 20%, of the car's total fuel consumption is used to actually "move" it to overcome the aerodynamic and rolling losses and to slow it down when required.

Based on the figure, several paths of vehicle energy consumption and emissions performance enhancements can be identified:

- Reducing the car's weight by using aluminum or carbon fiber materials rather than steel.

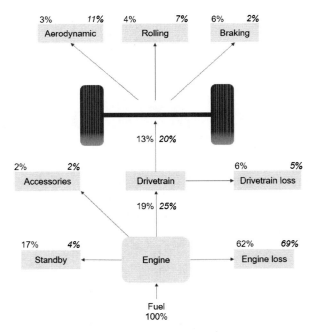

Urban driving / *Highway driving*

Figure 5.   Car consumptions — highway and urban driving / © *HBP and YB*

- Better efficiency of the ICE itself. As illustrated in an insert below, while improvements are continuously being implemented, significant progress has already been achieved since the first internal combustion engine.
- Recent "stop and go" technologies seek to address the standby losses.
- Automatic transmissions, initially proposed to enhance the driver comfort, are increasingly implemented to enhance the overall vehicle performance.
- Aerodynamic drag coefficient, C_x, reduction. While the C_x of 1950s sedans were in the 0.50 range, it can be below 0.30 for the latest models. Compromises between passenger comfort and low drag continue to be a major challenge for automobile designers.
- To achieve low rolling resistance, very hard and narrow tires should be used while good road adherence dictates wide tires made of soft

compounds with good road adherence leading to suitable compromises while also ensuring suitably long tire lifetimes.

* Braking losses are unavoidable. However, as will be seen later, by using various hybrid propulsion systems, the braking energy can be at least partially recovered to generate electricity that can be stored and reused for propulsion purposes.

Comfort enhancements have become key features for modern consumers: electric windows, air-conditioning, embarked entertainment systems, telecommunications, on-board and increasingly sophisticated navigational systems. However, they all increase the vehicle's fuel consumption. Safety-related on-board electronic systems, such as collision avoidance systems and cruise control, also add to the on-board energy consumption. Modern cars are increasingly becoming autonomous electric systems.

The latest trend is toward "autonomous vehicles" where the operator's role is evolving from being an actual driver to a system monitor/ supervisor.

ICE performance progress — from Etienne Lenoir to Formula 1 — a factor of 600

The first internal combustion engine car was constructed in 1860 by a French engineer, **Etienne Lenoir**. It was powered by an engine with a displacement of 2.5 liters, which produced 1.12 kW of mechanical output power, or 0.45 kW/l.

The 3.6 liter displacement, V8 engine (eight cylinders organized in two rows of four cylinders in a V configuration) produced by Ford in 1932, had a rated mechanical power of 48.5 kW, or 13.6 kW/l.

The 2019 Formula 1 specifications called for V6 engines with a maximum displacement of 1.6 liters; they are typically capable of 448 kW (600 HP; 1 HP = 746 W) or 280 kW/l, at 15 000 RPM. The total gasoline consumption is restricted to 100 kg for a 300 km race, i.e., 44 l/100 km.

From Lenoir's engine to the 2019 Formula 1 engine, the ICE performance has been multiplied by 600, in terms of kW/l, over 150 years while the underlying physical/chemical process involved remains unchanged.

Trucks and Buses

As seen above, the transport of goods using trucks only represents some 20% of the total ton·km. However, it is vital. Indeed, the transport of almost any goods from its origination to its final destination relies on trucks to and/or from the train station, harbor or airport. Over 85% of the road freight tonnage is carried over distances less than 150 km while only less than 1% is carried over distances exceeding 1 000 km [*iru.org*].

When propelled only using an internal combustion engine, modern trucks (referred to as lorries in some Anglo-Saxon countries) and buses overwhelmingly feature diesel engines. Among the main reasons for selecting a diesel engine are:

- Their larger torque capabilities due to their higher compression ratios. This is an important consideration for trucks when starting to haul heavy loads and for city buses during in-route stops and starts.
- Their generally lower fuel consumption; a crucial operational cost consideration.
- The higher fire safety of diesel fuel, which is less flammable than gasoline. This is an important consideration for buses and trucks because they both carry large amounts of fuels while operating in cities.
- The relative ease of producing biodiesel fuels from various waste products, including cooking oil.

Modern diesel truck engine displacements reach 16 liters with HP ratings well above 700 HP (515 kW) and torques exceeding 3 000 Nm.

Typical truck fuel consumption and CO_2 emissions

A tractor trailer hauling a payload of 25 tons over 100 km, i.e., 2 500 ton·km, typically consumes 35–40 liters of diesel fuel. Since the combustion of 1 liter of standard diesel fuel emits 2.5 kg of CO_2 (ref. "Units and Reference Values" annex), the emissions for the 100 km trip are: 87.5–100 $kgCO_2$.

(Continued)

(Continued)

For the transport of 100 ton·km by truck:

- Fuel consumption: 1.5 liters/100 ton·km
- Emissions: 3.75 kgCO$_2$/100 ton·km, rounded off.

This information can be compared to the values for a car transporting four passengers, each weighing 75 kg: typically 7.8 liters of fuel and 20 kgCO$_2$ for 100 km, i.e., 26 l/100 ton·km and 66.7 kgCO$_2$/100 ton·km.

Bus passenger transport is either used in an urban context or for long-distance travel. Compressed natural gas engines are making in-roads, primarily within large cities.

Comparing fuel consumptions between bus and private car travel

In 2014, Greyhound [*greyhound.com*] transported 18 million passengers for a total load of 8.8 billion passenger·km. Therefore, the average distance traveled by each passenger was 490 km. The average fuel consumption of a typical Greyhound bus is: 1 liter per 77.9 passenger·km. Thus, in 2014, the related total fuel consumption was 113 million liters.

As per the US Transportation Energy Data Book [*tedb.ornl.gov*], in the US, each car carries an average of 1.55 passengers. To transport the 18 million passengers, one would thus need 11.6 million cars, each traveling 490 km, as determined above. Assuming a fuel consumption rate of 7.8 l/100 km (30 miles per gallon), the total fuel consumption would be: (11.6 10^6 · 490 km) · (7.8 l/100 km) = 443 million liters.

The transport by bus requires 113/443 = 25% of the fuel required for the transport of the 18 million passengers over 490 km by car, thus again confirming the benefits of public transport!

Electric and Hybrid Propulsion

The efficiency of electric vehicle propulsion motors typically reach 90%, i.e., almost three times higher than ICEs.[7] The development of ICEs and

---

[7] The first car to reach the 100 km/h speed, in 1899, was built in France; it was electric.

the limited energy storage capacities of early lead-acid batteries, which seriously hampered the range of electric vehicles, led to their decline during the first half of the 20th century.

The energy density of any fuel or energy storage technology can be measured in kWh/kg, the amount of energy that can be stored per unit weight or in kWh/liter, the amount of energy that can be stored per unit volume. Clearly, one seeks to store the maximum amount of energy in as little weight and volume as possible.

Energy densities of gasoline, diesel and various batteries

For each teachnology, as of 2019, the high-end performance is given.

Batteries	kWh/kg	kWh/liter	
Lead-acid battery	0.047	0.16	Standard battery for ICE starters
Nickel metal hydride	0.08	0.29	First commercially available electric vehicles
Lithium-ion	0.22	0.56	Modern model electric and hybrid vehicles

To be compared with:

	kWh/kg	kWh/liter
Gasoline	12.0	9.0
Diesel	12.6	10.2

A modern intermediate-size car has a fuel tank holding 50 liters. If the car uses gasoline, the corresponding energy content is $50 \, l \cdot 9.0 \, \text{kWh/l} = 450 \, \text{kWh}$; the corresponding weight of the gasoline is $450 \, \text{kWh} / 12 \, \text{kWh/kg} = 37.5 \, \text{kg}$.

Using high-performance lithium-ion batteries to store the same amount of energy would require $450 \, \text{kWh} / 0.22 \, \text{kWh/kg} = 2\,045 \, \text{kg}$ in a volume of $450 \, \text{kWh} / 0.56 \, \text{kWh/l} = 800 \, \text{liters}$.

Even though electric motors have far better efficiencies than ICEs, as illustrated by these numbers, the development of new and highly reliable batteries remains at the core toward broad electrification of road transport.

When assessing the environmental impact, global and local emissions and noise for all-electric or hybrid-electric vehicles, be it for road or rail transport, it is important to separate the localized impacts, at the electric power plants, from the distributed impacts caused by each vehicle. Indeed, while an all-electric car or electric locomotive/subway may not cause any pollution by itself, their environmental impacts are, de facto, "outsourced" to the electricity supply system. Hence, electric propulsion has an increasingly positive environmental impact as the carbon content of the local electricity production mix decreases.

One can categorize electric and hybrid-electric propulsion systems as follows:

- All-electric and autonomous, where the propulsion motors are electric and the energy required is either stored electrochemically using on-board batteries, Battery Electric Vehicle (BEV), or supplied by a fuel cell (ref. "Multi-Energy Systems" chapter) supplied from an on-board hydrogen tank. Battery recharging or the hydrogen fill-up occur when the vehicle is stationary.
- All-electric and connected to an electric distribution system, where the propulsion motors are also electric but the electricity is provided by direct connection to a local distribution system using either overhead catenaries (electric locomotive, tramway or trolley) or various ground connection methods (metro).
- Hybrid propulsion systems feature both an electric motor and an ICE. As shown later, a number of configurations are possible. Hybrid propulsion systems bring many emissions and performance advantages compared to all-electric or ICE vehicles. However, they are also more complex, which explains their generally higher acquisition costs when compared to standard ICE propulsion. Plug-in configurations of both series and parallel hybrid vehicles provide for their battery recharging when stationary, thus extending all-electric range capabilities as on-board battery capacities are increased.

On-board power electronic controls enable regenerative braking to supplement standard brakes such that the electric motor is operated as a generator that either recharges the on-board batteries or reinjects energy onto the local distribution system. This energy-saving system has become standard for most electric and hybrid propulsion systems.

Battery Electric Vehicles (BEVs)

A schematic layout of a BEV, also referred to as Full Electric Vehicle (FEV), is shown in Fig. 6(a). Several variations are commonly used such as rear-wheel or four-wheel propulsion. In some cases, each wheel has a motor — referred to as a wheel, or hub, motor.

Airports, harbors, large industrial complexes and delivery vehicles,[8] where range limitations are of a lesser concern, increasingly rely on BEVs. On-going battery developments are leading the way to a broader variety of regular applications as the range limitations are gradually overcome. The low noise pollution of BEVs makes them particularly attractive for urban applications, as evidenced by the increasing number of city electric shared-mobility systems.

A detailed comparison of the key parameters — overall trip duration, consumption and emissions — of a 200-km and 520-km trip using an ICE powered car and a BEV is provided under *"Vehicle Refueling Versus Recharging Comparisons"* in the Companion Document. As the study suggests, the outcomes depend on a number of parameters

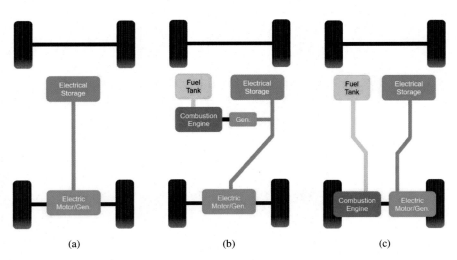

       (a)   (b)   (c)

Figure 6.   Electric vehicle configurations / © *HBP and YB*

---

[8]The deployment of such vehicles is increasing rapidly, bringing both noise and emission reductions for city centers.

including the travel distance and the battery capacity, in kWh, along with the installed charging capacities, in kW, of both the vehicle itself and the charging station, which both directly impact the recharging duration and hence the overall trip duration. Clearly, the carbon content of the electricity supply system is key to the emission reductions made possible by the EV.

Series Hybrid Electric Vehicles

A schematic layout of a series hybrid electric vehicle is shown in Fig. 6(b). Again, several propulsion configurations are possible as for BEVs.

In this case, an ICE is mechanically connected to a generator that helps to recharge the battery. The actual vehicle propulsion only relies on the electric motor. The ICE always operates at its optimal RPM, only providing for the recharge of the battery, hence the "range extender" designation that is sometimes used.

Parallel Hybrid Electric Vehicles

Figure 6(c) illustrates the basic parallel hybrid-electric configuration. The key difference compared to the series hybrid configuration is that the ICE contributes to the vehicle propulsion. Several alternate configurations are routinely implemented, among them: (a) the electric motor is connected to the front wheels and the ICE to the rear wheels or the other way around, and (b) the electric motor is connected to all four wheels.

Recent parallel hybrid electric vehicles, designated as "plug-in", provide for a connection to a regular or fast charging station such that the battery can be recharged while the vehicle is stationary. Depending on the size of the battery, plug-in parallel hybrid vehicles are capable of electricity-alone ranges of several tens of kilometers, thereby providing for an all-electric drive for most urban applications. Other control strategies favor enhanced vehicle performances, especially in terms of acceleration beyond what an ICE alone could provide.

Hydrogen Vehicles

The electric battery can either be entirely replaced or supplemented by way of a fuel cell (ref. "Multi-Energy Systems" chapter), which is supplied from an on-board hydrogen storage tank. Two hydrogen vehicle configurations are illustrated in Fig. 7. To extend the vehicle's range, the hydrogen is stored at very high pressures: typically 250–300 bars[9] for heavy vehicles, such as buses and trucks, and 700–800 bars for lighter vehicles, such as personal cars. As for EVs, hydrogen vehicles are low noise and non-polluting. The hydrogen production process is key to the emission reductions made possible by hydrogen vehicles.

Hydrogen buses are increasingly deployed in regular public transportation routes under a number of pilot programs in several cities in Canada, China, Europe, Latin America and the US.

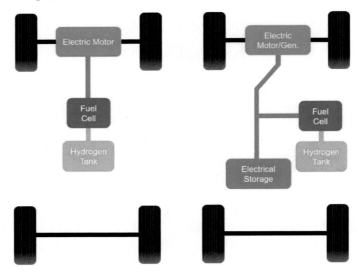

Figure 7.   Hybrid vehicles configurations / © *HBP and YB*

---

[9]The air pressure at sea level is close to 1 bar.

Fuel cell passenger bus

Among the leading selection criteria in favor of hydrogen buses are: (a) zero-emissions when operating, (b) short refueling durations, typically some 10 minutes and (c) ranges which can exceed 300 km.

Most buses feature hybrid configurations with a fuel cell providing electricity to electric motors. Electric batteries provide additional power when needed; regenerative braking is typically used to recharge them. Under very light load conditions, the batteries can also be recharged from the fuel cell.

An 18-ton, 12-m-long, 70-passenger bus with a combined electric motor rating at 180 kW, with a fuel cell system rated at 150 kW and batteries rated at 100–150 kW and 25–30 kWh, consumes 8–15 kgH_2/100 km.

Rail Transport

Rail transport not only includes classical railroads but also streetcars/tramways and subways.

Worldwide Rail Statistics [*uic.org*][10]

In 2015, railway passenger transport represented 6.7% of all passenger·km worldwide; it was 6.9% for all freight ton·km. Their combined CO_2 emissions were 4.2% of the overall transport sector emissions. Typically, passengers and freight are transported using separate trains but use common rail bed infrastructures. High-speed trains generally use dedicated rail bed infrastructures.

The total railroad passenger traffic worldwide was 3 100 billion passenger·km in 2015,[11] compared to 1 900 in 1990. At 1 200 billion passenger·km, India represented 39% of the world's railway passenger transport. Overall, the worldwide railroad freight activity is

---

[10]Railway Handbook, 2017, the most recent version when the book was written.

[11]Which corresponds to a worldwide average of some 400 km/cap·yr; in Switzerland, it was 2 500 km/cap·yr, the highest in the world.

declining; the US had the highest freight ton·km worldwide followed by Russia.

Very few railroad systems still use steam locomotives. Indeed, coal only represented 4.8% of the overall railway sector energy use in 2015 versus almost 25% in 1990. During that same time span, electricity's share increased from 17% to almost 39%.[12] Diesel fuel remains dominant in 2015 at 56%.

Worldwide, new rail beds are under construction, especially in developing countries, while others are or have been removed, especially in industrialized countries; overall, a net decrease is taking place to reach 1.4 million km of rail beds in 2015. The proportion of electrified tracks is increasing, having reached over 25%, also in 2015.

Trains

Locomotives, sometimes referred to as engines, of today's trains overwhelmingly are either full-electric or hybrid-electric.

Full-electric trains

The propulsion electric motor is supplied using an external source, by way of a pantograph sliding along an overhead catenary or by way of a supply incorporated in the rail bed. In most configurations, the main rails carry the return current.[13] The development of electric trains allowed for long-distance travels between end stations with no or short stops in-between; this makes battery trains impractical except for specific applications.

While full-electric trains do not cause any local emissions along their tracks, the overall associated emissions depend on the local electricity supply system. Their main drawback is the associated supply infrastructure along the track, which is costly in terms of initial investment and maintenance.

---

[12] Underground/subway systems are fully electrified; the same is true for urban streetcar systems.

[13] Some underground metro trains, such as the ones found in Paris, run on rubber wheels to provide a smoother ride; in this case, special provisions need to be made for the return current.

The typical consumption of full-electric trains is 2.25 kWh/100 ton·km [*uic.org*].

The emissions vary depending on the electricity production mix, for example, between 100 and 600 gCO_2/kWh, i.e., between 0.225 and 1.35 $kgCO_2$/100 ton·km.

Diesel-electric trains

As stated above, 75% of all rail tracks are not yet electrified. Aside from steam locomotives, the locomotive's traction wheels are driven by electric motors. In turn, the electric motors are either supplied using an on-board diesel engine driving an electric generator or using a fuel cell supplied from an on-board hydrogen tank, which is still rare but increasing steadily for commercial applications.

Diesel-electric locomotives cause emissions along the tracks they use. While hydrogen hybrid-electric trains do not cause emissions along the tracks, their overall emissions depend on the process used to produce and transport the hydrogen they consume. In either case, their main advantage is the lack of the need for any electricity supply infrastructure along the tracks, but regenerative network braking is not possible.

The typical consumption of diesel-electric trains is 4.5 kWh/100 ton·km [*uic.org*].

Since the combustion of diesel fuel leads to emissions of 250 gCO_2/kWh (ref. "Units and Reference Values" annex), the typical emissions are 1.125 $kgCO_2$/100 ton·km (4.5 · 0.250).

Rail freight transport in the US — fuel consumption and CO_2 emissions

CSX [*csx.com*] is a major freight rail transport corporation in the US. In 2018, using only diesel-electric locomotives, it transported 208 712 million ton·miles (333 940 million ton·km). The total diesel fuel consumption was 424 million gallons (1 603 million liters).

For the transport of 100 ton·km by diesel-electric train:

- Fuel consumption: 0.48 l/100 ton·km. With 9 kWh/l for the consumption of diesel fuel (ref. "Units and Reference Values" annex), one gets 4.32 kWh/100 ton·km, which confirms the value indicated above.

(Continued)

(Continued)

- Emissions: 1.08 kgCO$_2$/100 ton·km, which also confirms the value indicated above.

    To be compared with the fuel consumption of 1.5 l/100 ton·km and the emissions at 3.75 kgCO$_2$/100 ton·km computed above for the transport by truck.

    Comparisons with maritime and airfreight are provided later in dedicated inserts.

High-Speed Train Systems

High-speed trains generally use dedicated railway tracks and are capable of sustained speeds above 250 km/h. Typically, these trains only transport passengers. The Japanese Shinkansen, launched in 1964, is generally thought of as being the first truly dedicated high-speed train system. The proper contact continuity between the overhead catenary and the locomotive's pantograph has to be ensured for the high-speed trains, some of which have installed capacities reaching 10 MW or more.

As of 2019, there was slightly more than 50 000 km of high-speed railroad tracks worldwide; more tracks are rapidly added on all continents [*uic.org*]. It is noteworthy that by 2019, China, where high-speed railway lines did not exist at the beginning of the 21st century, had two-third of the high-speed tracks worldwide. The share of high-speed train travel is increasing as the share of slower trains is decreasing.

While the per passenger·km cost of high-speed trains is often thought of as being higher than for "classical" trains, this heavily depends on the number of stops between two terminal stations.

Maritime Transport

As presented above, while the maritime[14] transport's share for passengers is negligible, it represents 72% of the worldwide ton·km freight.

---

[14] Transport on oceans, seas, lakes, canals and rivers are included under the maritime transportation designation.

Excluding private pleasure boats, one can classify ships as follows:

- Freight: Bulk carriers, which carry grains, coal, for example; container ships; tankers, which carry various types of liquids; Liquid Natural Gas (LNG) ships; refrigerated ships; roll-on/roll-off ships, designed to carry wheeled cargo such as trains, trucks and cars; or other types of specialized ships such as coastal trading ships, barges, tugs, etc.
- Mixed passenger and freight, mostly ferries designed to carry both passengers and wheeled vehicles such as trucks and cars.
- Passenger ships, which presently overwhelmingly are cruise ships as opposed to ocean liners traveling on set scheduled routes.

Freight ships

As of early 2015, 6 000 freight ships were active on liner trades [*alphaliner. com*], i.e., large ocean-going freight ships operating on fixed schedules between major harbors. Roughly 80% of them were container ships whose capacity is generally given in TEU.[15] A ship's Dead Weight Tonnage (DWT) is the total load the ship can carry, including cargo, fuel and fresh water, passengers, crew, provisions and ballast, but excluding the ship's own weight. In 2014, the total worldwide container ship fleet capacity was close to 20 million TEUs with a DWT of 250 million tons, which corresponds to the weight of 25 000 Eiffel Towers in Paris.

Container ship fuel consumption and CO_2 emissions

As can be seen from Fig. 8, the daily fuel consumption of a 12 500 TEU container ship cruising at 20 knots is 175 tons,[16] which corresponds to 175 000 liters/day with a specific weight of marine diesel fuel at 1 kg/liter, i.e., heavier

(Continued)

---

[15]Twenty-foot equivalent container measuring 20' by 8' by 8'6", or 6.1 by 2.6 by 2.4 m. The corresponding volume is 38 m³.

[16]Container ships are designed, in terms of hull shape and propulsion systems, for "normal" cruising speeds between 20 and 25 knots. To save on fuels and reduce pollution, the trend is toward slower cruising speeds.

A speed of one nautical mile per hour is designated as one knot. One nautical mile = 1 852 m.

(Continued)

than regular diesel fuel. At 20 knots, the container ship travels 890 km/day. Assuming that each TEU weighs an average of 15 tons, the container ship transports (15 tons) · 12 500 · (890 km/day) = 167 million ton·km/day. The combustion of heavier marine diesel results in the emissions of 3.0 kgCO$_2$/kWh rather than 2.25 kgCO$_2$/kWh for regular diesel fuel.

As a result, for container ship transport:

- Fuel consumption:
  175 10^3 liter/day / 167 10^6 ton·km/day = 0.105 l/100 ton·km
- Emissions: 315 gCO$_2$/100 ton·km

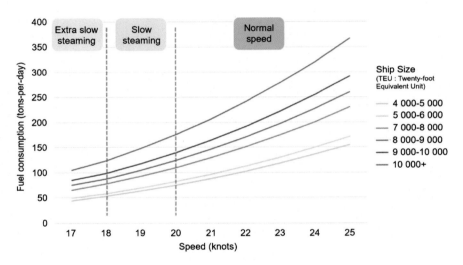

Figure 8. Container ships fuel consumption versus speed / © *sites.hofstra.edu/ jean-paul-rodrigue*

Propulsion Systems

Present-day ships, be they passenger or freight, either use diesel engines that directly drive the propellers or, for larger ships, a diesel-electric hybrid system where the propellers are driven by electric motors supplied from what has become on-board electricity production and distribution systems. Modern passenger ships feature so-called "pods" that can rotate in any direction to provide full steering flexibility, not only at sea but also with full intra-harbor maneuverability.

Maritime Fuels, Environmental Footprint and Water Ballast Concerns

Traditionally, large maritime diesel engines have been burning what is referred to as Heavy Fuel Oil (HFO), which can have sulfur contents reaching or exceeding 4% in some cases. The combustion of HFO produces sulfur oxides, as does the combustion of all oil products but more so for HFOs, which contribute to acid rain; it also produces nitrate oxides and fine particles. These emissions occur when the ships are at sea or in port and using their own energy supply rather than the, hopefully low-carbon, electricity provided by the harbor infrastructures.

Maritime transport is not directly included in the emissions targets set by the 2015 Paris Agreement, primarily since the related emissions cannot readily be assigned to any one country. The International Maritime Organization (IMO) [*imo.org*] is the United Nations specialized agency with responsibility for the safety and security of shipping and the prevention of maritime pollution. The IMO is setting increasingly lower limits for the sulfur content of diesel fuels, especially in the Emissions Controls Area (ECA), i.e., closer to main shorelines.[17] The IMO has also recently adopted a road map to decrease the shipping industry's CO_2 emissions by 40% by 2030, and by 70% by 2050, compared to 2008; overall greenhouse gas (GHG) emissions are also to be reduced by 50% by 2050.

Several measures are being implemented to reduce the shipping industry's environmental footprint, including:

- Using anti-fouling techniques to reduce hull-ocean friction.
- Improved propulsion system efficiencies, including waste heat recuperation.
- Using on-board renewable energy sources to cover ancillary consumptions.

---

[17] The ECA includes: the West and East coasts of Canada and the US, the Baltic Sea, the North Sea, and the Channel between England and the coast of Europe.

- Speed-optimization and/or reduction.
- Using electricity supplied by local port authorities, hopefully low-carbon, when at berth rather than the ship's own.
- Full electric propulsion for ships traveling shorter distances on set routes, such as point-to-point ferries, with recharging occurring during loading and off-loading operations.

For well over a hundred years, ships, especially freight ships, have been using water as ballast to stabilize them when running at light load. Increasingly stringent regulations are being enacted worldwide to limit the pollution and the impact on marine life as the associated large amounts of seawater is pumped in and out of the cargo holds.

Air Transport

Just one century after the first commercial flight, we now take for granted boarding a 10 000-km flight lasting over 12 hours with over 500 fellow passengers. Sending an express parcel between continents on a 48-hour delivery timeline has mostly become routine.

At present, air transport serves two main functions: passenger transport and freight. While most regularly scheduled passenger flights also carry freight, which represents significant revenues for the airlines concerned, a large number of flights are freight-only using specially equipped aircrafts.

In 2018, 4.2 billion passengers were carried by close to 1 400 commercial airlines using 25 000 airplanes that serviced almost 4 000 airports [*atag.org*]; this represents an increase of 610% in passengers carried compared to 1980. Also in 2018, 221 billion ton·km were transported worldwide, an increase of 800% over 1980 [*data.worldbank.org*].

The Wonder of Flying — Air Wing Lift and Drag

In 1748, **Daniel Bernouilli**[18] published the paper "Hydrodynamica" where he laid down the fundamental laws of fluid dynamics. In 1810,

---

[18] Swiss physician, physicist and mathematician.

Sir George Cayley[19] published his paper "On aerial navigation", which marked the beginning of the study of aerodynamics, i.e., for compressible fluids.

The basic idea behind the notions of lift and drag can be illustrated by way of Fig. 2. As an aircraft wing moves through air, its airspeed is lower on the side with the lower curvature, i.e., the wing's lower side, compared to the airspeed on its upper side, which has the higher curvature. Indeed, the distance the airflow travels on the lower side is shorter than on the upper side. As per the law of conservation of energy, the lower side of the wing is subject to a higher pressure, creating an upward force called lift. If the speed of the wing through the air increases, due to an increased thrust, so too will the lift, eventually exceeding the downward force due to the aircraft's weight, which allows it to lift off the ground.

Aircraft Propulsion Systems

Whereas rotational torque is key to moving a car or a train, thrust is key to moving an aircraft. Thrust is a force created by way of various types of propulsion systems; it is created when a volume of air is "forced" through an engine. The magnitude of thrust depends on the volume of air through the engine and by how much the air is accelerated by it. Three main propulsion systems are commonly used:

- Propeller engines, which are used primarily for lower speed aircraft (a few hundred km/h). The basic principle is to move large volumes of air through the engine while imparting a lower acceleration on it, compared to the jet engine. The propeller's driving torque can be provided by ICEs or by a gas turbine: the "turbo-prop" configuration.[20]
- Jet engines, where air is accelerated by way of combustion in a gas turbine. The basic principle is that of moving smaller amounts of air through the engine while imparting a higher acceleration on it. Jet

---

[19] British physicist, engineer and politician.

[20] Electric propeller propulsion systems are under development and demonstration.

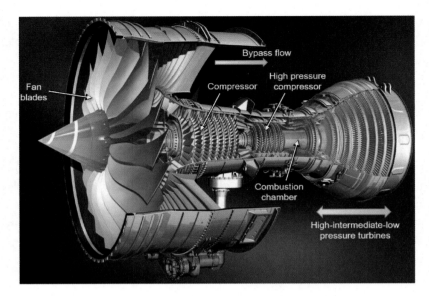

Figure 9.   Turbofan layout / © *2020 Rolls-Royce plc*

engines are used for higher speed aircraft, below or above the sound barrier, including modern military aircraft.

- Turbofan engines, shown in Fig. 9, which combine both systems mentioned above. These engines are predominantly used for civilian aircrafts as well as by some larger military transport planes. In a turbofan, the turbine not only accelerates the air through it, thus providing thrust, but also drives a large fan in front of the engine acting as a propeller, providing additional thrust. Some of the entering airflow "bypasses" the gas turbine; one speaks of a "by-pass ratio" that increases for the larger engines required for the largest passenger planes.

As of 2020, the combined maximum thrust of the two turbofan engines used on the largest twin-engine passenger aircraft in the world was 1 000 kN. For comparison, this is 1 400 times more than the force exercised on Earth by a person weighing 75 kg (75 · 9.81 = 735 N)!

Cargo plane fuel consumption and CO_2 emissions

The Boeing 747F is a commonly used civilian cargo aircraft. It can typically transport a useful load of 110 tons. Over an 11 000-km flight, it will consume some 150 000 liters of jet fuel, which corresponds to a weight of 120 tons (at 800 g/l, jet fuel is slightly heavier than gasoline). The related emissions are 2.25 kCO_2/l; ref. "Units and Reference Values" annex.

As a result, during the 11 000 km air cargo flight:

- Fuel consumption: 12.4 l/100 ton·km
- Emissions: 12.4 l/100 ton·km · 2.25 kCO_2/l = 27.9 $kgCO_2$/ton·km

This compares to:

- Truck: 1.5 l/100 ton·km and 3.75 $kgCO_2$/100 ton·km
- Diesel-electric train: 0.5 l/100 ton·km and 1.125 $kgCO_2$/100 ton·km
- Container ship: 0.105/100 ton·km and 0.315 $kgCO_2$/100 ton·km

Environmental Considerations

At 10.9% of the 2015 overall transport-related CO_2 emissions, which is almost identical to that of maritime transport, further attention is required to reduce the air transport sector's CO_2 emissions. It is the end-use sector for which it is the most difficult to implement significant emission reductions. Indeed, the weight of the on-board batteries required to transport several hundred passengers and/or several tens of tons of cargo for a transcontinental trip will remain prohibitive during several decades. The energy content, both per unit volume and unit weight, of carbonated fuels, presently fossil fuel, is such that they remain indispensable for air transport.

Air transport is thus the end-use sector for which biofuels are to be favored as long as their production is not in conflict with food production and does not otherwise negatively impact the environment. Universally agreed upon and adhered to biofuel standards are particularly important in this context, since an aircraft takes on fuel at airports worldwide.

As for maritime transport, air transport is not directly included in the emissions goals set by the 2015 Paris Agreement, primarily because the related emissions cannot readily be allocated to any one country. The

International Civil Aviation Organization [*icao.int*], which is the United Nations' specialized agency for the administration and governance of the Convention on International Civil Aviation (Chicago Convention), has decided to subscribe to the Paris Agreement terms by limiting future civil aviation emissions to their 2020 level through four groups of measures: (a) improved aircraft performances, (b) improved air traffic control technologies, (c) development of biofuels, and (d) creation of a compensation market with other end-use sectors to handle emissions exceeding their assigned quotas for participating airlines.

Comparative Freight Shipment Fuel Consumptions and CO_2 Emissions

The following table summarizes the earlier findings — the data is given for 100 ton·km.

Shipment Mode	Fuel Consumption — l	CO_2 Emission — kg
Diesel tractor-trailer	1.5	3.75
Diesel-electric train	0.5	1.125
Electric train[21]	2.25 kWh	0.225–1.35
Diesel container ship	0.105	0.315
Cargo aircraft	12.4	27.9

Case study: Coffee shipment from deep in Brazil to Flora, Energia

The Parana region in Brazil is one of the southern-most coffee growing regions in the world; it is well known for the quality of its beans.

We will follow a 15-ton shipment of coffee from Jacarezinho in Parana, Brazil, to Flora, Energia, where it will be redistributed toward several local roasting plants. The shipment took place in 2015.

While the detailed computations of the consumption and emissions resulting from each leg during the transport can be found under "*Multimodal Shipment*" in the Companion Document, the next table provides a summary.

(Continued)

---

[21] To facilitate the comparisons, 2.25 kWh, corresponds to the combustion of 0.25 liter of diesel fuel; ref. "Units and Reference Values" annex.

(Continued)

Leg	Mode	Distance km	Distance %	CO$_2$ %	CO$_2$ kg
Jacarezinho — Ourinhos	Diesel tractor-trailer	28	0.3	2.0	14
Ourinhos — Port Santos	Diesel-electric train	440	3.5	10.6	74
Port Santos — Portum	Diesel container ship	11 600	92.7	78.4	548
Portum — Flora	Electric train	440	3.5	9.0	63
Total		12 508	100%	100%	699

From the table above, it can be observed that while the emissions corresponding to the maritime leg are the highest in absolute value, they are also the lowest when computed on a per km basis. The exact contrary can be observed for the leg using the diesel tractor-trailer.

Industry and Agriculture

Producing Goods and Food

An industrial process consists of a succession of elementary processes. For example, the process from coffee trees to coffee beans for the over one billion cups of coffee that are drunk daily primarily consists of: (1) harvesting the red cherries, shown in Fig. 1, from the coffee tree, manually or mechanically, (2) extracting the two green grains contained in each cherry by removing the pulp using friction, followed by cleaning, (3) drying the grains, either using sunshine or a hot airflow, (4) removing the skin from the grains using a mechanical process, (5) roasting the grains followed by rapid cooling, to enhance and conserve the aroma by way of a chemical transformation resulting in the familiar brown color, also shown in Fig. 1, and (6) packing the grains under vacuum to prevent oxidation and humidity. Each stage relies on elementary thermal, mechanical or chemical processes that consume energy. The transportation energy between

(a)   (b)

Figure 1.   (a) Coffee cherries / © *C sa Bum, Shutterstock*, (b) Coffee beans / © *Kankitti Chupayoong, Shutterstock*

115

production and consumption sites must also be taken into account. The overall energy required by a 200-ml cup of American coffee is some 1 kWh.

Thus describing elementary process technologies allows for the description of the related energy consumptions; indeed, most are common to many industrial processes. New processes are continuously implemented, which can broaden application possibilities and occasionally lead to breakthroughs, such as three-dimensional printing for the manufacturing industry.

The same object can often be manufactured in several different ways. The choice depends on the materials at hand, the ease of handling them, the energy resources available, and economic considerations, etc. For example, tubes and pipes can be made using cement, copper or plastic materials; they can be molded, rolled and then welded or extruded.[1] Toys manufactured from wood a hundred years ago are now made of plastic.

Industry consumes energy to:

- Heat: To enable physical transformations and/or chemical reactions, to compress or release stresses, to mold or deform, to liquefy or to vaporize, to dry, weld or glue materials.
- Cool: To interrupt or slow down physical transformations and/or chemical reactions, to loosen up, to solidify or liquefy.
- Mechanically move, position, assemble, to deform, break, crush, cut, mix or separate, to compress or loosen up, to pump or ventilate, to polish, etc.
- Electrically or magnetically separate, to depose thin films, etc.

Statistics

In 2018, the final energy consumption of the industry and agriculture sector was 35 600 TWh,[2] i.e., 30.8% of the worldwide final energy consumption. The 2018 decomposition was:

---

[1] Toothpaste is dispensed by extrusion from the tube containing it.
[2] Including 2 579 TWh for agriculture and fishing.

Source/Vector	%
Coal	26.5
Oil and oil products	13.4
Natural gas	19.9
Wind, solar PV, other non-hydro renewables	0.1
Bioenergy and waste	7.1
Electricity	28.2
Network heat	4.8

The 2018 industry and agriculture electricity consumption was 10 054 TWh, which represented 45.1% of the worldwide electricity consumption.

In 2017, the sector caused emissions of 12 030 $MtCO_2$, i.e., 36.6% of the worldwide CO_2 emissions [*iea.org*].[3]

In 2015, the most energy-intensive industries were: refining, chemicals[4] and steel, which combined represented more than 50%, and then the cement and glass industries, followed by agro-food and paper. These percentages vary considerably from one country to the next; websites such as [*ec.europe.eu/growth*] provide listings of professional national or international associations where more detailed information can be found regarding the industries they represent.

Heating

Heating is the highest energy usage in the industry; it is thus the main focus in this chapter. A material or an object can be heated using a number of methods. Directly using sunshine remains a solution in some cases. Heating energy usages can be categorized depending on:

- The energy source/vector used:
  - ○ Fossil fuel, biomass, or even waste materials.
  - ○ Electricity, using, for example, the Joule effect or heat pumps; ref. "Housing" chapter.

---

[3] IEA CO_2 emissions from fuel combustion, Highlights, 2019.

[4] The chemical industry manufactures products such as soda, chlorine, plastics, paints, etc.

- The method by which the heat is added to the process:
  - *Around* the object or mixture, which is heated using thermal conduction and/or convection and/or radiation. Gas or electric ovens, plasma torches or lasers are examples of this method.
  - *Inside* the object or mixture itself, using combustion or using electric conduction by way of the Joule effect in a conducting material. In this latter case, the electric current is obtained by creating an electric circuit, by way of induction or an electric arc.
- Whether the process is continuous or not. Some processes cannot be interrupted without serious risk of damage to the installation, such as a tank containing a liquid metal, which therefore requires a high-reliability energy supply.

In a number of cases, heat enables a chemical reaction which breaks up stable molecules to replace them with other stable molecules. The reaction itself can be exothermic, i.e., providing heat that must be properly evacuated or valorized, or endothermic, i.e., requiring heat.

A few examples of heating processes are provided next, starting with the oldest ones, i.e., ovens.

Ovens: The Production of Bread, Bricks, Glass, Cement, etc.

Ovens are enclosures providing heat. Industrial ovens primarily use natural gas or electricity. The shape of the oven, the number and layout of the heat sources (burners, resistances), the characteristics of the walls, the use or not of forced air circulation, the absence of air, the humidity level, the characteristics and position of the load itself inside the oven, all influence the oven's operation. Depending on the particular process, thermal convection or conduction between the support and the load or radiation is of primary importance. The relative importance of heat transmission modes may change during the various phases of the process.

The oven may be a closed enclosure, even airtight; it may operate under controlled pressure and environment. It may operate without oxygen to avoid combustions, which is then referred to as pyrolysis; coke, which is frequently used in industry due to its properties, is obtained using pyrolysis to purify and dry coal.[5] Continuous ovens, or tunnels, are open at both ends. They can be from a few meters to several hundred meters long; the loads move inside at speeds properly adapted to the process, which can last from a few minutes to several days.

An oven's thermal efficiency is the ratio between the thermal energy provided inside the enclosure and the energy supplied by the gas or electricity; the lower the losses through the walls and due to the doors, the higher the efficiency becomes. The thermal efficiency of a closed electric oven is generally at the 95% level. The efficiency of an entire cycle (heating, cooling) is the ratio between the energy stored in the loads and the thermal energy provided inside the oven. The efficiency for an entire process is obtained by multiplying both efficiencies; however, it is the average over several cycles that is important.

Three examples of common products manufactured using ovens are provided below:

- *Bread* is manufactured starting from dough containing primarily flour and water as well as yeast and salt. The baking of the dough occurs after it is kneaded and after the fermentation has started; this typically lasts 20 to 30 minutes and represents some two-third of the energy used in a bakery. The temperature reaches 200 to 250°C. Water steam is required (using a mist producer, which may or may not be part of the oven) to create a suitable crust.[6]

---

[5] Pyrolysis is used to clean modern house ovens by drying/solidifying accumulated grease.
[6] The complex chemical reactions involved are not discussed here. The brown color, similar to coffee roasting and many other common bakings, is the result of the transformation of sugar, a reaction named after **Louis Camille Maillard**, a French chemist who described it in 1911. The mastery of the bread crust is part of the baker's art!

Six billion baguettes eaten yearly by 66 million French residents

A bakery's electric or natural gas oven typically has four levels with a total baking area of 5.5 m²; its installed capacity typically is 40 kW. The oven is preheated, at full power, for an hour before a batch of 72 dough pieces can be introduced. Figure 2 illustrates the many reactions taking place during the 30 minutes of baking as the power is maintained at 20 kW. The losses through the walls at 200°C are 6.5 kW while losses through the doors depend on how long they are opened. For four successive baking sessions to produce 288 baguettes of 250 g, the energy required is: (40 kW · 1 hr) + (20 kW · 0.5 hr · 4) = 80 kWh, i.e., 0.3 kWh per 250 g baguette and 60 g of CO_2 are emitted (ref. "Units and Reference Values" annex) using a natural gas oven. The emissions using an electric oven depend on the local electricity production mix, 20 gCO_2 per baguette in France. The energy consumption would be higher should interruptions occur between bakings.

Baking the six billion baguettes consumes around 2 TWh yearly; i,e., some 20% of the yearly production of a modern 1 250 MW nuclear power plant.

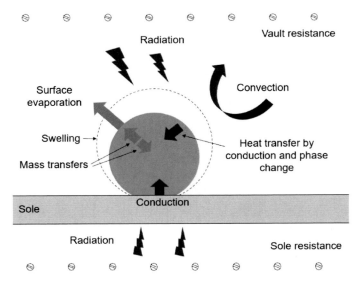

Figure 2.   Physical phenomena during bread baking / © *HBP and YB*

- *Bricks*, an essential construction material in many countries, are manufactured starting from a paste, primarily consisting of clay and water. The paste is first crushed and mixed to make it homogeneous. The bricks are then shaped, generally using extrusion, followed by cutting. To avoid cracks, the bricks are dried before baking, generally in a tunnel oven. The baking can last from 12 to 48 hours at temperatures reaching 800 to 1 200°C, depending on the paste's characteristics and desired outcomes.

Some 1.5 MWh is required to manufacture one ton of bricks [epa. gov], slightly less if the bricks are hollow or for tiles.

- *Common glass*, hollow for bottles and flat for windows, is produced from a mixture essentially consisting of white sand,[7] the main component, as well as lime, from limestone, and soda carbonate, produced, for example, from salt and chalk. This mixture is brought to fusion and homogenized at temperatures above 1 500°C and then suddenly quenched to ensure that the glass has the desired properties.[8] Industrial manufacturing processes use oven tunnels, most often using natural gas; the process can last several days.

Some 2 MWh are required to manufacture one ton of "primary" glass. Roughly 600 kg of CO_2 are emitted, two-third of which come from the natural gas combustion in the oven and one-third from the chemical transformations of the materials used [icglass.org].

Recycled glass, which has been purified and crushed, is referred to as cullet. To achieve a glass of a specific color, a homogeneous cullet is needed, such as bottles of that color, which requires careful sorting.[9]

---

[7] Sand is not only required to manufacture glass. It is also the core material for most semiconductors used in microelectronic circuits and the most commonly deployed photovoltaic solar cells. Sand is mixed with cement to make concrete. Affordable and low environmental impact procurement of sand is reaching strategic importance.

[8] When homogeneous, common glass is transparent for visible light but not so for ultraviolet light.

[9] Beyond common glasses, other types have been developed, such as: glass wool, fiberglass, glasses for mobile phone touch panels, and the "Skywalk" glass above the Colorado Grand Canyon.

Since the chemical bindings resulting from the fusion already exist in the cullet, a temperature of 1 000°C is sufficient.

The required energy is roughly two-third of that required for "primary" glass, confirming the importance of recycling.

Electric Heating: Induction Heating and Electric Arc

Electricity, at power levels from a few kW to several MWs, can be used to heat materials by way of:

- *Induction heating*, based on the same principle as for induction cook-tops (ref. "Housing" chapter); an electrically conducting material is increasingly used to: (a) weld or glue thin objects together, for example in the aeronautical and automobile industries and (b) thermally "staple" metal and plastics pieces together, the latter being melted in the process. Since temperatures above 900°C can readily be reached, one can heat pieces to improve their characteristics.[10]
- *An electric arc*, which is a spark or flash produced by Man, voluntary or not.[11] When imparting a sufficiently high electric voltage between two electrodes, in air or gas, electrons are "torn away" from molecules, thus creating ions. The gas then contains electrically charged particles and becomes conducting. This principle is used in arc furnaces; temperatures from 1 800 to 3 600°C can be reached, which is sufficient to melt most metals, such as scrap iron, as further presented later.

Other processes are also used, such as plasma torch or heating/liquefying using lasers, for example in three-dimensional printing which is further described later.

---

[10] Metals often go through structural changes at these temperatures, which are below their fusion temperatures. Following their manufacturing process, metal pieces may feature internal stresses that can be harmful in the long run and result in cracks. Heating a piece followed by a suitable cooling can reduce such stresses.

[11] An electric arc may occur when interrupting an electric current, which can be quite dangerous.

Heating to Separate: Refining

When extracted from a well, crude oil contains a number of different hydrocarbons,[12] C_nH_p. Its quality depends on its precise composition, including its density as well as the presence of other elements such as sulfur and phosphor; it depends on the particular extraction region.[13]

Starting from raw oil, the industry seeks to elaborate:

- *Energy products*, useful for transportation (gasoline and diesel for road vehicles, kerosene for planes, etc.), heating or industry (butane, heating oil, etc.).
- *Non-energy products* such as lubricants, bitumen, or other products useful in the chemical industry to manufacture plastics, etc.

The purpose of oil refining is to obtain products containing hydrocarbons that are well suited for specific applications, in particular the ratio between carbon and hydrogen atoms.

The two basic refining processes are:

- *Distillation,* i.e., heating the oil to some 370°C and then separating it into various sub-products with similar characteristics while relying on the different vaporization temperatures of hydrocarbons; the lighter ones vaporizing before the heavier ones. A distillation column, illustrated in Fig. 3, can be as high as 60 m. When the heated oil is introduced from the bottom of the column, the various products, transformed into gases, are recuperated at selected levels and then transferred to other treatments (further distillation, desulfurization, etc.) in other units.
- *Cracking,* which consists of breaking down heavier molecules into lighter ones. Several methods are used; some remove carbon while others add hydrogen.

---

[12] Hydrocarbon for hydrogen and carbon.

[13] The most commonly mentioned raw oil categories are: (a) WTI, from Texas, and (b) Brent, from the North Sea. Oil from the Middle East is heavier and thus more difficult to refine than the other two.

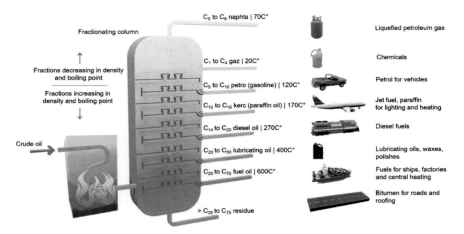

Figure 3.   Petroleum distillation column / © *HBP and YB*

In 2015, the petroleum industry used roughly 15% of the energy of its products for its overall operations. This means that for the production of seven liters of gasoline from crude oil, one liter is required for its production. This implies that the petroleum industry itself emits CO₂.

Heating for Metallurgy: Steel Production

Steel is fully recyclable and can be reused indefinitely, which is one of its advantages.

When mined, iron is found as iron oxides.[14] Some 1.7 billion tons of steel were manufactured in 2017 [*worldsteel.org*]. The two main industrial processes used are: blast oven, for 70% of the total production, and electric arc, for the remaining 30%, primarily used for recycling; both are briefly described below.

Blast furnace and converter steel making

Using the blast oven approach, the iron ore's oxygen is removed by melting at temperatures exceeding 1 200°C. As illustrated on the left side of

---

[14] Primarily Fe_2O_3, designated as hematites, and Fe_3O_4, designated as magnetites.

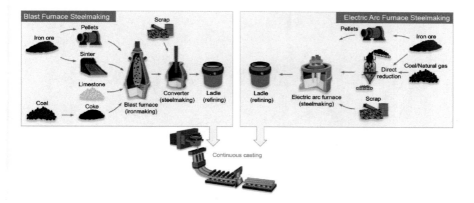

Figure 4. Steel making process / © *Worldsteel Association*

Fig. 4, iron ore, in the form of pellets and sinter (iron ore recuperated from iron extracting), and coke[15] (obtained from the pyrolysis of coal) are layered into the blast furnace along with limestone. Hot air is blown into the bottom of the blast furnace. The iron ore's oxygen reacts with carbon from the coke to form CO_2.[16] The melted iron, referred to a pig iron, is too brittle to be used directly since its carbon content, between 3 and 5%, is too high whereas less than 2% is required for steel.

As also illustrated on the left side of Fig. 4, the pig iron is poured into the converter, also referred to as Basic Oven Furnace (BOF), into which oxygen is blown. Almost all of the carbon and remaining undesirable elements, such as sulfur and silicon, are oxidized. The converter step is exothermic and results in liquid steel at 1 600°C, which is then poured into ingots, as shown in Fig. 5, and subsequently laminated, rolled, etc.[17]

The process is CO_2 emissions intensive due to: (a) the coke production, (b) the energy required to heat the blast oven's air and (c) the reactions in the blast oven and the converter.

---

[15] Coke is used rather than coal due to its high calorific content and its ability to withstand high pressures in the blast oven.

[16] $Fe_2O_3 + 2\ C + \frac{1}{2}\ O_2 \rightarrow 2\ Fe + 2\ CO_2$ and $Fe_3O_4 + 2\ C \rightarrow 3\ Fe + 2\ CO_2$

[17] Several variations of the process are used to adapt it to the characteristics of the ore used, the need to eliminate undesirable elements or to add elements, such as nickel or chrome to achieve selected steel specifications.

Figure 5.   Hot steel pouring in a steel plant / © *Yermolov, Shutterstock*

The blast furnace steel making process requires some 7 MWh per ton of steel produced;[18] *roughly 2.5 tCO$_2$ is emitted.*

The largest blast furnace in Europe, in Duisburg, Germany

The key dimensions of this blast furnace, which produces 12 000 tons of pig iron per day, are: 90 m high and 15 m in diameter. It has 42 hot air injectors. Along with its converter, it, therefore, consumes 12 000 tons · 7 000 kWh/ton per day, which corresponds to an average power of 3 500 MW, i.e., the installed capacity of a modern multi-unit thermal power plant.

Direct reduction of iron and electric arc steel making

Illustrated on the right side of Fig. 4, using direct reduction of iron (DRI), the iron ore's oxygen is removed without melting, as in a blast furnace. The reduction happens either by reaction with hydrogen, resulting in water vapor, or with carbon monoxide (CO) resulting in CO$_2$ emissions. The iron obtained from the DRI is purified in an electric arc furnace.

---

[18]This value can vary by ±50% depending on whether the energy required from the mine to the plant is included or only that of the steel making itself; the amount of scrap iron included also affects the energy consumption as further detailed later.

The electric arc steel making process requires some 3 MWh per ton of steel produced, including 640 kWh of electricity. The CO_2 emissions depend on the carbon content of the electricity production mix.

Recycling: From scrap to steel

Recycling of scrap metals from manufacturing plants, household waste using selective recycling, and scrapped vehicles as well as from deconstructed buildings is increasingly done. As illustrated in Fig. 4, scrap metal can be recycled using either of the two processes described above:

- Blast furnace. On average, 1.37 ton of iron ore and 0.13 ton of scrap iron are used per ton of steel produced.
- Electric arc furnace. On average, 0.6 ton of iron ore and 0.7 ton of scrap iron are used per ton of steel produced.

Recycling scrap iron is primarily done using the electric arc furnace approach in view of its significantly lower energy consumption and thus CO_2 emissions, especially as lower carbon content electricity is used.

Recycling provides over 30% of the total steel used worldwide; even more so in industrialized countries. Since most of the steel recycling relies on steel that is over 20 years old, this percentage is increasing in emerging regions. Overall, the worldwide energy consumption and associated emissions per ton of steel produced have decreased by close to 50% during the last 50 years. This progress is made possible by the increased impact of recycling and the implementation of new technologies in both processes, including heat recuperation from hot gases [*worldsteel.org*].

Electrolysis and Electrochemistry

Electrolysis is a chemical reaction process in a fluid using an electric current flowing through it.[19]

---

[19] The oldest example is the electrolysis of water, which was first achieved in 1800 by two British chemists, **William Nicholson** and **Sir Anthony Carlisle**.

Water Electrolysis

Water, H_2O, is decomposed into hydrogen and oxygen when it is placed in a recipient with two electrodes connected to a DC source, with an electrolyte, generally salt, added to facilitate the current circulation. Provided the carbon content of the electricity is low, the CO_2 emissions resulting from electrolysis are also low. A more detailed description of water electrolysis can be found in the "Multi-Energy Systems" chapter.

Some 5 kWh of electricity is required to produce 90 g of hydrogen, i.e., 1 m^3, at 20°C and atmospheric pressure.

Hydrogen production using steam reforming of natural gas[20] is significantly less costly than electrolysis; it represents some 95% of the worldwide production. However, 11 kg of CO_2 is emitted per kg of hydrogen produced.

Aluminum Electrolysis

Electrolysis is also used to manufacture other products such as aluminum. The process consists of several stages:

- Extraction of bauxite ore, consisting of small red pebbles containing over 50% of alumina, Al_2O_3.
- Refining, which results in alumina, a white powder. This process requires temperatures reaching 1 100°C. The residue is red sludge, which requires suitable handling from an environmental viewpoint.
- Electrolysis of the alumina to separate the aluminum from the oxygen.[21] It takes place in a special tank where alumina is brought to fusion at 970°C using the Joule effect. The alumina is decomposed into aluminum and CO_2.

Roughly 5.5 tons of bauxite ore is required to produce 2 tons of alumina and then to extract 1 ton of aluminum from it; the related electricity consumption is some 14 MWh [world-aluminium.org].

---

[20] $CH_4 + O_2 \rightarrow CO_2 + 2 H_2$.

[21] Independently invented in 1886 by **Charles Martin Hall** in the US and by **Paul Héroult** in France. The electrodes are made of carbon, hence the CO_2 emissions. If electrodes without carbon could be used, the CO_2 emissions could be replaced by oxygen.

As illustrated by the following insert, the process' electricity consumption is intense, which explains why aluminum manufacturing plants have traditionally been located close to electric power plants, generally hydro plants.

Recycling aluminum requires less than 5% of the energy required for the manufacturing of primary aluminum.

Indeed, it generally only involves a fusion without chemical transformations; it also avoids the need for new ore. This highlights the importance of recycling worldwide.

Alma aluminum electrolysis factory in Canada: 660 MW

This factory, located in Québec, occupies a 95-ha plot. In 2013, it comprised 432 tanks operating in parallel. Electricity was supplied at 161 kV from hydro plants belonging to the same company and from Hydro Québec. Also, in 2013, it produced 440 000 tons of aluminum while drawing 660 MW continuously, which means that the energy required was 13.1 MWh per ton of aluminum produced [*riotinto.com*].

Cooling

Cooling is frequently used to:

- Suitably lower the temperature at the correct moment after a heating process. For example, bread is naturally cooled using preferably dry air to keep it crusty and to further evacuate water. Liquid glass, steel and aluminum are cooled using air in processes adapted to the desired product shape; subsequent reheating may be implemented to relieve any internal stresses resulting from the previous heating and/or mechanical processes.
- Avoid degradation of pieces heated up in an industrial process due to motions (friction) or transformations. For example, the walls of blast furnaces are coated to withstand the material in fusion and the gas pressures as well as the temperature fluctuations. To avoid premature degradations, the coating itself has often to be cooled down using water.

- Liquefy gases to transport them or to be used as coolants.
- Refrigerate at a temperature below the ambient temperature, for example, deep-frozen food at −18°C.

Cooling technologies do not actually create "cold"; their purpose is to remove heat from the object to be cooled and then transmitting it to the surrounding space, either directly or indirectly using a coolant that can be air and/or water or a liquid selected for the particular purpose. To evacuate any given amount of heat, using evaporation-based cooling requires far less fluid than using the fluid in direct contact with the object to be cooled before it is returned to its original medium having been heated up.[22]

The selection of a cooling system depends on the industrial process in which it is to be used, in particular on the temperature and pressure ranges, the rate at which cooling is to occur, on the power available, and on the location, for example, proximity to a river.

- For powers below a few MW, between −30°C and +90°C, heat pumps can be used.
- For power above a few MW, cooling towers are used such as for thermal power plants; ref. "Thermal Combustion Power Plants" chapter.

The deep frozen food industry

Deep freezing is an industrial process involving fast cooling (from a few minutes up to 30 minutes) of foods and then preserving them at −18°C. Fast processing allows better conservation of the food's texture by reducing the number of water-containing cells damaged by crystallization.

Some of the processes used to deep freeze foods include:

- For thin slices of food: Passage between two plates where a coolant is circulating at −30 to −40°C.
- For small pieces of food, such as peas: Passage through a very cold air stream.
- For thick foods: Once hermetically sealed, immersion in an inert fluid, such as liquid nitrogen at −196°C.

---

[22] This is the way humans evacuate heat by perspiring.

Variations of the technologies described above, as well as other technologies, are presented on several sites such as [*iifiir.org*].

Water Treatment and Desalination

The processes used to collect, treat and desalinate, distribute and discharge water are illustrated in Fig. 6. The use of desalination is increasing rapidly to cover the growing need for potable water. In 2017, over 18 700 desalination plants of various sizes were in operation worldwide, producing over 87 million cubic meters of fresh water daily, thus satisfying the needs of 300 million people in over 100 countries on all continents [*idadesal.org*]. The two main processes, distillation and reverse osmosis, are described under *"Desalination"* in the Companion Document.

The total amount of electricity needed for tap water is around 2 kWh/m^3 without desalination and 6 kWh/m^3 with desalination using reverse osmosis. This data illustrates that it is preferable to treat used water before consumption rather than relying on desalination.

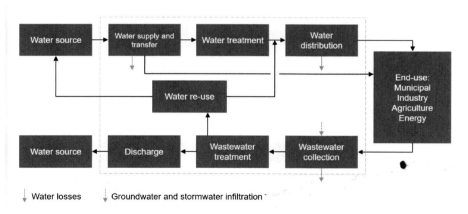

Figure 6. Water treatment processes / © *IEA*

Selected Other Energy Uses in Industry

Moving and Positioning

Moving objects and materials within a factory contributes to industry energy consumptions. In addition to manual movements, internal transports using specialized vehicles and conveyor belts are required to operate, for example:

- Conveyors in the automobile industry, for solar panel and electronic circuit manufacturing,
- Automatic sorting systems such as mail sorting.

The recognition of an object to be moved, followed by its retrieval and positioning using a number of rotations and linear motions, is increasingly done using robots. Moving from stationary to moving robots is one of the present challenges. Another challenge is the deployment of cobots, which are robots that can collaborate with the operator to assist with particularly heavy manipulations or with very small and precise ones.

> ### Reduction of robots' energy consumptions in automobile manufacturing plants
>
> A high-volume automobile manufacturing plant producing 1 000 vehicles per day typically consumes 2 400 MWh daily, i.e., 2.4 MWh per car and an average of 100 MW continuously over 24 hours. Over two-third of the consumption is due to electrical motors driving conveyor belts, pumps and robots.
>
> Generally, robots make jerky motions with high accelerations and decelerations with precisions at the mm level. Studies show that using software to smoothen the motions could result in energy savings of 10 to 20% (or more) while still maintaining the same cadences and inter robot synchronizations.

Three-dimensional Printing/Additive Manufacturing

These processes, which have been in on-going development for the past 30 years, do not remove materials from an object (as for many traditional manufacturing processes) but add materials to it. The three key elements are: (a) a three-dimensional digital description of the object's geometry, in "slices" of 20 to 100 microns, (b) a material in the form of a powder (metallic, plastic, etc.) and (c) a printer to deposit the powder in layers before melting and/or polymerizing, often done under a controlled atmosphere. It is therefore an electro-thermal process. There is an increasing number of variants.

Initially used to build one object or a small series, additive manufacturing is increasingly used to manufacture series of thousands of identical objects. Energy consumption audits, not yet performed, will thus become important; less materials waste should be possible compared to traditional manufacturing, and lighter objects reduce energy consumptions. However, the manufacturing of the powders consumes energy.

Agriculture: Energy and Emissions

In 2018, agriculture represented 2.2% of the worldwide final energy consumption. In 2017, it produced 13% of the overall emissions. These values vary considerably from one country to another.

Some two-third of the emissions were directly tied to the type of farming pursued and to the fertilizers used.[23] For example, burning agriculture fields and combusting waste result in CO_2 emissions.

One-third of the emissions depend on the local climate and topography as well as on the local hydrology and habits, which in turn, vary significantly depending on the type of farming pursued.

The energy consumptions relate to:

- Mechanical energy requirements:
  - Mobility: To plow, seed, harvest, using agriculture equipment, such as tractors, combined harvesters, which generally consume fossil fuels, and to transport animals and products.
  - Stationary: To irrigate, grind, etc.

---

[23] The impacts of various farming methods on greenhouse gas emissions, including methane (CH_4) and nitrogen oxides (NO_x), are beyond the scope of the book as are the discussions related to the eating habits, food chains, losses and waste utilization.

- Heat and cold requirements:
  - ○ To cool and heat greenhouses as well as animal barns.
  - ○ To heat, boil and cool products.

Replacing liquid fuels for mobile energy requirements is presently difficult, given the power levels involved; as a result, the availability of biofuels is crucial to reduce the related emissions. While the other requirements are often still covered using fossil fuels, they tend to be replaced by electricity when available at an affordable cost.

Tertiary and Services

At the Office and in Public Spaces

To define this sector, it is easier to observe that it does not include the residential, transportation or industry and agriculture sectors. It, therefore, regroups entities offering a broad range of services, mainly:

- Shops and shopping centers.
- Administration and business offices, including call centers and postal offices.
- Teaching institutions, from schools to universities.
- Hotels and restaurants.
- Hospitals and health institutions.
- Sports and leisure facilities.
- Community life centers, such as retirement homes, student housings and jails.
- Places of worship.
- Train stations, airports and ports.
- Computing centers.
- R&D centers and laboratories.

This list is not exhaustive if one seeks to include all energy consumptions; indeed, how does one include small enterprises, such as a bakery or a convenience store located on the ground floor of an apartment building? Furthermore, some office buildings are located on industrial sites. Generally, university or industrial campuses include multi-purpose buildings. Train stations, airports and harbors are increasingly becoming

135

shopping centers. These examples illustrate the difficulties in describing the tertiary and services sector for which a universal definition does not exist. As a result, obtaining precise statistical information is often a challenge.

One can consider tertiary and services establishments, more specifically their buildings, along several energy consumption characteristics:

- Whether the establishments receive visitors or not. Their energy consumption not only depends on the behavior of their staff[1] and service providers but also on that of their visitors; for example, related to the frozen food displays doors in a grocery shop. It may be difficult to assess the energy required for some services.[2]
- Establishments operating 24/7 all year round or only on part-time schedules.
- Establishments for which energy consumptions only depend on the presence of occupants (HVAC[3] hot water, lighting, office machines and communication) or having specific requirements such as refrigeration (food storage), cooling (computer center), heating (cooking), or air quality (hospitals). Security and resilience (redundant electric supplies) are also important.

Particular energy requirements are quite different, in proportion and quantity, from one establishment to the next. Aside from controlled air quality constraints, the needs are those of the residential sector but at different scales. While the technologies are generally the same, larger-scale applications provide for improvements and performances difficult to reach for residential applications. Some technologies can only be deployed within the tertiary and services sector for economic reasons.

Having presented selected energy performance improvements in the tertiary and services sector as well as related to some technologies which

---

[1] Some shop operators leave their entrance doors open during winter or summer to encourage shoppers to enter, thus increasing their heating or cooling requirements.

[2] For example, when an email is sent, energy is consumed by the sending computer, network, servers and the receiving computer.

[3] Heating/venting/air conditioning.

are becoming more prevalent, we have deliberately opted to focus on selected examples of energy issues in shops, office buildings, educational institutions, hotels, hospitals and computing centers.

Energy consumptions and the related emissions are described in this chapter; the potential for local production are presented in related chapters.

Statistics [*iea.org*]

The 2018 overall final energy consumption of the tertiary and services sector was 9 404 TWh, i.e., a share of 8.1%. The decomposition was:

Source/Vector	%
Coal	3.8
Oil and oil products	9.8
Natural gas	25.6
Wind, solar PV, other non-hydro renewables	1.1
Bioenergy and waste	3.8
Electricity	51.0
Network heat	4.9

In 2017, in the European Union (EU) 28 it was 12.9% and 13.6% in the United States (US). This information somewhat hides significant disparities; it was 17.7% in Switzerland but only 2.5% in Nigeria, for example.

The 2018 tertiary and services sector electricity consumption was 4 799 TWh, which represented 21.5% of the worldwide total electricity consumption. At 51.0%, the electricity share in the sector's final energy consumption is generally higher than for the housing sector where a broader range of energy vectors is used for heating and where air-conditioning is less prevalent.

Final energy consumption distributions by activities or by end-use categories are rarely provided.[4]

Indicators such that the energy consumption related to the age and types of buildings ([*bpie.eu*] for Europe and [*eia.gov*] for the US) or alternatively, by m^2 in offices or shops, are sometimes provided.

CO_2 emissions can be computed from the final energy consumptions. In the US, for example, in 2017, the tertiary and services sector was responsible for 17% of the overall emissions — 73% from the use of electricity and the balance primarily due to natural gas [*eia.gov*]. In most other countries, the emissions proportion was lower since the sector was less developed, for example, it was 15% in Europe.

This type of information is often cumbersome and time-consuming to assemble. Periodic update surveys are implemented in some countries to facilitate the choices as to which types of buildings should be favored for renovations toward more rational energy end-uses and CO_2 abatement.

Energy Performance Improvements in the Tertiary and Services Sector

The final energy consumption depends on:

- The workspace itself — HVAC and lighting.
- The equipment used, including computing installations — workstations, printers, networks, etc.
- The occupants and their presence. The air quality in an office or meeting room can be degraded by the CO_2, by body and other smells perceived when entering the space if ventilation is insufficient. Local regulations generally rely on the CO_2 density as the global air quality indicator; they generally stipulate that it should be less than 1 000 ppm (parts per million).[5] From this value, the required airflow can be determined.

---

[4] For example, the 2018 data for the United Kingdom was [*gov.uk/government/statistics*]:
- Offices (18.3%), shops (16.9%), hotels and restaurants (15.6%), health services (15.1%), teaching institutions (11.4%), leisure and culture establishments (9.8%) and the balance (12.9%).
- Heating (47.1%), restaurants (10.6%), lighting (9.0%), hot water (6.7%), ventilation and air conditioning (5.5%), IT (4.1%), and the balance (17.0%).

[5] The air in rural regions contains some 350 ppm whereas it reaches 600 to 700 ppm in cities, sometimes more. The average CO_2 density on Earth reached 400 ppm in 2015.

Elevators: Daily public transport for many persons

Elevators have become ubiquitous since, **Elisha Otis**, an American engineer, invented a safety brake mechanism in 1852, which rendered high-rise buildings possible. An elevator's main energy consumption relates to its motion. The elevator cabin is generally connected, using cables and pulleys, to a counterweight whose weight is that of the cabin itself and its passengers when it is half full,[6] thus making the weight to overcome smaller on average. Variable speed motors are presently used to ensure comfortable transitions to the cruising speed, which can reach 15 m/s, i.e., 54 km/h, in modern skyscrapers, and to ensure smooth stops at the floors. Modern elevators also feature energy recovery systems: when the cabin goes up (or down) at less (or more) than half of its capacity, braking is required thus providing for energy recuperation systems. Thus, from a strictly energy consumption limitation perspective, it is best if the elevator goes up empty and comes down full!

The second energy consumption is due to the lighting, only when occupied, as well as the monitoring and safety systems, which must always be available including for the doors. Proper ventilation, when required, consumes energy and induces losses since the elevator shaft cannot be airtight.

High occupancy buildings generally have several elevators. Their overall operation needs to be optimized not only to reduce passenger waits but also to limit the number of intermediate stops, thus reducing energy consumption and avoiding excessive installation wear and tear. As a result, in modern elevators, users are asked which floor they wish to go to when calling for the elevator, thus providing for enhanced optimization possibilities; however, changing floor requests are not possible once the elevator is underway.

Elevator electric energy consumptions are rarely measured separately but represent a significant load of high-rise common services. For an average nine-floor building with 80 residents, the yearly elevator electric energy consumption can be estimated at 2.5 MWh, on average. Assuming that each resident uses the elevator six times daily, it leads to an average consumption of 15 Wh per user and per use. A 75 kg resident taking the stairs to the 5th floor, i.e., 13.5 m, performs work of: 75 kg $\cdot$ 9.81 m/s^2 $\cdot$ 13.5 m $\approx$ 10 kJ or 2.8 Wh, or 2.4 kcal, i.e., roughly five times less than using the elevator. Not only is it good for our health to take the stairs but also much more energy efficient! Of course, for the 50th floor, it is more difficult!

---

[6] For high-rise buildings, the weight of the cables becomes significant in comparison to the cabin itself and must be taken into account.

Aside from a building's energy performances and equipment along with the occupants' requirements and behavior, two tools are available to enhance rational energy end-uses:

- *Presence/absence detectors*, which can reduce the lighting and HVAC energy demands, especially in spaces such as offices, meeting rooms, corridors and restrooms.
  - Infrared (IR) detectors are the most commonly used and less expensive ones.
  - Ultrasound (US) detectors are also commonly used.
  - Depending on the particular application, sensors using a combination of both are used. In some cases detectors based on other physical principles are used.
  Most detectors draw less than 1 W when used in spaces of common dimensions. The return on investments is generally measured in months.
- *Centralized energy management systems*, combined with sensors and actuators, provide for overall systems monitoring and optimizations, such as automatic office heat reduction during weekends and increase in time to ensure a comfortable situation on Monday mornings. More sophisticated systems can control selected zones along with specifications as to which controls are left to the user.[7] Such systems, while increasingly installed in new buildings, may be difficult and expensive to install in older buildings since new cabling may be required; modern WiFi systems may circumvent this obstacle. Furthermore, existing equipment may not be easy to control remotely and may rely on a variety of technologies.

Shops and Shopping Centers

The small 20-m^2 grocery shop in a 1930s building in town and the 10 000-m^2 shopping mall built in 2010 in a suburb have little in common other than that they both need cooling for their food and to attract customers. These

---

[7] Homogeneous remote temperature control is not easy to implement since offices may have different exposures, shadows may come from surrounding buildings at different times of the day, and this also depends on weather conditions.

energy requirements, combined with long operational hours, result in yearly final energy consumptions that can readily reach 500 kWh/m^2 for a large air-conditioned grocery store in a mild climate region, i.e., five times more than for a house constructed during the 2010s; ref. "Housing" chapter. The three main energy consumptions are: food refrigeration, HVAC and lighting.

Corporate image, customer attraction and upgrades to abide by new standards lead to more frequent renovations compared to the residential sector; energy considerations are generally not the deciding factor leading to renovations.

Compared to apartment buildings, there are fewer different types of commercial buildings. Indeed, industrial construction technologies are more prevalent and large international corporations are more active in this construction market. Commercial buildings often feature metallic (or concrete or wood) core structures where the walls are not weight bearing. Using outside insulation is easier to implement, resulting in shorter returns on investment. Replacing these buildings on the same site rather than renovating them after some 30 years is often more cost effective.

Shop operators must ensure that customer and staff safety is maintained and that sanitation constraints are adhered to in terms of air quality, temperature and humidity, which are generally imposed by standards. Avoiding olfactory nuisances to the staff and clients are also important considerations. This implies ensuring sufficient air recycling to avoid excessive CO_2 concentrations while avoiding unpleasant drafts (less than 20 cm/s). Temperatures are to be maintained all year round such that they are neither too low nor too high even during peak occupancy. Grocery stores are energy intensive since proper food storage cooling must be maintained: at least $-18°C$ for frozen foods and $4°C$ for dairy products — cooling represents between 35 and 45% of their energy consumption.

To properly design the HVAC systems as well as their operational range, detailed thermal, humidity and CO_2 audits must be done, taking the opening hours and local seasonal conditions into account.

Selected systems design orders of magnitude

During peak hours, over 5 000 persons may be present in a 10 000-m^2 shopping center located on the bottom floor. Each occupant "brings" some 100 W of heat, 50 ml/h of humidity (from breathing and perspiration) and 36 l/h of CO_2. Practically, to design an air handling system, one starts from the number of occupants during each week, day and hour, on one hand, and the local climate conditions, by way of the local HDD and CDD (ref. "Housing" chapter), on the other hand, to compute the hourly heating/cooling, humidifying and CO_2 mitigation requirements.

Sophisticated heat pumps are increasingly used not only to meet specific cooling requirements but also for HVAC purposes. Grouping all systems in a rooftop configuration provides for the extraction of stale air and the injection of fresh air using ducts; such solutions, illustrated in Fig. 1, provide for maximum flexibility of the space inside since they can readily be reconfigured, especially for tertiary applications. Providing fresh air directly from outside (day time and/or nights, depending on the season), designated as "free cooling", to avoid using air conditioning can reduce energy consumptions. Another summer cooling method can consist in circulating the heating/cooling fluid used in a geothermal heat pump to heat the building during winter and cool it during summer, since the ground is then colder than the outside air. In addition to rooftop HVAC systems,

Figure 1.   Rooftop HVAC / © *Tupungato, Shutterstock*

shopping center roofs are increasingly used to install photovoltaic (PV) panels to cover some of the overall electricity consumption; ref. City Hall Tagum shopping center insert in the "Solar Energy" chapter.

To accommodate the changing loads of shops, especially since they are only sparingly used during nights, the air handling system must be designed to operate efficiently under such varying conditions,[8] i.e., at low power and not only at full power. One convenient solution is to install two systems, one operating alone at close to its maximum power when the air-handling load is moderate, and with both in operation under heavier load conditions. When using multiple systems, not only is the performance of each one important but also the coupling between them to reduce their combined energy consumption. An example is the use of the heat from a heat pump with a primary cooling objective to heat the building or to produce the hot water. While such couplings can be beneficial under normal operating conditions, they could be detrimental if the failure of one subsystem also causes the failure of the other.

Proper lighting is not only important for the overall ambiance but also to attract the customer's attention to the merchandise. The spatial optimization of the lighting system, coupled with the technology used, can have a significant impact on the energy consumption. For example, the transition to LED lamps (ref. "Housing" chapter), can divide the energy consumption by two compared to fluorescent tubes, which already consume three times less energy than incandescent bulbs.

Office Buildings and Hotels

Office Buildings

Energy costs generally represent less than 5% of the overall office buildings' operational and maintenance costs, which in turn, only

---

[8] The airflow of a fan, i.e., the air mass, m, coming through it per second, is proportional to the air speed, v. The kinetic energy of air is $0.5 \cdot m \cdot v^2$. The fan's airflow thus depends on v^3. As a result, when dividing the speed by two, and thus the air flow by two, the energy consumption is divided by eight (neglecting friction). This highlights the impact of fan controls. This v^3 relationship is the same for wind turbines, where the airflow through them imparts energy to them.

represent 10 to 30% of the personnel costs. Aside from an uncomfortable work situation for the staff, energy-saving actions have limited impacts on the overall productivity when compared to organizational or work method improvements. In spite of these ratios, the financial value of the potential cost savings may be sufficient to encourage corporations to move ahead, especially when an operations general services department exists. Corporate image may be another motivation, especially for "prestige" buildings such as corporate headquarters.

Hotels

Hotel energy issues are analogous to those of office buildings and teaching institutions since space is used intermittently: the rooms at night and the restaurants and conference rooms during the day. Energy costs can be significant in this growing sector — some 3 to 6% of the overall costs. The distributions between various energy consumption categories are significantly different for a 40-year-old 40-room tourist hotel or a 400-room modern hotel featuring a conference center along with a spa and pool. The occupation of the rooms by zones and decreasing the temperatures in the unused ones while providing for a quick increase when needed, are not only proper management tools but also reduce the hotel's energy consumptions!

Main hotel energy consumptions

The ranges indicated below were published in 2016 by an international hotel operator which includes a variety of hotel categories worldwide.

- HVAC, rooms, meeting rooms, etc.: 40–45%.
- Hot water: 10–20%.
- Electricity in the rooms: 5–10%.
- Kitchens, including cooking and refrigeration: 10–25%.
- Laundry: Less than 10%.
- Public space lighting, inside and outside: 3–10%.
- IT: 2–5%.

Hospitals

Hygiene and security to provide proper patient care and comfortable working conditions for the staff are primary considerations rather than energy consumption when designing and operating a hospital. Energy generally represents some 5 to 10% of a hospital's budget. Proper energy management must take several considerations into account:

- The temperature in the patients' rooms is to be maintained between 21 and 24°C.
- To maintain proper hygiene, only fresh air ventilation is used as opposed to circulation. The air is completely renewed at least 15 times per hour in surgery blocks to eliminate all contaminants.
- Most medical systems, such as a scanner, shown in Fig. 2, consume electricity and require a high-reliability supply. The power required for some of them, such as a magnetic resonance imaging (MRI) system can be quite high: 8 kW when not in operation and 20 kW when scanning.
- Back-up electricity supplies, generally on-site diesel generators, are required to maintain patient and staff safety along with equipment and fire protection in case of local network outages.

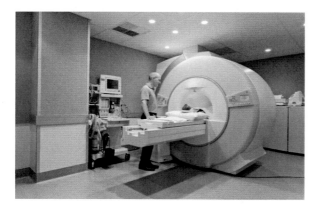

Figure 2. Scanner / © *Shutterstock*

Evolving medical practices and the associated medical technology developments, along with population evolutions may lead to renovations of existing hospitals or reconstructions on existing sites, when possible, or to entirely new constructions. In the latter situation, the comfort of both patients and staff can be better taken into account from the design of the facility. Separate air handling systems can be included to meet specific hygiene requirements as called for by modern health practices. Significant energy savings can then also be reached, especially regarding HVAC systems.

Worldwide Digital System

The deployment of digital systems is increasing worldwide and will continue to do so for decades to come, driven by evermore users, connected objects, applications and integrated services in all areas (health, finance, mobility, education, for example). This growth induces additional electricity consumption; however, deploying digital systems may, in turn, result in consumption reductions elsewhere, including, for example, improved process performances and decreased travel when replaced by teleconferencing.

The three main digital system subsystems are: (a) computers or data centers, (b) networks, and (c) peripherals (local area networks, personal computers (PCs), printers, etc.). From an energy point of view, the recent evolutions include [*itu.org*]:

- While the worldwide digital system energy consumption is difficult to accurately evaluate, it is broadly agreed that it is growing faster than the overall electricity consumption. Depending on the peripherals included, in 2017, the consumption was between 5 and 8% of the worldwide electricity consumption, i.e., between 1 000 and 1 600 TWh.[9]
- The peripherals represent roughly 45% of the total consumption, i.e., some 600 TWh of the median value from above; the networks represent 30% and the data centers 25%,

---

[9]The consumption related to the manufacturing of equipment may add another 50% which confirms the benefits of slower equipment replacements and recycling.

- By mid-2019, some 4.5 billion people were connected to the Internet. Based on the 600 TWh from above, the average yearly consumption per user is 130 kWh/yr; it is closer to 400 kWh/yr for an intensive user.

The electricity consumption per computer "elementary operation" keeps decreasing. However, the number of operations per second grows faster than the decrease in the per operation energy consumption; the number of computers and data centers also grows. As a result, the electricity consumption of computer servers is becoming a new challenge; this is especially true for large "supercomputers" such as those used, for example, to handle the large models required to study climate changes.

Super-computer energy consumptions

The Top 500 ranking of the 500 most powerful computers in the world [*top500.org*] not only gives the computing powers, measured in petaflops/s (one petaFlop is equal to 10^{15} Flop, floating point operation; in turn, a Flop designates the addition or multiplication of real numbers) but also the corresponding electricity they draw from the local supply, in MW. The computer's energy performance can thus be determined; it is the number of Flops per second and per Watt. For modern computers, it is given in GFlop/s·W; a GFlop is equal to 10^9 Flop.

In June 2020, the most powerful computer on the list was named Fugaku in Japan, with 415 PFlops/s when it draws 28.3 MW. The computer with the best energy performance was 393th on the list; it is also located in Japan and has a computing power of "only" 1.6 PFlop/s; its energy performance is 21.1 GFlop/s·W compared to 14.7 for the Fugaku.

The decrease in telecommunication costs allows for the re-localization of computer centers away from their users to take advantage of lower energy costs, which is illustrated in the following insert.

Swiss Scientific Computing Center

The center was constructed in Lugano, in 2012, to take advantage of the water in the Lake of Lugano; it is pumped some 2 km away from the center at a depth of 45 m where its temperature is 6°C all year round. A heat exchanger provides for the cooling of the water from the computing facilities using the lake's cold water. The heat from the computer center is also used to heat nearby office buildings. The center can draw up to 25 MW from the local electricity supply. In addition to other servers and data storage systems, the 2 000-m^2 computer room houses the CRAY "Piz Daint" Cray supercomputer, ranked 10th in the June Top 500 ranking; it draws 2.4 MW at a computing power of 21.2 PFlop/s. The backup supply is implemented by way of 960 batteries.

The Lugano location, far removed from the main users, primarily located in Zürich and Lausanne, was selected due to the some 30% electricity consumption reduction made possible by the lake's cold water; it also illustrates the high performance of the telecommunication networks.

How Electricity Is Produced

Introductory Remarks

Energy Sources

Since the end of the Second World War, Man has relied on three main sources of energy: renewables, fossils and nuclear.

Renewables

They emanate from the Sun's radiation, the Earth's sub-surface or from the Sun–Earth–Moon interactions.

Radiation from the Sun

With some 1 000 million TWh received yearly on the Earth's surface, it is almost 9 000 times larger than the 115 600 TWh worldwide final energy consumption in 2018. It is thus inexhaustible at Man's scale. However, it is not easy to use due to its variability, which can be significant over short time spans.

The Sun's radiation can be used on Earth:

- Directly, to use its heat,
- Indirectly:
  - To produce heat, which, in turn, can be used to produce electricity.
  - To produce electricity by:
    - Taking advantage of the natural water cycle.

- Using the photovoltaic effect.
- Extracting energy from the wind.

Biomass growth uses solar energy by way of photosynthesis. In turn, biomass can be used to produce heat, electricity or gas.

Geothermal energy

Geothermal energy emanates from the Earth's sub-surface and not from the Sun. It is also inexhaustible at Man's scale. However, its use can result in greenhouse gas and polluting emissions.

Marine energy

Currently, the most commonly used form is tidal energy that is due to the Earth's movements relative to the Moon and the Sun. While still largely untapped, tidal energy's potential is immense at Man's scale. While variable, its availability can be accurately predicted long in advance.

Fossil

Strictly speaking, fossil energies are also renewable but at time scales that are incompatible with those of Man, thus effectively making them non-renewable. For example, the transformation of vegetal sediments into coal takes several million years.

Proven and readily accessible reserves of coal were estimated at 1 055 Gt in 2018; they are broadly distributed worldwide. The production of coal was 7.8 Gt[1] in 2018 [*iea.org*] such that reserves are sufficient to cover our needs for at least 135 years. For both oil and natural gas, the proven reserves are sufficient to last at least 50 years. In addition, recent discoveries of new types of fossil resources, such as shale gas and shale oil, along with the development, at reasonable costs, of new extraction techniques of these resources, further increase the reserves.[2]

---

[1] China's production represented some 45%.
[2] The "peak oil" worry that was prevalent 10 years ago is no longer a concern.

Fossil resources are not infinite. Aside from the emissions caused by their combustion, it behooves us to leave as much as possible for future generations and thus primarily use them in petrochemical and pharmaceutical processes, for which carbon is indispensable, rather than burning them to produce heat.

Nuclear

Uranium and thorium reserves are also broadly distributed worldwide. While nuclear energy remains highly controversial, when in operation, nuclear power plants do not emit any CO_2. The proven and readily accessible reserves of uranium are sufficient to operate the existing nuclear power plants worldwide for some 85 years; the estimated resources are three times larger [*iea.org*]. In addition, present nuclear power technologies only enable the use of some 5% of the energy contained in the fuel; technologies under development would significantly enhance that proportion.

AC and DC Currents and Voltages

As further discussed in the "Electric Power Systems" chapter, the early investigations dealing with electricity and, as a result, its first applications were such that the currents and voltages involved did not vary with time. They are said to be DC, for direct current, similarly to the AAA and AA batteries we are all familiar with.

The power drawn by an electric load is the product of the voltage across it and the current through it. It follows that for a given power, P, drawn by a load, if the voltage across it doubles, the current through it is divided by two. Furthermore, the losses in the wires connecting the electricity source to the load depend on the square of the current through them such that if the current is divided by two, the losses are divided by four. Therefore, to reduce the losses, the voltages should be as high as possible. The main problems with the DC electrical systems first deployed in the late 1800s were that: (a) generators could only be built with limited voltage capabilities and (b) using electricity at higher voltages would be dangerous for the end-user — it would be ackward and not safe to use 2 000 V appliances!

What was needed is a "device", which would make it convenient to increase the generators' output voltages before supplying the electricity to the network bringing it to loads located at a distance and then reduce the voltages before it is consumed. That device is a transformer, which safely and efficiently changes the voltage levels from its input terminals to its output terminals[3]; it is further described in the "Electric Power Systems" chapter. Transformers only work if the voltages at its primary terminals vary with time, such that they reach a maximum and a minimum value a number of times per second; they are said to be alternating along with the currents, hence the designation AC. The frequency of AC currents and voltages is defined as the number of cycles per second; it is measured in Hz.[4]

The problem remained of designing and manufacturing generators that could produce AC electricity and motors which could operate under AC conditions. This was one of the major contributions from **Nikola Tesla**. He demonstrated that by using AC generators and motors, interconnected via long distance lines and transformers, electric power systems could reliably and efficiently be deployed over entire continents.

Present electrical power systems overwhelmingly rely on AC currents and voltages, primarily due to the ease of adapting voltage levels to various operating situations. While it is true that modern power electronics could once again make DC electric power systems technically feasible, the sheer size of worldwide legacy of AC systems is such that DC electric power systems are likely to be only used for specific applications for decades to come.

Electricity Production

As shown in Fig. 1, and in the following table, between 1973 and 2018:

- The electricity production increased by a factor of 4.34, which is consistent with the information regarding its consumption, as seen in the "Energy and Emissions — Where We Are" chapter.

---

[3]They can either "step up" the generator voltages or "step down" the load voltages.

[4]In most countries, the frequency of electric power systems is 50 Hz; it is 60 Hz in some regions of the world such as northern America, some countries in Latin America and the southern part of Japan.

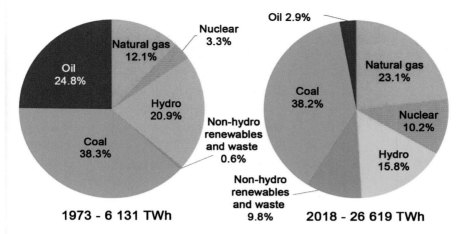

Figure 1. Electricity production by source, 1973 and 2018 / Data from the International Energy Agency

- Nevertheless, the production per capita increased only by slightly more than two, from 1.56 to 3.51 MWh/cap. Indeed, the worldwide population almost doubled during the same time span.
- Overall, the use of fossil fuels decreased from 75.2% to 64.2%, primarily due to the significant decrease in the use of oil. The use of coal increased by more than four times, while that of natural gas increased more than eight times.
- In 2018, hydro energy remained, by far, the principal renewable electricity production source.

Electricity Production By Source

	1973	6 131 TWh	Population	3.92 Billion	1.56 MWh/cap
	2018	26 619 TWh	Population	7.59 Billion	3.51 MWh/cap
Ratio		*4.34*		*1.94*	*2.24*

	1973		**2018**		
	%	**TWh**	**%**	**TWh**	**Growth**
Coal	38.3%	2 348	38.2%	10 168	4.33
Oil	24.8%	1 521	2.9%	772	0.51

(Continued)

(Continued)

Electricity Production By Source

	1973		2018		
	%	TWh	%	TWh	Growth
Nat. Gas	12.1%	742	23.1%	6 149	8.29
Nuclear	3.3%	202	10.2%	2 715	13.42
Hydro	20.9%	1 281	15.8%	4 206	3.28
Non-hydro Renewables and Waste	0.6%	37	9.8%	2 609	70.91
	100.0%	6 131	100.0%	26 619	

Furthermore, the 2 609 TWh under non-hydro renewables and waste were decomposed as follows:

Type	TWh	Percentage of Total Production
Solar PV	550 ⎫	2.1%
Solar Concentrating	12 ⎭	
Wind On-shore	1 195 ⎫	4.7%
Wind Off-shore	68 ⎭	
Geothermal	88	0.3%
Bioenergy	522	2.0%
Waste	174	0.7%
Total	**2 609**	**9.8%**

Largest Power Plants in the World

As of early 2019, the total worldwide installed generation capacity was some 7 000 GW; the largest ones include:

- The five power plants with the highest installed capacities in the world were all hydro, with the Three Gorges in China leading the way at 22 500 MW. Eleven of the 15 power plants with the highest installed capacities worldwide were hydroelectric; the remaining four were nuclear.

- The Kashiwazaki-Kariwa nuclear plant, in Japan, with an installed capacity of 7 900 MW, was the largest in the world before its operation was suspended after the Fukushima 2011 accident. The Bruce nuclear power plant, at 6 300 MW, in Canada, was the largest one in operation worldwide.
- The fossil power plant with the highest installed capacity in the world was Shoaiba, at 5 600 MW, in Saudi Arabia. It burns oil; although quite unusual, it is explained by the country's vast oil reserves. It is coupled with a desalination plant.
- The Surgut-2 power plant in Russia also had an installed capacity at 5 600 MW. The fuel is natural gas, which is explained by Russia's vast natural gas reserves.
- The coal-fired power plant with the highest installed capacity in the world was Taichung, with 5 500 MW, in Taiwan. It is burning mostly imported coal.
- The largest wind farms had installed capacities below 5 000 MW, and the largest solar power plants did not reach 1 000 MW.

Power Plant Efficiency

The overall efficiency of a power plant is the ratio between the plant's electric output energy and the input energy supplied — hydro, thermal, fossil or nuclear, solar, wind, tides, tidal current. This efficiency is crucial for the financial assessment of each power plant project; this is particularly the case given the long lifetimes planned for some types of power plants — 50 years and beyond for hydro plants.

Power Plant Capacity Factor

The capacity factor is an important index used to compare the operational possibilities of various power plants related to their underlying technologies, physical implementation site and operational restrictions, if any (limit on the yearly emissions, for example).

It is defined as the ratio between the electricity actually produced by the plant during a year and the electricity it could produce if operated at full capacity during an entire year, i.e., during 8 760 hours.

The capacity factor is typically expressed as:

- A percentage

or

- The number of hours per year the plant would have to operate at full power to produce the electricity it actually produced during a year.

For example, a capacity factor of 20% corresponds to 1 752 hours/year at full power.

The capacity factor of a power plant incorporates several key operational aspects, such as:

- The variations, intermittencies, of the raw energy availability during the year:
  - River water flow variations and yearly variations of water supplies into hydro dams.
  - Wind and solar energy variations.
- The need for refueling shutdowns as required for nuclear power plants.
- Regular, periodic, maintenance requirements.
- Failures of the plant itself, resulting in more or less prolonged outages.[5]

A particular power plant may be technically able to produce more energy during a year than indicated by its capacity factor. For example, a fossil power plant might not be fully used due to its higher operational costs or for environmental impact reasons. As a result, the capacity factor

---

[5] On occasion, the electric distribution or transmission network to which the plant is connected might not be available due to various events external to the plant. Such outages, when prolonged, are not included in the plant's capacity factor; for example, if the local network was not available 200 hours during a year, which would be quite unusual, the capacity factor would only be based in 8 560 hours for the year.

of a particular power plant may not only be indicative of its particular technology but also of its physical implementation and operational circumstances.

CO₂ Emissions

As seen in the "Energy and Emissions — Where We Are" chapter, in 2016, the CO_2 emissions represented 75% of all greenhouse gases, i.e., 34.6 $GtCO_{2e}$ [*wri.org*]. In Fig. 2, the IEA [*iea.org*] indicates that in 2018 the CO_2 emissions strictly from fuel combustion were 33 513 $MtCO_2$. As shown in the figure, between 1973 and 2018, the emissions due to:

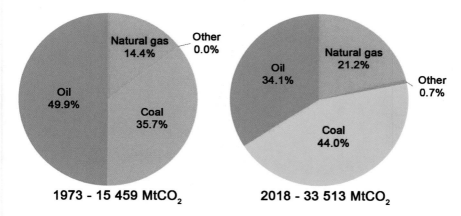

Figure 2. Fuel shares of CO_2 emissions from fuel combustion, 1973 and 2018 / *Data from the International Energy Agency*

- Natural gas went from 2 242 to 7 105 $MtCO_2$, i.e., they more than tripled.
- Coal went from 5 504 to 14 746 $MtCO_2$, i.e., they almost tripled.
- Oil went from 7 714 to 11 428 $MtCO_2$, i.e., they increased by 48%.

As to the emissions on a per capita basis, they went from 3.94 in 1973 to 4.42 tCO_2/cap in 2018, confirming that humanity has not yet reduced its emissions.

CO₂ Emissions from Electricity Production

The worldwide average CO_2 emissions from electricity production were 485 gCO_2/kWh in 2017 (the most recently available IEA data); it was 532 gCO_2/kWh in 1990 (the earliest IEA data).[6] The following table shows the evolution of CO_2 emissions due to electricity production worldwide as well as in some regions and countries.

Emissions (gCO_2/kWh)	1990	2000	2010	2017	Change 1990–2017 (%)
World	532	537	528	485	−8.8
OECD	507	495	441	384	−24.2
Africa	684	665	626	568	−17.0
Asia	642	670	663	610	−5.0
China	911	893	749	623	−33.0
Middle East	742	708	678	641	−13.6

Hydro, Wind and Thermal Turbines: Providing Rotating Energy to Electric Generators

Aside from photovoltaic electricity production, which is a direct, static, process, all other electric power plants presented in the six electricity production chapters rely on one type of turbine, or the other.

While hydro, wind and thermal turbines all provide rotating energy to electric generators, they do so relying on different principles:

- *Hydro turbines*, ref. "Hydro Energy" chapter, convert potential or kinetic energy from water, either stored in dams or flowing in rivers or due to tides.
- *Wind turbines*, ref. "Wind Energy" chapter, convert the kinetic energy in the blowing wind.

---

[6] IEA — CO_2 Emissions from Fuel Combustion, 2019.

- Thermal turbines are of two main types:
  - *Steam turbines*, where hot steam expands as it flows through the various turbine stages and their properly shaped blades. The heat production occurs outside of the turbine itself. The hot steam can:
    - Be produced by running water through pipes heated in a boiler burning coal, oil, natural gas, or biomass; ref. "Combustion Thermal Power Plants" chapter.
    - Come from heat exchangers or steam generators where the primary heat is provided by nuclear fission reactions; ref. "Nuclear Thermal Power Plants" chapter or by solar thermal power plants; ref. "Solar Energy" chapter.
    - Come from geothermal sources; ref. "Geothermal Energy" chapter.
    - Be wasted steam from industrial processes.
  - *Gas turbines*, where the combustion occurs inside the turbine itself. As the heated exhaust gases exit the combustion chamber, they expand thus driving the rotating turbine rotor blades. The shaft not only drives the generator but also the compressor which forces air into the combustion chamber. For power plant applications, the combustion fuel used is predominantly natural gas and only rarely oil.

Power Plant Indicators

The operating and environmental impact factors for each of the six electric power plant technologies presented next can be found under "Power Plant Indicators" in the Companion Document.

Hydro Energy

Using Water

Man has always collected rainwater for drinking and irrigation purposes; for centuries, he has also used the water cycle to assist with tasks such as grinding or water pumping. For over a century, the main application of hydro energy has been electricity production by way of hydro turbines. The world's first hydro power plant was installed in 1882 at Appleton on Wisconsin's Fox River, United States (US), with an installed capacity of 25 kW. The first hydro power plants with installed capacities of over 100 MW were constructed in Western Europe and North America over a century ago.

As electric networks became ubiquitous, smaller and often artisanal installations fell by the wayside in favor of larger plants. Concurrently, the technical and economic feasibility of larger power plants led to, and was enabled by the development of higher voltage,[1] longer distance and interconnected networks.

Before the Second World War, hydro power plants provided the bulk of the electricity production in most industrialized countries. As available sites for new major hydro plants became increasingly difficult to identify, particularly in Europe, large bulk fossil and nuclear power plants came on line while the role of selected storage hydro plants, presented later, evolved toward providing for load variations while thermal power plants provide the baseload.[2] As will be further discussed in the chapter dealing with the

---

[1] To limit the losses to a few percent of the energy delivered; ref. "Electric Power Systems" chapter.

[2] A power plant is said to be "base loaded" if it is primarily operating at or close to its installed capacity, with outages primarily due to maintenance or breakdowns.

overall electric power system, the expanding role of alternative renewable energy sources, such as solar and wind, will make hydro power plants even more important, providing for energy storage between changing loads and varying availabilities of these new sources of electricity.

A renaissance of smaller hydro power plants is underway to make enhanced and local use of the energy resources available in drinking water and water treatment systems, for example. Moreover, emerging technologies seek to use ocean waves, marine currents and ocean thermal energy, in addition to a few tidal power plants already in operation.

While the potential for new major hydro power plants is limited in OECD countries, significant development possibilities remain elsewhere; indeed, the worldwide installed capacity of hydro power plants continues to grow. Although their operation is CO_2-free, the main limiting factors to build new major hydro power plants generally are acceptance by local populations, environmental impact issues, and in some instances, installation costs.

Statistics

Hydro energy remains, by far, the largest electricity renewable source in the world with a 15.8% contribution in 2018 [*iea.org*] compared with 64.2% for fossil fuel-based production, 10.2% for nuclear energy and 9.8% for other sources, including non-hydro renewables.

As of 2019, the total worldwide installed hydroelectric capacity had reached 1 308 GW; the hydro production was 4 206 TWh versus 1 281 TWh in 1973. Also, in 2019, with an installed capacity of 356 GW and a production of 1 302 TWh, China had the largest hydro installed capacity and production in the world [*hydropower.org*]. At 95.7%, Norway had the largest contribution of hydro energy to its overall electricity production.

As of 2019, the largest power plant in the world, all production technologies included, is the Three Gorges in China, with an installed capacity of 22 500 MW [*ctg.com.cn/en*]. The Itaipu power plant, on the Parana River along the Brazil-Paraguay border, is the second-largest one at 14 000 MW [*itaipu.gov.br/en*]. Also, in 2019, the ten largest hydroelectric power plants in the world all had installed capacities of above 6 000 MW, the 60 largest ones were above 2 000 MW and over 50 000 dams were said to be large, i.e., higher than 15 m, primarily for irrigation and electricity production purposes [*icold-cigb.org*].

The Natural Water Cycle

Water is primarily stored in oceans and lakes as well as in mountain glaciers, the Arctic and Antarctica. As shown in Fig. 1, energy from the Sun provides for: (a) evaporation from oceans and lakes, (b) evapotranspiration from the ground and trees, and (c) direct sublimation.[3] The water is condensed as drops or ice, creating clouds and is precipitated back to ground as rain, snow or hail. The cycle closes as rain runs off into oceans and lakes or transits through groundwater and as glaciers melt under the influence of the Sun. The annual natural precipitation over the entire surface of the Earth is some 500 000 km³, or 500 Ttons [*unep.org*]. This

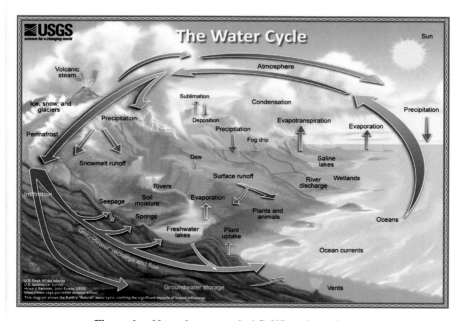

Figure 1.   Natural water cycle / © *US geological survey*

---

[3] I.e., the direct transformation from snow or ice into vapor without a transition through the water phase.

amount of precipitation would cover the 48 contiguous states of the US with 60 m of water or the European Union with 110 m.

In tropical regions, the water cycle is often repeated daily. In colder climates, it typically evolves yearly with the seasons. Long-term climate evolutions will impact the water cycle and/or rain distribution patterns, which, in turn, will affect local hydroelectric production patterns.

Hydro Power Plants — Dams

Two main types of hydro power plants are primarily used to harvest energy from the water cycle: run-of-the-river and storage.

Run-of-the-River Plants

They take advantage of a natural drop in a riverbed. Since the water simply runs through them, they do not disrupt the river's flow; its temperature is also unaffected. The most common types are:

- Using a dam built directly across the river to collect all or part of its flow; they were often initially built for flood control purposes with electricity production as an additional benefit. The power house, with the turbine-generators, is typically located inside the dam itself. Even though the dams' heights typically do not exceed a few meters or tens of meters, artificial upstream lakes results, thus providing for some flow control. Navigation can be provided for by way of a lock, aside from the dam itself.
- Also using a dam, part of the river's flow is diverted from the main flow, either by way of open channels or closed conduits, to a power house located further downstream, after which the water is returned to the main river. This approach provides for a higher fall, thus higher electricity production; an artificial upstream lake is not created. On the other hand, the river flow is reduced between the water intake and release points.

Several run-of-the-river power plants may be implemented along one same river. Since only limited or no local water storage capacity is provided for, the electricity production is entirely dependent on the natural river flow as it evolves with the seasons during the year. Electricity production is lost if the available water flow is not fully taken advantage of when available.

The Bonneville Dam run-of-the-river power plant, US

Built with the initial main intent of flood control of the Columbia River, it is located on the border between the Oregon and Washington States. As shown in Fig. 2, it features two power houses, the first one with ten turbines and the second one with eight, with combined installed capacities of 600 and 642 MW, respectively. The flood gates, used to release excess water, are in the middle. The navigation lock can be seen at the right of the figure. All 18 turbines are Kaplan types, described below. The plant has a fish ladder to protect the local fisheries.

In 2009, for example, electricity production was 4 466 GWh. The capacity factor thus was 41%,[4] which confirms that the Columbia River flow changes with the seasons.

Figure 2. Bonneville run-of-the-river dam (United States) / © *US Corps of Engineers*

---

[4] Capacity factor; ref. "Introductory Remarks" chapter under "How Electricity Is Produced". In this case: 4 466 000 MWh / (8 760 h · (600 + 642) MW) = 41% or 8 760 h/yr · 41% = 3 592 h/yr at full power.

Storage Plants

They may feature sometimes very large storage capacities to accommodate daily, weekly or even seasonal fluctuations of the water inflows. The three main types are:

- *Gravity dams*, which hold the water back by their sheer weight. They are dimensioned such that they neither slide nor tilt over under the pressure of the water stored.

Grande Dixence gravity dam, Swiss Alps

Completed in 1961, it is shown in Fig. 3. At 285 m, as of 2019, it was the highest gravity dam in the world. It is 695 m long at the top and 200 m thick at its base. Fifteen million tons of concrete were poured during its construction. Its water intake basin covers 420 km^2, including 35 glaciers and 80 water intake points from several valleys. It can store up to 400 million m^3 of water.

The Bieudron power plant, supplied by this dam, and featuring Pelton turbines, is described later in this chapter.

Figure 3.   Grande Dixence dam (Switzerland) / © *Grande Dixence SA Photo: essencedesign.com — www.dpicard.ch*

• *Arch dams*, where the water is held back by way of its pressure on a carefully computed shape such that the pressure is transferred to the valley walls onto which the dam is solidly anchored. The configuration of the valley itself is critical — shape and rock resistance. The actual shape of the arch is computed similarly to cathedrals and arched bridges, which transfer their weights to their surrounding walls. The dam is typically roughly 10 meters wide at the top and no more than a few tens of meters thick at the base, thus requiring far less concrete than gravity dams. Figure 4 shows the Hoover Dam in the US, which was completed in 1936.

Figure 4.   Hoover dam (United States) / © *US Department of Agriculture/NRCS*

Jinping-I arch dam, China

Completed in 2015 and located on the Yalong River in the Sichuan Province of China, it is the tallest arch dam in the world at 305 m. The artificial lake has a capacity of 7.7 billion m^3, of which 4.9 billion m^3 can be used as active storage while the other 2.8 billion m^3 must remain in the artificial lake to maintain the pressure required to keep the dam solidly anchored to the valley rock formation.

Electricity is produced using six 600 MW Francis turbines, which are described later in this chapter.

Figure 5.   Serre-Ponçon earthfill dam (France) / © *Christophe BEN*

- *Earth and rock-fill embankment dams*, which represent over 70% of all large dams; they are also gravity dams but built using rocks, dirt and waterproof clay. Figure 5 shows the Serre-Ponçon dam in France. Most of them were constructed for flood control and irrigation purposes, which explains why several are over a hundred years old. Some were later updated to include electricity production; others were constructed with electricity production as the main purpose.

Tarbela earth-fill embankment dam, Pakistan

Located in the Pakhtunkhwa region, as of 2017, it was the largest earth-fill embankment dam in the world. Consisting of a main earth-filled dam and two shorter concrete gravity dams, it dams the Indus River. It is 148 m high and spans 2 740 m at its crest. The total reservoir capacity is 14.3 billion m^3, of which 12 billion m^3 can be used as active storage while 2.3 billion m^3 provide irrigation reserves. Since these are gravity dams, water pressure behind them is not required for structural integrity. Four main tunnels were dug through the main earth-fill embankment dam: one for irrigation purposes and three to supply energy to 14 Francis turbines with an installed capacity of 3.5 GW.

Spillways and Penstocks

For all types of dams, emergency water evacuation, such as in case of floods, is provided by spillways, an example of which is shown in Fig. 6.

Figure 6.   Spillway, Tarbela dam / © *Akhmad Dody Firmansyah, Shutterstock*

Figure 7.   Srinakarin hydro power plant, Thailand — penstocks and power house / © *OrapanCh, Shutterstock*

Water is brought from the dam to the power house and the turbine-generators by way of penstocks as shown in Fig. 7 along with the power house; they can also be imbedded in the dam itself with the power house or be underground.

Water Hammer and Surge Tanks

Under emergencies, it may be necessary to rapidly interrupt the water flow to the turbine to avoid severe damage to the turbine-generator assembly. Such rapid interruptions may create a shockwave traveling back up through the water trapped in the penstock. Commonly referred to as a water hammer, it may, in turn, cause severe damage to the penstock itself.[5] To avoid water hammer damages, a vertical surge tank is added reaching from the end of the penstock, i.e., close to the emergency valve, to the highest level reached in the storage dam; they can thus be several hundred meters high.

Hydro Power Plants — Turbines

Hydraulic Power

Intuitively, the hydraulic power delivered to a hydro turbine directly depends on:

- The water flow rate, q, in m^3/s, along with the density, ρ, of water, i.e., 1 000 kg/m^3,
- The pressure, i.e., the head, h, in meters, between the water intake and the power house,
- The gravity, g, on Earth, i.e., 9.81 m/s^2.

Neglecting the frictional and other losses between the water intake and the turbine, the hydraulic power, P_h, in W, can thus be expressed as:

$$P_h = q \cdot \rho \cdot h \cdot g$$

This means that the hydraulic power obtained from a flow of 100 m^3/s and a head of 500 m is roughly 500 MW.[6] The same power would be obtained with a flow of 500 m^3/s and a head of 100 m.

Based on experience and laboratory experiments as well as simulation techniques, a number of turbine designs have been perfected for a broad range of operational conditions — primarily head and flow.

---

[5] In older houses, the pipes may make a rattling sound when a faucet is rapidly turned off — the hammer effect at work!

[6] 100 m^3/s · 1 000 kg/m^3 · 500 m · 9.81 m/s^2 = 490.5 10^6 kg·m^2/s^3 = 490.5 MW.

Turbine and Generator Losses

The proper design of the turbine wheel, or runner, is such that the hydraulic energy delivered to the turbine is converted into mechanical rotational shaft energy with losses that are as small as possible, typically a few percent, such that the efficiencies of larger modern turbines reach 90%. In turn, the rotational shaft energy is converted into electric energy in the generator, sharing the same shaft with the turbine, with efficiencies which routinely reach 95% for large modern machines. The combined turbine-generator efficiencies are thus in the 85–90% range for the larger power plants built since the later part of the 20th century.

Pelton Turbine[7]

The Pelton turbine, shown in Fig. 8, was developed starting from the water wheel. It can have a vertical or horizontal axis. It has buckets around its periphery, which are "hit" by one or more high-speed water jets around its periphery; it is referred to as an *impulse* turbine. The properly shaped buckets ensure that the incoming water jet's kinetic energy is almost entirely transferred to the turbine.

Bieudron power plant Pelton turbines, Switzerland

The Bieudron hydroelectric power plant is supplied from the Grande Dixence dam presented above. The head between the water intake and the power house is 1 883 m, which was the highest head of any hydroelectric power plant in the world, as of 2019. The power plants' installed capacity is 1 269 MW, i.e., equivalent to a modern nuclear reactor. Each of the three Pelton turbines is rated at 423 MW, making them the largest Pelton turbines in the world. Shown in the figure below, the diameter of each one is 4.6 m. Each is supplied by five injectors with a combined water flow rate of 25 m^3/s; the diameter of each jet is 193 mm when it hits the buckets at 192 m/s, i.e., 690 km/h.

Pelton turbines are used for high to very high water heads, up to 1 800 m and for low water flows. They can be designed to have a mechanical output power of several hundreds of MW per wheel. Some turbines feature several injectors around their periphery to increase the water flow, as shown in the figure.

---

[7] Named after the American inventor **Lester Alan Pelton**, who perfected it during the 1870s.

Figure 8.   Bieudron plant Pelton runner / © *Alpiq-Grande Dixence SA_ Heinz Preisig*

Francis Turbine[8]

Francis turbines, shown in Fig. 9, are primarily used for heads of a few tens to several hundred meters. They are the most commonly used turbines; they can accommodate very high flows. Their rated power can exceed 700 MW; the efficiency of modern ones can be above 90%. They are referred to as *reaction* machines since they are completely immersed in the water with which they react to create rotational motion.

The turbine assembly consists of:

- The *spiral casing*, which ensures that the incoming water flow is evenly distributed around the entire periphery of the turbine. To this end, its cross-section decreases around the periphery.
- The stationary *guide vanes*, which convert the pressure from the incoming water into kinetic energy before reaching the runner blades.

---

[8] Named after the British-American engineer **James Francis,** who perfected it during the 1850s.

- The set geometry *runner*, with carefully designed blades to optimize the energy conversion from the incoming water flow kinetic energy to the rotational energy at the shaft.
- *The draft tube*, through which the water leaves the turbine assembly with the lowest speed possible.

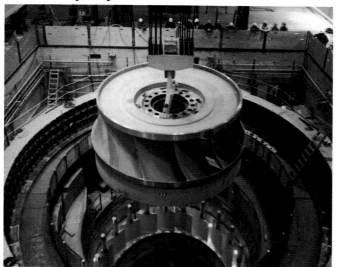

Figure 9. Francis turbine — runner and guide vanes / *Pierre Henry, Turbomachines hydrauliques, Presses polytechniques et universitaires romandes, 1992*

Three Gorges hydroelectric power plant, China

The Three Gorges power plant, located on the Yangtze River, was the largest in the world as of 2019. Fully functional as of 2012, it serves three purposes: electricity production, flood control and navigation improvement. Its gravity dam is 180 m high and 2 335 m long; it is 40 m wide at its crest. Its combined installed capacity is 22 500 MW consisting of 32 Francis turbines, each rated at 700 MW and two 50 MW Francis turbines to supply the plant's auxiliaries. The average capacity factor is 45%, i.e., 3 940 h/yr at full power. In 2014, it set a new world record for annual electricity production for one single power plant at 98.8 TWh.[9]

(Continued)

[9]As of 2016, the record is 103.2 TWh at the Itaipu power plant in Brazil and Paraguay. Since its installed capacity is "only" 14 000 MW, the capacity factor was 84%, confirming that the Parana River's flow is steadier than the Yangtze River flow.

(Continued)

The power plant also provides for enhanced navigation between Shanghai and Chongqing, i.e., 2 400 km, thanks to two five-stage lock systems (one for upstream and one for downstream navigation).

Kaplan Turbine[10]

An evolution of the Francis turbine, it is derived from a maritime propulsion propeller. It is also a *reaction* turbine. It is well suited for low heads, below 100 m, and a broad range of water flow intensities. The ratings typically range from a few tens of kW up to 200 MW.

As for a Francis turbine, a Kaplan turbine assembly features a spiral casing to evenly distribute the water flow around its periphery. The water flows into the turbine by way of adjustable wicket gates, which can regulate the water flow and hence the power output. The blade pitches of more sophisticated Kaplan turbines can also be adjusted to allow for a double regulation along with the wicket gates, thus providing for the turbine's performance optimization over a broad range of operational conditions.

The two main installation types of Kaplan turbines are:

- Vertical axis machines are used for higher heads and higher powers; the turbine's diameter can reach 10 m.
- Horizontal axis machines, shown in Fig. 10, sometimes referred to as "bulb" machines. In this case, the wicket gates, as well as the generator, are underwater with the turbine. They are typically installed directly in a riverbed, thus accommodating for lower heads but larger flows; they are frequently used for run-of-the-river plants as well as for tidal power machines.

---

[10] Named after the Austrian engineer **Viktor Kaplan** (1876–1934), who developed it in the early 1900s.

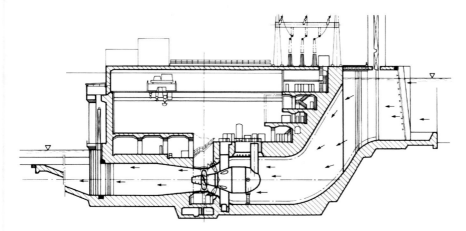

Figure 10. Kaplan bulb turbine / *Pierre Henry, Turbomachines hydrauliques, Presses polytechniques et universitaires romandes, 1992*

Pump-storage Power Plants

As further discussed in the "Energy Storage" chapter, pump-storage hydro plants, sometimes referred to as pump-turbine, are essential for the large-scale deployment of renewable energy sources such as solar and wind. As shown in Fig. 11, similarly to "standard" hydro storage power plants, energy is produced when water flows from an upper reservoir; however, in this case, the water is released into a lower reservoir, which can be natural, such as a lake or even the sea, or an artificial lake.[11] Daily load variations combined with the availability of excess generation, renewable or not, can occasionally make it economically advantageous to pump water back up into the upper reservoir to make it once again available for production when necessary.

It is important to highlight that, aside from the natural water inflows, pump-storage hydro plants do not produce energy; the energy outcome of a turbine-pump cycle is a loss.

Modern installations reach efficiencies above 80% for an entire turbine-pump cycle. Aside from Pelton turbines, most turbines can also be used as pumps either by reversing their rotational direction or by inverting

---

[11] The arch dam forming the lower lake of the pump-storage plant shown in Fig. 11 is visible to the lower left of the figure.

their blade pitches, as for Kaplan machines. Recent developments of electronic adjustable speed drives have significantly contributed to the operational flexibility of pump-storage power plants.

Figure 11.　Cortes La Muela pump-storage power plant (Spain) / © *Burakyalcin, Shutterstock*

The pump-storage power plant with the highest rated power, Bath County, US

Located in Virginia, it was originally completed in 1985. Upon the replacement of its six vertical axis Francis turbines, completed in 2009, it became the most powerful pump-storage power plant in the world. The installed capacity is 3 000 MW in the turbine (production) mode and 2 880 MW when pumping. The overall efficiency turbine — pump cycle reaches 78%; the full rotational speed, 257 RPM, can be reached in 6 minutes upon startup.

Both the upper and lower reservoirs are earth-fill embankment dams. The upper reservoir, with an area of 1.23 km² when full, can store up to 44 million m³ of water; the lower one, with an area of 1 km² when full, 380 m below, can store up to 34 million m³. The highest turbine flow rate is 850 m³/s, and the highest pumping rate is 800 m³/s. This means that the upper reservoir could be completely emptied in 14 hours and the lower one in 12 hours, thereby confirming the intended daily operational cycle of the plant. Given the relatively small surface of the two lakes, their levels can change quite significantly during turbine and pump sequences.

By 2019, the total installed capacity of pump-storage plants worldwide was 158 GW [*hydropower.org*]; it continues to increase. The turbine and pumping installed power capacities, the storage capacities of the upper and lower reservoirs as well as the rates by which they are allowed to change, set the operational limits of these facilities.

Mini Hydro Power Plants

While a worldwide agreed-upon definition of mini hydro plants does not exist, they are generally rated below 1 or 10 MW. As their smaller sizes do not justify highly sophisticated design implementations, the inevitable mechanical and electrical losses become more important relative to the installed capacity. Their efficiencies generally are in the 40 to 60% range. Even though mini hydro power plants generally have higher installation as well as higher operational and maintenance costs, they are increasingly built to harvest the potential energy available in urban water treatment and distribution systems.

Marine Energies

Harvesting the enormous marine energy potential remains a challenge. While accurate estimates are difficult, it is generally agreed that using all forms of marine energy, the theoretical installed capacity could reach several thousands of GW, and the annual energy produced could approach 100 000 TWh, i.e., of the same order of magnitude as the total final energy consumption worldwide.

The three main marine energy paths pursued are:

- Tidal or marine currents power plants, with dams (tidal power plants) or without dams (in-stream machines), which are further discussed later.
- Wave power plants; on-going full-scale prototype testing are necessary before these technologies are ready for broad implementation in what are generally hostile physical environments.
- Ocean Thermal Energy Conversion (OTEC), which aims to leverage the temperature differences between the bottom and at the surface of oceans. While several experimental OTEC systems have been built

since 1880, the design of industrial-sized systems remains at the conceptual level.

Tidal and In-stream Power Plants

Tides are primarily influenced by the position of the Moon; the tides are higher when it is closest to the Earth, i.e., when its attraction is the strongest. The Sun's tidal contribution is roughly half of that of the Moon; it can either increase or decrease tidal amplitudes. Coastline and local ocean bottom configurations also influence the tide levels. Despite the many factors influencing both the periodicity and level of each tide at a particular location, it can be accurately computed very long in advance. Highly detailed tables are available, indicating the time and amplitude of tides around the world; these tables are essential not only for navigational purposes but also to properly integrate the electricity production of tidal power plants.

The main types of tidal power plants are:

Tidal dam power plants

Tidal dam power plants essentially are run-of-the-river plants taking advantage of tidal rather than river flows. Kaplan machines are generally used; since they are reversible, depending on the local circumstances, energy can be produced when the tide is coming in or ebbing. Furthermore, depending on local circumstances, the Kaplan machines can also be used as pumps; the tidal height and timing can thus be adjusted to optimize the production cycle, environmental constraints allowing.

The Rance River tidal power plant, France

Inaugurated in 1966 and as of 2019, the largest reversible tidal power plant in the world, it is located at the estuary of the Rance River in Brittany, where the tidal amplitudes reach 13.5 m, one of the highest in the world. Figure 12 shows the entire power plant with the Rance River (upper right) and the British Channel (lower left) along with the navigational sluice at the bottom of the picture. Twenty-four Kaplan bulb turbine-generator assemblies (each rated at 10 MW) are located in the dam itself and can be seen to the left of the lock. In the figure, the water is flowing from the river to the Channel.

In-stream tidal power plants

Tides also induce significant currents in ocean straights, which can be used to produce electricity; currents in sufficiently deep rivers could also be valorized. Several turbine configurations have been proposed, primarily derived from the Kaplan turbine. Industrial-sized in-stream tidal turbines, rated at a few MW, while having significant potential, are at the dawn of their development. Aside from the initial construction and installation costs, the actual maintenance and, if required, repair of in-stream tidal machines remain problematic. The impact on the local underwater fauna also needs to be considered.

Figure 12. La Rance tidal power plant, France / © *Benoit MAREMBERT/ www.survoldefrance.fr*

Accidents

A broad range of safety measures is incorporated during the design and construction of hydro power plants. Their operating status is continuously monitored using several types of sensor networks. Despite these precautions, a few serious accidents have occurred; a brief description of four such accidents can be found under *"Hydro Power Plant Accidents"* in the Companion Document. It is important to keep in mind that in 2019, well over 8 000 hydro plants rated above 1 MW were safely operating worldwide.

Hydro Power Plants in their Environmental and Societal Contexts

Oppositions to the construction and operation of hydro plants, when voiced, primarily relate to:

- The potential flooding of villages and towns as reservoirs are built.
- Visual and environmental impacts on often pristine natural sites.
- Impacts on fish migrations. Fish ladders are added to mitigate this impact. Minimum water flows are often required to protect downstream fisheries.
- Downstream sediment flows may affect agriculture, positively or negatively as the case may be.

Hydro power plant operational and impact factors can be found under "Power Plant Indicators" in the Companion Document.

Hydro power plants remain an important renewable energy source. Considerable development potential remains to be leveraged in developing regions. The role of hydro energy will remain crucial for the proper and broad integration of variable renewable energy sources.

Combustion Thermal Power Plants

Burning Fossil Fuels

Thomas Edison (1847–1931) launched the Edison Electric Illuminating Company of New York in September 1882. The Pearl Street power plant supplied electricity to 82 customers and their 400 light bulbs. Four boilers burning coal supplied steam to six piston steam engines, which, in turn, drove six so-designated "Jumbo" generators, each rated at 100 kW and weighing 27 tons. By 1884, the Pearl Street power plant was supplying over 500 customers and their 10 000 bulbs. The plant was destroyed by fire in 1890. Despite its short life, the Pearl Street power plant set the stage for combustion thermal power plants, which presently produce over two-third of all electricity worldwide.

Progress of both the capabilities and reliability of electric transmission and distribution systems have made it technically and economically feasible to locate power plants away from urban regions. This is essential for coal-fired plants since it is not unusual that the mines and/or harbors where coal is delivered are a distance away from electricity consumption centers. A careful balance thus needs to be found between the transport of coal to power plants closer to load centers, on one hand, and long electricity transmission systems, on the other.

Statistics

As of 2018, 38.2% of all electricity produced worldwide was from power plants burning coal, which is abundant and broadly available; natural gas plants provided 23.1% of all electricity. Aside from diesel generators, primarily used for isolated villages and islands or as emergency back-up systems, oil is hardly used anymore to produce electricity; its share has declined to reach only 2.9% in 2018.

The total worldwide installed capacity of all thermal combustion power plants was 4 150 GW in 2017 [*iea.org*].

Main Types of Combustion Thermal Power Plants

Combustion thermal power plants can be broadly classified as:

- *Steam power plants*, illustrated by Fig. 1(a). Coal is primarily used as fuel, or more rarely, oil, natural gas, waste or biomass.[1]
- *Gas turbine power plants*, illustrated by Fig. 1(b). Natural gas is primarily used as fuel.
- *Combined cycle power plants*, illustrated by Fig. 1(c), which, for improved overall efficiencies, feature a steam turbine valorizing the waste heat from the exhaust of a gas turbine.
- *Combined heat and power (CHP) plants* that produce both electricity and heat, burning gas or, rarely, coal or oil.
- *Diesel generators*, where a diesel internal combustion engine directly drives an electric generator. In some cases, the exhaust is used as a heat source by way of a heat exchanger.

Steam and Gas Power Plant Efficiencies

The efficiency of steam and gas power plants, η, is expressed as:

$$\eta = \text{Electricity produced}^2 \text{ / Heating value of the fuel used}$$

---

[1] Strictly speaking, waste and biomass are not fossil fuels. However, the combustion technologies are essentially the same as for fossil fuels and their combustion emits CO_2. As seen in the "Introductory Remarks" chapter of "How Electricity Is Produced", the 2018 contribution of biomass represented 2.0% of the overall electricity production while that of waste incineration represented 0.7%.

[2] A distinction between gross and net efficiency is often made: the first relates to the output at the generator terminals and the second to the output actually delivered to the local network, i.e., after the electricity consumption of the power plant's auxiliary services such as conveyor belts, circulation pumps, etc., have been taken into account.

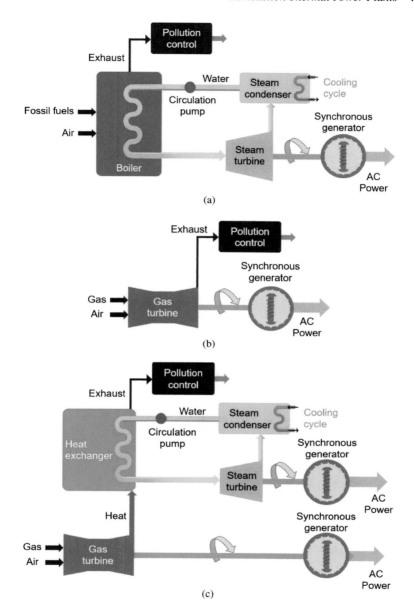

Figure 1. (a) Steam fossil fuel plant block diagram, (b) Gas turbine power plant block diagram, and (c) Combined cycle power plant block diagram / © *HBP and YB*

It is computed at constant output power over short time periods, typically hourly. The efficiency varies as the plant's output power varies due to the thermal and rotational losses.[3] The plant's rated efficiency corresponds to its rated electric power output.

If the plant is producing electricity and heat, its overall efficiency is:

η = Electricity and heat produced / Heating value of the fuel used.

Steam Thermal Power Plants

As of 2017, some 200 steam power plants with installed capacities above 2 000 MW were in operation or under construction in 19 countries; over half in China, 20 in the United States, 15 in India and 10 in South Africa.

The unit size for modern steam power plants can reach 800 MW or more.

As mentioned above, steam power plants are mostly coal-fired. Several types of coal, with different energy densities, are used which impacts the design as well as the operation and emissions of the plants.

Different types of coal

Several types of coal are used in power plants; their calorific value and moisture content are:

- Lignite, also referred to as brown coal: 4.0–5.5 kWh/kg and 40–50% moisture.
- Sub-bituminous: 5.5–8.4 kWh/kg and 5–16% moisture.
- Bituminous: 8.4–9.1 kWh/kg and 2.5–15% moisture.
- Anthracite, also referred to as black coal: above 9.1 kWh/kg and less than 15% moisture.

Better Efficiencies — Toward Higher Steam Pressures and Temperatures

Referring to the *"Thermodynamic Cycles"* presentation in the Companion Document, the overall efficiency of steam power plants increases with steam temperature and pressure. Recent steel alloy developments have

---

[3] The rotational losses, that remain constant since the rotational speed of the turbine-generator assembly does not change (ref. "Power Systems" chapter) with the plant's output levels, become relatively more important at low plant outputs levels.

enabled the construction of turbines capable of operating at pressures well beyond 200 bar[4] and temperatures above 600°C. Referring to the *"Phase Shifts"* presentation in the Companion Document and applying it to water, one can distinguish three types of steam power plants:

- *Subcritical*, operating at pressures well below 200 bar. Most steam power plants presently in operation are subcritical.
- *Supercritical*, operating at pressures in the 250–260 bar range and at temperatures reaching 600°C.
- *Ultra-supercritical*, operating at pressures above 270 bar.

Multi-Stage Steam Turbines

Modern steam power plants generally feature multi-stage turbine configurations as illustrated by Fig. 2. High pressure and high-temperature steam supplied by the boiler enters the high-pressure turbine. After some of the heat is converted to rotating energy by way of the turbine blades, the steam is generally reheated in the boiler to increase its energy content before entering the medium/intermediate turbine; the reheating is not explicitly shown in the figure. The steam is then supplied from the medium/intermediate turbine to the low-pressure turbine. Figure 3 illustrates the size of a multi-stage steam turbine. Each turbine is carefully designed to operate within specific pressure/temperature ranges. Larger steam turbine assemblies may feature two intermediate pressure and three low-pressure turbines. As the steam flows from one turbine to the next, its pressure decreases and its

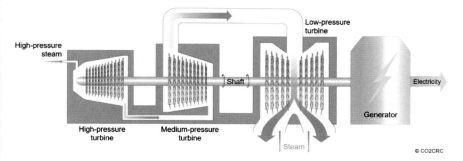

Figure 2. Multistage steam turbine configuration / © *CO2CRC*

---

[4] Ambient air pressure at sea level is at one bar; ref: "Units and Reference Values" annex.

Figure 3.   Low pressure turbine / © *EDF_Photo: Pierre MERAT*

volume increases, as do the diameters of the successive turbines; steam turbines are thus sometimes referred to as "expanders".

The high-pressure turbine, intermediate-pressure turbine(s) and the low-pressure turbine(s) share the same shaft with the generator to convert the kinetic rotational energy into electricity. The weight of large turbine-generator assemblies with ratings at 800 MW may reach 1 000 tons.

Technological Challenges

Steam turbines are subjected to three types of stresses as the plant's loading conditions change:

- *Centrifugal forces*. As illustrated by the Lünen power plant insert, the peripheral speed of low-pressure turbine blades can reach 1 500 km/h, thus creating significant traction stresses.
- *Blade deformations*. Highly specialized steel alloys are used for the turbine blades to minimize steady state and/or transient high temperature and pressure-induced deformations and stresses.

- *Shaft deformations*. The overall length between supporting bearings of a multi-turbine and generator shared shaft can reach over 50 m. Two types of shaft deformations must thus be avoided:
  - *Longitudinal*. Power plant load changes will induce turbine operating temperature changes, which, in turn, can result in longitudinal expansion and contraction effects.
  - *Bending*. Even though the shaft is supported by several bearings, which themselves require careful design and operational attention, during plant maintenance/repair operations bending shaft deformations are mitigated by slow rotations using a dedicated motor.

Coal Power Plant Layout

The plant's main subsystems are:

- The coal supply system, including conveyor belts either directly from a storage pile at an adjoining mine, or at a dedicated train station or harbor dock.
- The boiler, which provides steam to the high-pressure turbine and then to the intermediate-pressure turbines after reheat.
- The turbines sharing the same shaft as the generator.
- The cooling system, including the cooling tower, in which the steam is condensed back to water after it exits the low-pressure turbines.
- The generator, which is connected to the electric grid by way of a transformer.

A detailed description and layout of a coal power plant can be found under "*Coal Power Plant Layout*" in the Companion Document.

Increasing or decreasing the power output of coal-fired power plants inherently implies increasing or decreasing the coal supply to the boiler, i.e., affecting the speed of the long and often complex conveyor systems, which can only take place slowly. This explains why, when in operation, steam power plants are most often operated in a constant production mode.

Belchatow Coal Power Plant, Poland

Rated at 5 392 MW, it is the second-largest coal-fired power plant in the world. It occupies an area of 11 km^2. The lignite coal is provided directly from an adjoining open pit mine visible in the foreground of Fig. 4.

Belchatow I, commissioned in 1982 and recently renovated, consists of 12 units with a total installed capacity of 4 534 MW. Figure 4 shows the 12 units, the two exhaust chimneys and the six cooling towers (four are hidden behind the units; only their water vapors are visible in the background).

Belchatow II, commissioned in 2011 and visible to the left with its larger cooling tower, consists of one 858 MW unit with an efficiency of 42%.

In 2015, Belchatow's overall electricity production was 32 TWh, which represented 21.4% of Poland's total electricity production. As a result, its overall capacity factor was 67.7%.[5] The plant's emissions are 1 000 gCO$_2$/kWh; its emissions thus are 32 million tons of CO$_2$/year. The yearly consumption of lignite coal is 42 million tons, i.e., 115 000 tons/day, which equals the weight of 10 Eiffel towers!

Figure 4.   Belchatow coal power plant with open pit coal mine forefront (Poland) / @ *NV77, Shutterstock*

---

[5] Capacity factor; ref. "Introductory Remarks" chapter of "How Electricity Is Produced". In this case, 32 000 000 MWh / (8 760 h · 5 392 MW) = 67.7% or 8 760 h/yr · 67.7% = 5 935 h/yr at full power.

The Lünen high efficiency coal power plant, Germany

Located in the Ruhr region and shown in Fig. 5, the Lünen 750 MW ultra-supercritical steam power plant was put in service in 2013. It is intended to operate up to 7 000 hours per year.

The steam exits the superheater of the 100 m high boiler at 280 bars and 600°C before entering the high-pressure turbine. Upon reheat of up to 610°C, the steam is returned to the intermediate pressure turbine and from there to two low-pressure turbines. The residual heat from the low-pressure turbines is used for district heating by way of a heat exchanger before it is condensed using a cooling tower.

The plant's overall efficiency is 46%, which makes it one of the highest efficiency coal power plants in the world. It burns average quality coal with a heating value of 7 kWh/kg and features a number of pollution mitigation measures. Its emissions are 750 gCO_2/kWh, i.e., roughly 25% below the emissions from older coal power plants generally emitting 1 000 gCO_2/kWh.

The longest low-pressure turbine blades measure 1.15 m and the turbine diameter reaches 2.8 m. Lünen is a 50 Hz power plant; its turbine-generator thus rotates at 50 RPS, i.e., 3 000 RPM. The peripheral speed of the longest blades is: 50 1/s · 2.8 m · π = 440 m/s or 1 600 km/h, thus inducing high longitudinal stresses.

Figure 5. Lünen coal power plant (Germany) / © *Trianel Gmbh*

Gas Power Plants

The first deployment of a single cycle gas turbine power plant took place in 1939 in the City of Neuchâtel, Switzerland. It was rated 4 MW at the generator terminals and had an efficiency of 17.4%. It was retired from service in 2002. As of 2019, over 50 gas power plants rated over 1 000 MW were in operation in 20 countries worldwide — 19 of them in Japan and 10 in China.

Stationary gas turbine[6] performances have increased significantly during the last decades, primarily driven by progress in materials. Gas turbines overwhelming burn natural gas, a hydrocarbon[7] gas extracted from ground wells and then transported via dedicated pipelines or, more recently, as liquid natural gas (LNG) transported by ships over long distances. During the last few years, shale gas has significantly augmented the available/proven reserves of natural gas.

A presentation of the various types of gases and their uses and processing can be found under *"Types of Gases and Their Processing"* in the Companion Document.

Gas Turbine

Combustion occurs internally in gas turbines as in internal combustion engines (ICEs); however, contrary to ICEs, the combustion is continuous. Referring to the *"Thermodynamic Cycles"* presentation in the Companion Document, the operating principle of gas turbines is based on the Brayton cycle. Efficiencies of modern single-stage stationary gas turbines can reach 40%. Figure 6 shows the main stationary gas turbine parts: compressor, combustion chamber and turbine, which share the same shaft.

- The air drawn into the turbine assembly is gradually compressed in successive stages, similarly to fan jet engines; ref. "Transport and Travel" chapter. However, in this case, the entire compressor output flow is fed

---

[6] The stationary designation serves as a reminder that gas turbines are also used for mobile applications in transport, such as for aircrafts.

[7] Hydrocarbon for hydrogen and carbon.

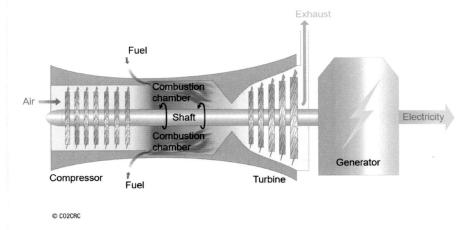

Figure 6.   Stationary gas turbine / © *CO2CRC*

to the combustion chamber. Although no heat is added in the compressor, the air temperature increases as it is rapidly compressed.

- Using carefully designed injectors, the compressed hot air is mixed with fuel and injected into the combustion chamber where the mixture ignites, bringing the combustion gas temperatures up to well above 1 200°C, reaching 1 600°C for some configurations.
- The hot combustion gases are fed to the turbine where they expand as the turbine blades convert their thermal energy into mechanical rotating energy. The exhaust temperatures can exceed 600°C for larger units, which suggests further valorization of the output heat.

Combined Cycle Gas Turbine (CCGT) and Combined Heat and Power (CHP) Plants

As mentioned above, gas turbines exhaust temperature readily reach 600°C or above. As illustrated in Fig. 7, rather than releasing the hot exhaust fumes, they are used in a Heat Recovery Steam Generator where the heat is used to create steam, which, in turn, is supplied to a steam turbine. Both the gas turbine and steam turbine contribute to electricity production. These types of gas power plants are designated as combined cycle (CCGT), in

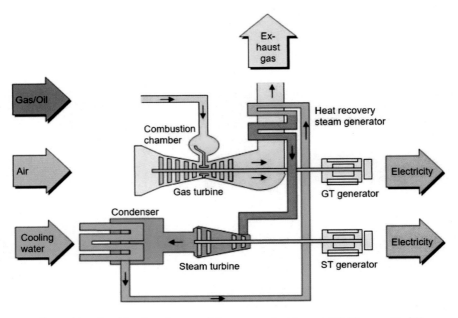

Figure 7.   Combined cycle gas turbine power plant layout / © *Siemens GmbH*

contrast to single-cycle plants only featuring a gas turbine. Modern CCGT power plants can reach efficiencies above 60%.

Even using a CCGT configuration, significant heat energy remains to be valorized after the steam turbine. This heat can be further used either for district heating or various industrial processes requiring heat energy; these systems are referred to as combined heat and power (CHP) systems.

Two main categories of CHPs are presently in operation around the world:

- Electricity production is the main purpose of the plant while heat is a byproduct. Most CHP plants are of this type.[8]
- Heat production is the main purpose while electricity is a byproduct. These types of installations can be found within large industrial sites, such as cement factories, refineries and chemical plants.

---

[8]They can either be single-cycle, using a steam or gas turbine, or combined cycle. Some residual heat from a few tens of nuclear power plants, which inherently are single cycle steam power plants, is valorized by way of heat energy provided to close-by industrial processes or by way of district heating networks.

From a technical feasibility viewpoint, CHP capabilities must be included already at the plant's construction; retrofits are generally cost-prohibitive.

The proper controls of CHPs can be a major challenge, especially when electricity production is the primary purpose. Indeed, if the heat production is tied to a district heating system, the plant's overall control settings may have to be adjusted according to the seasons as well as to daily meteorological situations, which may affect its overall efficiency. Furthermore, back-up systems for the district heating system must be planned for in case of power plant failures.

Fortuna CCGT and the Düsseldorf district heating power plant, Germany

Located in the port of Düsseldorf, the Fortuna combined cycle power plant was commissioned in January 2016. The electricity production rating is 600 MW using a gas turbine and steam turbine combination; the maximum efficiency is 61.5%, a world record at the time. From a hot-start, the maximum power may be achieved in 25 minutes.

In addition, the power plant can also produce 300 MW of heat for the local district heating system. When the heat is fully used, the plant's overall efficiency reaches 85%.

Thermal Power Plant Cooling Systems

Thermal power plants, be they fossil or nuclear, require a cooling system to properly evacuate the heat that cannot be transformed into mechanical energy or be used for other purposes. The overall objective is to achieve a temperature that is as low as possible at the exit of the power plant's condenser, where it is cooled to achieve high efficiency, while limiting the heat's environmental impact.

Depending on the local water availability and climate, three cooling systems can be used:

- *Open circuit*, used when the plant is located on a sea or lakeshore or along a high flow river. A small portion of the available water is pumped into the condenser pipes using one or more water intakes. The cooling water and the steam coming from the turbines exchange heat by conduction between the tubes or the plates of the condenser. As the steam is condensed, the cooling water temperature increases. The cooling water is entirely returned to the sea, lake or river. Since the cooling water volume is small compared to the available water, the temperature increase is also small. The cooling water intake and outlets are typically located far apart. A cooling tower is not required.

- *Semi-open*, used when the plant is located along a river with lesser flows such that directly releasing the cooling water back into it would result in excessive river temperature increases that could potentially be harmful for the local fauna and/or the flora. A cooling tower is then generally used to take advantage of the higher energy required by evaporation, i.e., lower temperature resulting from evaporation. The basic principle is illustrated in Fig. 8. Similarly to open cooling, water from the river is pumped through the condenser where its temperature

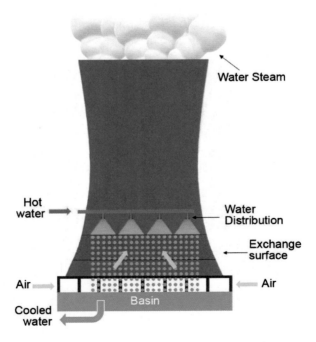

Figure 8.   Cooling tower layout / © *HBP and YB*

increases. The hot water from the condenser is then sprayed over an exchange surface, at about one-third up the cooling tower, where it encounters cold air rising through the tower. It releases one-fifth to one-fourth of its heat by convection and the rest by the evaporation. The rising, clean water vapor exits at the top of the cooling tower. Most of the water is returned to the river. Cooling towers can be up to 200 m high and 150 m in diameter at their base; their hyperbolic shape is designed to enhance the free draft airflow from the bottom to the top. The cooling tower itself does not consume energy. Since some of the cooling water is evaporated, it is not entirely returned to the river.

- *Closed circuit*, used when neither open nor semi-open cooling is feasible for lack of a proper water resource such as when the available river flow is insufficient or when the river's temperature is such that any additional heat rejection would be harmful to the local fauna and/ or flora. The cooling occurs by forcing cold air, using a draft fan, through an assembly of pipes carrying water from the condenser. Closed circuit cooling, while not requiring high cooling towers, is less effective than the other two methods described above. It also requires more energy for the draft fan and is used only when it is the sole feasible option. Closed circuit cooling provides for a "dry" plant, i.e., not requiring any cooling water.

During particularly hot and/or low river flow periods, the output of thermal power plants may have to be reduced for lack of proper cooling, except for closed circuit cooling systems.

CO$_2$ Emissions Mitigation

CO$_2$ emissions occur during any combustion of a carbonated fuel using air as the oxidant. From this simple observation, the key engineering[9] avenues to reduce emissions can be put forward:

---

[9] Economic, societal and political considerations should also be taken into consideration; ref. "Public Policy and Carbon Fee" chapter under "Where We Should Be Heading".

- Using fuels combusting with low CO_2 emissions, which depends on their chemical compositions. Using natural gas is better than oil or coal:
  - Among fossil fuels, burning coal results in the highest emissions per kWh produced; ref. "Units and Reference Values" annex.
  - In addition, natural gas power plants have better efficiencies than other combustion power plants since gas turbines operate at higher temperatures.
- Whenever technically possible, using biomass and/or waste products.
- Developing both energy production and end-use technologies featuring ever higher efficiencies.
- Deploying combustion technologies resulting in lower emissions.
- Whenever CO_2 emissions are concentrated at specific sites, such as power plants and some industry sites, for example, cement plants or oil refineries, trying to capture the CO_2 and either storing it or, better, valorizing part or all of it for other useful purposes. CO_2 capture from diffuse emission sources, such as transportation uses — cars, trucks, buses, diesel trains, aircrafts — remains impractical.
- Preserving and developing forests that naturally store the CO_2 emissions as they grow.[10]

Some of the avenues mentioned above are illustrated in Fig. 9. While the figure shows only coal and biomass as fuels, the pathways shown also apply to natural gas and petroleum products.

In addition to other pollutants and dust, CO_2 emissions are the major drawback of the intensive use of fossil fuels to produce electricity. To mitigate the emissions, the CO_2 can be captured and then used or stored, either locally or after transportation. Under the Carbon Capture and Sequestration (CCS) designation, CO_2 capture is first discussed, followed by CO_2 reuse and storage. The dimension of the challenge justifies CCS efforts; however, as of 2020, broad-scale deployment remains elusive for economic reasons.

While the next presentation focuses on combustion thermal power plant applications, several technologies and systems that are described can be and sometimes are applied to selected industrial processes such as cement manufacturing, oil refining, petrochemical production, to name a few.

---

[10] CO_2 is permanently stored as we use wood for building construction or furniture, for example. It is returned to the atmosphere when we burn wood or as it decays in forests.

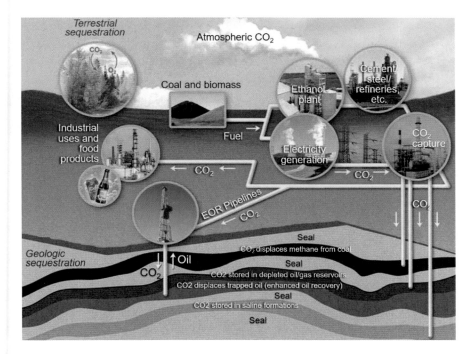

Figure 9. CO_2 emissions mitigation pathways / © *US Department of Energy*

CO_2 Capture

The main processes available to reduce the CO_2 impact of combustion thermal power plants are:

- *Post-combustion*, also referred to as "CO_2 scrubbing", is the least complex one. Having been cooled, the exhaust, i.e., post-combustion, flues are made to flow through chemical solvents that absorb the CO_2 while the cleaned exhaust gas is released. The solvent and CO_2 are separated. The regenerated solvent, once cooled, is reused while the CO_2 is stored or, if feasible, valorized as discussed next. The energy required is generally bled from the steam turbine's supply which decreases the plant's overall efficiency by typically 20 to 30%. The post-combustion process can be technically retrofitted to existing coal power plants, provided sufficient space is available on-site.

- *Pre-combustion*, in which the fuel is in the form of a gas, i.e., converted, using chemical reactions, into a mixture of hydrogen and CO_2 which are subsequently separated. The hydrogen is then combusted in the power plant boiler while the CO_2 can be stored or valorized, similarly to the post-combustion process. Pre-combustion is typically more complex to implement, thus more costly. It remains at the pilot plant level.
- *Oxy-fuel combustion*, in which oxygen is used as an oxidant rather than air. A key requirement is the availability of sufficient pure oxygen supply. The result is a flame of higher temperature, which explains why oxy-fuel combustion has been used for many years in welding and steel cutting applications. It also remains at the pilot plant level.

CO₂ Valorization

As illustrated by the previous figure, several avenues are being pursued, such as:

- Enhanced oil recovery (EOR). Only 20 to 40% of the oil in a natural reservoir can be recovered using standard extraction technologies. Three main EOR techniques are used: thermal, chemical or mechanical by gas injections. CO_2 injection EOR is the most commonly used; an additional 30 to 50% of the reservoir's original oil volume can be extracted using it.[11] Two fossil power plant CO_2 capture and EOR integrated facilities are described under *"Integrated CCS Projects"* in the Companion Document.
- A number of chemical manufacturing processes involve intensive uses of carbon, which could be extracted from stored CO_2. Examples include the manufacturing of synthetic fuels and fertilizers.
- CO_2 can be added to water and other beverages to make them "fizzy". Dry ice, i.e., frozen CO_2, is used in food processing. Pressurized liquid CO_2 is used in fire extinguishers for specific applications.

---

[11] It is primarily used in North America where over 7 700 km of CO_2 pipelines have been built for this purpose. At the present time, the CO_2 is mainly of natural origin due to the still insufficient CCS CO_2 supplies.

The combined CO_2 volume valorized by these applications remains quite small compared to the overall CO_2 emitted by power plants worldwide.

Carbon Storage

Injecting CO_2 into oceans, while seriously considered in the past, is no longer an option since it exacerbates the problem of ocean acidification; it has been declared illegal in several countries.

The two main approaches to carbon storage are:

- *Geological*, where the CO_2, is generally stored in critical state,[12] i.e., pumped into depleted oil or gas fields, into inoperative coal mines or saline formations. Injection into old coal mines will result in methane release as the CO_2 combines with coal; if recovered, the methane can be further used as a fuel. Since detailed knowledge of saline formations is often lacking, CO_2 injections into them are generally not the preferred approach. The Sleipner CO_2 storage facility described in the following insert is an example of what remain few industrial projects.

Sleipner — the world's first large-scale CO_2 storage, Norway

The Sleipner off-shore gas development in Norway's North Sea consists of two fields: Sleipner East and Sleipner West, which came on line in 1993 and 1996, respectively. The European Union market specifications dictate that the CO_2 content of natural gas cannot exceed 2.5%, which the Sleipner East field satisfies, but which the Sleipner West does not with a CO_2 at around 9%. In 1991, Norway introduced a CO_2 tax that has been gradually increased to its present level of US\$65 per ton of CO_2. To avoid the tax, Statoil, the Sleipner operator, decided to capture the CO_2 at the West field using post-combustion amine separation. The CO_2 is brought to the critical state before injection into a sandstone reservoir, which is located 800–1 000 m under the sea and is 250 m thick. A 700 m thick gas-tight caprock above the reservoir provides a natural seal. The CO_2 injection rate is 0.85 ton/year; the total amount injected since the 1996 project inception has reached more than 16 $MtCO_2$.

---

[12] At ambient temperature and pressure, CO_2 behaves like a gas. When contained, it reaches its critical state at 31.1°C and 72.9 bar. It then expands like a gas to fill an enclosure but with a density like a liquid; ref. *"Phase Shifts"* in the Companion Document.

- *Mineral storage*, where the CO_2 enters into an exothermal reaction, i.e., a heat-producing with naturally available metal oxides. The results are stable carbonates such as calcites and magnesites. Great amounts of surface limestone result from such reactions. However, the process takes several years; research is underway to accelerate it to make it economically attractive without requiring excessive amounts of energy.

Environmental Impacts and Accidents

The main consequences of the use of fossil fuels to produce electricity are: greenhouse gas emissions, mainly CO_2, air pollutants and accidents, primarily related to the supply chain of fossil fuels for the power plants.

CO_2 Emissions

As orders of magnitude, the CO_2 emissions from coal and natural gas power plants are:

- For a coal-fired steam power plant with a typical efficiency of 38%: 360 gCO_2[13] for 0.38 kWh (38% of 1 kWh) = 950 gCO_2/kWh.
- For a natural gas power plant with a typical efficiency of 42%: 200 gCO_2[13] for 0.42 kWh (42% of 1 kWh) = 475 gCO_2/kWh.
- For a modern combined cycle power plant with a typical efficiency of 60%: 200 gCO_2[13] for 0.60 kWh (60% of 1 kWh) = 333 gCO_2/kWh.

Plant efficiencies play a crucial role in reducing CO_2 emissions along with fuel quality.

Air Pollutants

The main air pollutants due to combustion thermal power plants are: carbon monoxide (CO), nitrogen oxides (NO_x), and sulfur oxides (SO_x). Pollution monitoring and abatement programs are in place in industrialized countries and are being expanded in emerging regions of the world.

---

[13] Average emissions from the combustion of coal, respectively, natural gas; ref. "Units and Reference Values" annex.

While combustion thermal power plants play an important role in air pollution, they are not the only cause. Indeed, transportation and travel, as well as industry, also contribute to air pollution. Increasingly strict air pollution regulations are driving electric industries to implement mitigation measures or to switch from coal to natural gas or to shutting plants down.

Accidents

Accidents involving the combustion section of combustion thermal power plants are very rare and are mostly due to operational errors. Accidents involving the electricity-producing section, be they combustion or nuclear, also occur very rarely and primarily relate to transformer oil fires or explosions or mechanical failures due to rotor damages, primarily turbine blade failures that can have catastrophic consequences given the rotational speed and size of the blades. Deaths or serious personnel injuries from combustion thermal power plant accidents are extremely rare.

Accidents involving coal mining as well as oil or natural gas exploration and extraction do, however, occur more frequently and may result in deaths or serious injuries. Serious accidents may also result from the transportation of fossil fuels along complex and multi-stage supply chains. Four examples of such accidents can be found under *"Fossil Fuel Supply Chain Accidents"* in the Companion Document.

Combustion thermal power plant operational and impact factors can be found under "Power Plant Indicators" in the Companion Document.

The decarbonization of electricity production requires that the dominant contribution of combustion thermal power plants be reduced to eliminate them as soon as possible. In the meantime, should it become economically feasible on a broad scale, CCS can accelerate the decarbonization. A broad range of initiatives are underway in selected countries of the world to replace coal with natural gas as fuel, to enhance plant efficiencies, and to reduce emissions whenever possible. However, the pace of change remains insufficient.

Nuclear Thermal Power Plants

Splitting the Atom

Medical nuclear applications and the nuclear energy industry both emanate from physics developments during the first half of the 20th century, similar to military applications, starting with the two nuclear bombs of 1945. While several hundred nuclear power plants have been producing electricity for several decades without major problems and CO_2 emissions, the risks inherent to radioactivity and the severe accidents at Three Mile Island, Chernobyl and Fukushima continue to feed debates and controversy around the nuclear energy industry.

The Experimental Breeder Reactor was the first nuclear reactor worldwide to produce heat transformed into electricity. Rated a few kW, it started operations in 1951 at the National Reactors Testing Center in the State of Idaho, United States (US). In 1954, the first connection of a nuclear reactor, rated 5 MW, to a local network occurred in Obninsk, the Soviet Union. France and the United Kingdom (UK) started building nuclear reactors during the 1950s. The different types of reactors presently in operation were developed during that same period; the main ones are presented in this chapter. Presently, nuclear reactors can reach installed capacities of over 1 500 MW; several reactors may be located on the same site.

Statistics

According to the International Atomic Energy Agency[1] [*iaea.org*], by the end of 2019:

- 450 nuclear reactors, with a combined installed capacity of 399 GW, were operable in 31 countries.[2]
- 178 reactors were in permanent shutdown.
- 55 reactors were under construction, including 10 in China, 7 in India and 5 in Russia.
- The highest numbers of reactors were located in the US (96), France (58), China (48) and Russia (37).

In 2018, the worldwide nuclear electricity production was 2 715 TWh, i.e., a share of 10.2%. This situation evolves as new plants under construction, or in a planning phase, are put into service or taken out of service, either for technical or economic reasons (aging, safety) or following governmental decisions. For example, Germany decided to shut down all of its remaining nuclear power plants by 2022. In Switzerland, it was decided to gradually phase out its five nuclear reactors, which together produced 38% of the electricity in 2015, thus making Switzerland the country with the highest nuclear contribution percentage to have reached such a decision.

Main Reactor Types

Nuclear reactors can be categorized by:

- The construction period:
  - *Generation I*, for 1950s and 1960s reactors.
  - *Generation II*, for the reactors constructed after 1970 and still in service.

---

[1] Created in 1957 by the General Assembly of the United Nations and headquartered in Vienna, Austria.

[2] Including the reactors in Japan presently not in operation as well as those in prolonged outages, but which are intended to be put back in service. Only electricity-producing reactors are included but not the several hundred research reactors used for the production of various medical-purpose materials, or industrial application reactors.

- ○ *Generation III*, for the reactors developed since 1990, incorporating the experience gained from the construction, operation and maintenance of Generation 2 reactors as well as from accidents. The first Generation 3 reactors are in service.
  - ○ *Generation IV*, for reactors still at the R&D stage; large-scale operation is not expected before 2035–2040.
- The installed power range: "Small" below 300 MW, "medium" between 300 and 700 MW, and "large" above 700 MW.
- The coolant, i.e., the liquid transporting the heat. Pressurized Water Reactors (PWRs) and Boiling Water Reactors (BWRs) use "ordinary" water as coolant. Other reactors use heavy water, liquid sodium or a gas, such as helium or CO_2.
- The nuclear fuel[3] used: Uranium oxide, uranium and plutonium oxides mix, or, potentially thorium.
- How the neutrons are "moderated":
  - ○ Slowed down using a moderator: Liquid water under pressure in PWRs, boiling water in BWRs, or graphite or heavy water, for example.
  - ○ Not slowed down such as in a Fast Breeder Reactor (FBR).

The selection of a nuclear reactor type is inherently tied to the nuclear fuel and its cycle; its characteristics and manufacturing are conceived in coherence with the reactor type as well as with its processing after usage.

As of early 2020, the main types of nuclear reactors in operation were:

- PWRs represented 67% of the worldwide nuclear power plant fleet.
- BWRs with 15% of the worldwide nuclear power plant fleet.
- Heavy water reactors, CANDU, with 11%, which use natural uranium, i.e., non-enriched, contrary to PWRs and BWRs, which use enriched uranium.
- Other types with 7%, including liquid sodium fast neutron reactors that ensure that nuclear fuel will be available for several hundred years, if not more.

---

[3] This is the term normally used in this context even though it is not referring to a combustionable fuel, chemically speaking.

Most presently deployed reactors are PWRs and since most plants under construction or planning are also PWRs, it is used as the basis for the presentation hereafter; the other main technologies are briefly presented under "*Alternative Nuclear Reactors*" in the Companion Document.

From Radioactivity to Chain Reactions

Nuclear reactors use heat obtained from nuclear fission, i.e., from the energy produced by atomic ruptures. Atoms are composed of a nucleus, in turn composed of neutrons and protons, with electrons spinning around it. Neutrons and protons have almost the same mass; the mass of the electrons can be neglected as a first approximation. Neutrons are electrically neutral. A proton's charge is positive. An electron also has a charge, equal in absolute value to that of the proton but negative. An atom has the same number of protons and electrons, Z, which is its atomic number; as a result, it is electrically neutral. For carbon: Z = 6, for example. The total number of neutrons and protons, A, is the atomic mass. Atoms having the same atomic number but different numbers of neutrons are referred to as isotopes and have the same name. For example, carbon, Z = 6, has several isotopes with A = 12 (with 6 neutrons), the most common one, but also A = 14 (with 8 neutrons). **Mendeleyev's**[4] table provides a systematic element classification.

Almost all atoms on Earth are stable. Others are unstable; they can spontaneously decompose after a more or less long time and are said to be fissile or radioactive. This is referred to as the spontaneous radioactivity discovered by **Henri Becquerel** and **Marie** and **Pierre Curie**.[5] A radioactive element is characterized by its period, also referred to as its half-lifetime, which is the time required for half of its atoms to be transformed by spontaneous fission into other atoms; after 10 periods, less than one-thousandth of the initial atoms remain in their original composition. Depending on the particular radioactive elements, their periods range from a fraction of a second to several hundred million years. Aside from

---

[4] **Dimitri Mendeleyev** was a Russian chemist; he published the first version of his table in 1869.

[5] French physicists who shared the 1903 Nobel Prize in Physics for this discovery.

fissile nuclei continuously created in nature due to cosmic rays, such as carbon 14, the principal fissile element still existing in the Earth's crust is an isotope of uranium, U235 (Z = 92 and A = 235), having a period of 703.8 million years.

Other atoms, designated as fertile, while more stable than fissile atoms, can still be decomposed when hit by a neutron. In nature, this is the case for U235 and U238, which is the most abundant uranium isotope (99.3% and only 0.7% for U235). This is also the case for thorium (Z = 90 and A = 232). Fission itself can produce neutrons, which, in turn, can produce more fissions and so on. This is the chain reaction principle first implemented in the US by **Léo Szilard** and **Enrico Fermi** in 1942.[6] The chain reaction's multiplication factor, k, is the average number of neutrons emitted by way of fission. To enable a chain reaction, it may be necessary to slow the neutrons down to avoid them passing by atoms too fast to cause fissions; the neutrons are then said to be moderated.

Fission produces a lot of energy. Nuclear reactors are designed to recuperate the energy produced during chain reactions and transform it into heat. This is one of a number of applications of radioactivity, which is also used for therapeutic purposes, food sterilization, or dating of remains from the deep past. As further discussed later, above certain levels, it can also be dangerous for humans.

Further information is provided under *"Radioactivity and Chain Reactions"* in the Companion Document.

Pressurized Water Reactor: Description and Operation

Overall Description

The main difference between nuclear thermal power plants and combustion thermal power plants is that heat emanates from a nuclear reaction (i.e., the splitting of atoms) rather than from a chemical reaction,

---

[6]**Léo Szilard** was a Hungarian-American physicist. **Enrico Fermi** was an Italian-American physicist; he received the 1938 Nobel Prize in Physics.

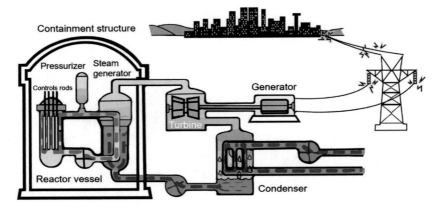

Figure 1.   PWR layout / © *NRC, United States*

combustion (i.e., atom bonding). As seen in Fig. 1, a PWR has three separate loops.

- The closed primary loop provides the heat. The primary loop water is heated as it flows through the reactor vessel, the "nuclear furnace", where the reactor is located. It flows through steam generators before returning to the reactor vessel; the pressurizer ensures a sufficient pressure to maintain the water at the liquid state, hence the reactor's name.
- The closed secondary loop converts the primary circuit heat into electricity. The secondary loop water recuperates the primary loop heat as it vaporizes by conduction in the steam generator. The steam is supplied to the steam turbine and then follows the same path as for steam power plants.[7]
- The tertiary loop, i.e., the cooling system, can be open or closed depending on the type and cooling source availabilities (sea, river, air).

Figure 2 shows the features of a nuclear power plant with four PWRs on the same site:

---

[7] Some 50 nuclear power plants worldwide are operated in cogeneration — a small part of the heat produced is used for district heating or industrial process purposes.

Figure 2.   Paluel nuclear power plant (France) / © *EDF_Photo: Marc Didier/PWP*

- The four nuclear fuel storage buildings next to the reactor buildings (right hand side).
- To the left of the fuel storage buildings, the four containment vessels, i.e., the four concrete cylindrical reactor buildings and their domed covers. Each one contains the entire primary loop, including the reactor and its reactor vessel, as well as the steam generators and associated primary pumps.
- To the left of the containment vessels, the rectangular buildings house the turbines and generators.
- The sea, which is the primary cooling source (left hand side); as a result, cooling towers are not needed.

The ratio between the electric power output and the thermal power produced by the reactor, i.e., the plant's efficiency, is approximately 33 to 35% for PWRs in operation.[8]

---

[8]The efficiency is lower than for combustion thermal power plants since the irradiation of the materials requires that lower operating temperatures be maintained to limit their degradation.

From the Nuclear Fuel to Fuel Assemblies to the Reactor

Nuclear fuel most often consists of uranium oxides (UO_2). Starting from uranium ore, the U235 content is enriched to some 3–4.5% to allow for a chain reaction; the fuel is said to be lightly enriched. This enrichment level is far below that required for atomic bombs, which is above 80%; this is why a nuclear reactor cannot physically become a nuclear bomb.[9]

After manufacturing, nuclear fuel is shaped as cylindrical pellets, stacked into fuel rods subsequently organized into fuel assemblies. The pellets' diameter and height are some 1 cm. They are placed in cylindrical zirconium[10] sheaths. The fuel rods (Fig. 3, right) are 4 m long and closed at each end to contain the fissile products, i.e, the uranium "descendants".

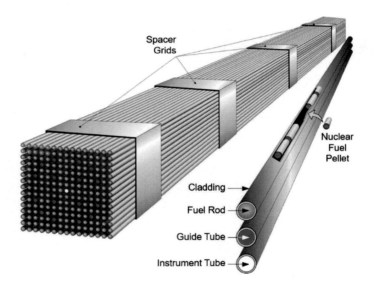

Figure 3.   Fuel rod and fuel assembly / © *US Department of Energy*

---

[9] In a nuclear bomb, the density and thus the proximity of U235 atoms ensures that the multiplication factor reaches a level of 3 as the neutron travel distances are very short. The explosive destruction level is thus reached in less than a millionth of a second.

[10] Zirconium is selected because it is a good heat conductor and allows neutrons to flow through while also resisting the hot water under pressure flowing around the sheaths.

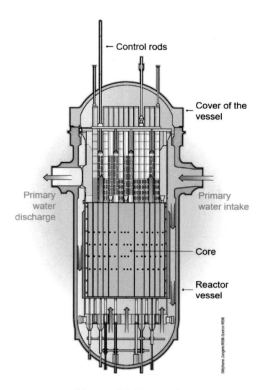

Figure 4. Nuclear reactor vessel layout / © *France, Research Institute on nuclear safety, IRSN*

As shown in Fig. 3 (left), the fuel rods are grouped in square grid assemblies, which also include guiding tubes for control rods made of neutron-absorbing materials; they can be moved up and down vertically to moderate the chain reaction as needed for operational purposes or even to rapidly stop it.[11]

As seen in Fig. 4, the fuel assemblies are placed next to each other; they constitute the reactor core located inside the reactor vessel, which is made of special steel and shown in Fig. 5 illustrating its size compared to the nearby operator. The reactor vessel and fuel configuration provide for

---

[11] Nuclear power plant operators can also moderate the chain reaction in a PWR, independently from the control rods, by injecting neutron-absorbing boric acid in the primary loop.

Figure 5.   Handling a nuclear reactor vessel / © *EDF_Photo: Alexis Morin*

the primary loop water to freely flow between the fuel rods from the bottom to the top while serving two purposes: (a) as coolant to evacuate the heat and (b) as moderator to slow down the fission neutrons.[12]

Depending on the reactor power, the heat is evacuated using three or four loops, each including a primary pump and a steam generator, as shown in Fig. 6. The rest of the plant, i.e., the turbine and condenser, is similar to other steam power plants but generally of larger size.

Contrary to combustion thermal power plants, the fuel is not supplied continuously, since the energy available in nuclear fuel is approximately 100 000 larger than that available in fossil fuel of equal weight. During the four years a 10 g nuclear oxide pellet typically remains inside the reactor core, it produces approximately 10 MWh of heat, i.e., slightly more than that produced by the combustion of one ton of coal, 9 MWh; ref. "Units and Reference Values" annex.

---

[12]The speed reduction is from some 20 000 km/s to 2 km/s, i.e., 7 200 km/hr.

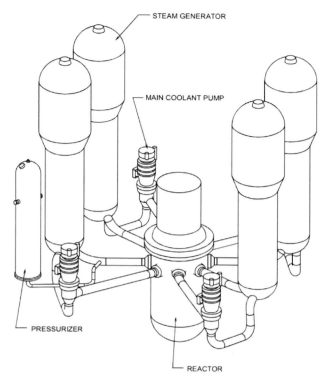

STEAM GENERATOR

MAIN COOLANT PUMP

PRESSURIZER

REACTOR

Figure 6. Four-loop nuclear power plant primary circuit / © *Framatome*

Characteristics of a 1 450 MW French PWR reactor

The core and its components

- Each pellet is 13.5 mm high with a diameter of 8.5 mm and weighs 10.5 g. It has a thermal power of roughly 290 W, i.e., with an annual thermal production of 290 W · 8 760 h = 2.5 MWh.
- Each fuel rod contains 272 pellets and thus represents some 80 kW.
- Each 17 by 17 fuel assembly grid contains 264 fuel rods as well as 24 tubes in which the neutron-absorbing control rods are inserted; one additional tube is reserved for measurement and monitoring purposes.
- The core contains 205 fuel assemblies; as a result, it contains 54 120 fuel rods, i.e., 14 720 640 pellets!

(Continued)

(Continued)

Efficiency
- Rated thermal power: 4 250 MW.
- Rated electrical power: 1 450 MW.
- Efficiency at rated power: 34%.

A few dimensions and characteristics:
- Reactor vessel: 4.5 m diameter, 13.6 m high, 22.5 cm thick (at the core level) and an empty weight of 460 t, of which 85 t is for the cover alone.
- Each steam generator is 22 m high and weighs 420 tons. It contains 5 610 heat exchange tubes, each 1.1 mm thick with a diameter of 19 mm. The total heat exchange surface is 7 300 m^2.
- Each primary circulation pump: 6.6 MW, flow of 24 500 m^3/h, 8.5 m high and a weight of 116 t.
- The concrete containment vessel is 1.75 m thick; its inside diameter is 43.8 m and its inside height is 57.5 m.

Load Following

Nuclear power plants are normally used for base production. However, production adjustments are possible between 20 and 100% of the installed capacity at a rate of 40 MW/mn during two-third of a fuel campaign but generally no more than twice per day to limit temperature variation effects.

Nuclear Fuel Cycle, Waste Treatment and Dismantling

Nuclear Fuel Cycle

The nuclear fuel cycle — from mine to final storage — is illustrated in Fig. 7; the main stages are:

- In nature, uranium ore is found as oxides, with rather high densities. It is purified before transportation to processing centers; the result is a yellow material, the yellow cake.

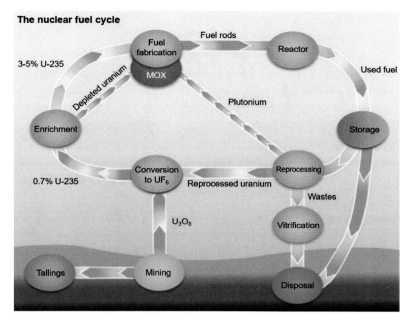

Figure 7.　Nuclear fuel cycle overview / © *World Nuclear Association*

- The uranium ore is chemically converted into gaseous uranium hex-afluoride (UF_6). Using centrifugation, the U235 concentration is increased while that of U238 is decreased. The mass difference between the two isotopes, while small, is sufficient to implement the separation.[13]
- The enriched gas is then chemically transformed into UO_2, a solid material from which the nuclear fuel pellets are manufactured.
- Following its use in the reactor, the spent fuel is radioactive and hot. It contains U235, U238 and plutonium as well as other fission materials. The initial storage is temporary, generally in a pool at the power plant site. One can then:

---

[13] This technology potentially enables the proliferation of nuclear fuels with higher enrichment levels. This issue, while quite important, is beyond the scope of this book.

○ Consider it as waste, i.e., material not to be reused. In this case, the nuclear fuel cycle is said to be open, or

○ Reprocess it by separating the plutonium and the uranium, which represent more than 95% of the spent fuel, to use them later. In this case, the fuel cycle is said to be closed. The residual wastes must be properly processed.

Recycling spent fuel has the disadvantage of concentrating high activity radioactive waste, which is difficult to handle; it has the advantage of limiting the overall quantities of waste and reducing the need for uranium ore extraction. Plutonium is obtained during the recycling, which could potentially be used for military applications. It can also be used to manufacture the MOX nuclear fuel that is a mixture of uranium and plutonium oxides, which can be used in PWR and BWR power plants as well as in fast neutron reactors; ref. *"Alternative Nuclear Reactors"* in the Companion Document.

Nuclear Waste Treatment

Nuclear waste results from installations, primarily industrial, medical and R&D, where radioactive materials are used. Only some of this waste is radioactive. Aside from its origin and physical state, solid, liquid or gaseous, nuclear waste is categorized depending on:

- Level of activity: Low, intermediate or high.
- Period, which depends on the elements it contains: Short or long. The limit between the two depends on the particular regulation, generally around 30 years. Some waste has ultrashort periods, less than 100 days, which means that it is no longer dangerous after the intermediate storage.

The heat released, the reactivity to oxygen, for example, and the volatility are taken into account when selecting the treatment method. A chemical or physical transformation[14] is also possible. The waste

---

[14] A physical transformation such as a transmutation, i.e., the transformation of atoms into other atoms, may eventually reduce the quantity, the period and/or the level of activity of some waste with high activity–long periods.

conditioning and the storage site are interdependent along with the special precautions necessary at each step of the process and during transport. Aside from the dismantling, the most active nuclear power plant waste is the spent fuel, reprocessed or not. Some equipment that was replaced during the plant's lifetime may also be radioactive.

Some orders of magnitude for PWR wastes

The high activity and long period waste represents less than 3% of the overall waste volume but over 95% of the radioactivity. The low-level activity waste, on the contrary, represents more than 90% of the volume but less than 1% of the radioactivity.

In a PWR, after three years in the reactor core, one ton of uranium oxide fuel enriched at 4.5%, is typically transformed into: 955 kg of uranium still enriched at 0.9%, 10 kg of plutonium and 1 kg of minor actinides[15] as well as 34 kg of fission products.

Depending on its characteristics, nuclear waste can be encapsulated in concrete, metals, or in glass for high-activity waste. For high- or intermediate-activity waste with long half-lifetimes, the storage location must not only prevent any exposure to the public-at-large and any groundwater contamination, it must also prevent any inadvertent or deliberate intrusion. Storage in pools, above ground or subsurface, can thus only be considered as temporary. Countries operating nuclear power plants are considering deep underground storage, with or without pretreatments, in a geological structure durably reducing the risk of radioactivity percolation emanating from the waste. Research is still on going about this type of solution, which remains controversial. Finland is the first and, as of 2019, the only country in the world to construct a final storage facility for civilian spent nuclear fuel without reprocessing. Other countries have projects that have either been decided upon or are suspended; other countries operate underground laboratories in granite, clay or other rock formations.

---

[15] Uranium, plutonium and some other fission materials are referred to as actinides in Mendeleyev's table. Since the other fission materials only represent a small portion, they are referred to as "minor". However, they are quite active or have longer periods; as a result, they are to be handled with special care.

The final waste storage site at Onkalo, Finland

Since the 1990s, Finland has gained experience with nuclear power plant waste storage in granite at a depth of 100 m. After several decades of preparatory research, it decided, in 2015, to store spent nuclear fuel without reprocessing at an underground site close to the Olkiluoto nuclear power plant.

The installations required to encapsulate the fuel in copper canisters are on-site and above-ground. Tunnels and elevators allow for transits between the surface and galleries at depths between 450 and 500 m. The installation is designed to store nuclear waste starting in 2020 until 2120, at which time it will be permanently sealed. More information is available on the Finnish nuclear safety authority's website [*stuk.fi/web/en*] and on the operator's website [*posiva.fi/en*].

Nuclear Power Plant Dismantling

It implies removing all equipment, deconstructing the buildings and then returning the land to a status compatible with its future use. Once the nuclear fuel has been removed, part of the remaining equipment, primarily the reactor vessel and the equipment inside, contain radioactive elements due to the intense nuclear irradiation they have been subjected to. Several entities worldwide specialize in various dismantling stages, increasingly using robots; indeed, dismantling is not only required for nuclear power plants but also for experimental reactors of various sizes as well as for civilian and military laboratories.

Delaying the dismantling several decades significantly reduces the radioactivity of elements having short or intermediary periods, which are the most abundant ones aside from the nuclear fuel itself. The quantity of radioactive waste to be handled is thus reduced; however, the site remains unusable. Therefore, two possibilities exist: dismantle without delay or mothball the installation into a safe and monitored state and delay the final dismantling by one or more decades.

Further information about both options can be found under "*Nuclear Power Plant Dismantling*" in the Companion Document as well as on [*iaea.org*] and for nuclear plants in the US on [*nrc.gov*].

Environmental Impacts — PWR and BWR

A major feature of nuclear power plants is that, aside from water vapor, greenhouse gases are not emitted during their operation. If one includes the construction and dismantling as well as the nuclear fuel cycle, during a 40-year operation, the emissions for PWRs and BWRs are some 30 gCO_{2e}/kWh as per data provided by the US National Renewable Energy Laboratory (NREL) [*nrel.gov*].

The impact on the local water resources is similar to those of a combustion thermal power plant. For a PWR or a BWR using a cooling tower, the evaporation is some 2.5 l/kWh produced. As indicated under *"Power Plant Indicators"* in the Companion Document, on the basis of total kWh produced during an entire life span, nuclear power plants have the lowest land footprint among all power plants, roof-mounted photovoltaics aside.

Risks and Radiation Protection

Radioactivity-related risks must be prevented. Short or longer exposures significantly above the levels of natural radioactivity, which is present everywhere, can be dangerous or even deadly. Overall, the effect of radiations is the ionization of tissue atoms, which, if intense, can damage cells beyond repair by the organism. While a detailed explanation of these important and complex issues is beyond the scope of this book, a summary is provided below; additional information is provided by specialized websites cited below.

To quantify the impact of radiation on humans, the notion of effective dose, E, is introduced. Measured in sieverts, Sv,[16] it takes the type of radiation, its intensity and the organisms exposed into account. Since the sievert unit is large, commonly encountered doses, such as local natural radioactivity, are generally given in mSv. Any organism is exposed to natural radiation:

- External, from Earth and the Cosmos.
- Internal, from the body, water and food as well as from the inhalation of radon, the only heavy gas from natural radioactive elements, which can be found, for example, at the bottom of valleys or in basements.

---

[16] From the Swedish physicist **Rolf Sievert** (1896–1966).

Overall, the average natural radiation is 2.4 mSv/yr, with variations between 1.2 and 12 mSv/yr depending on the regions. Negative health effects due to natural radiation have not been observed. In addition, they vary from one person to the next. Overall, a particular dose received over a longer period has a lesser impact than if received over a shorter period. Following an accidental exposure:

- Negative effects have not been demonstrated for doses below 100 mSv; however, low-dose exposures remain a controversial issue.
- Doses above 700 mSv are always harmful, including cancers.
- Accidental doses at the 10 000 mSv level result in rapid death.

A number of medical studies have been completed regarding the effects of accidental or planned radiations; information can be found on the website of the World Health Organization [*who.int*] or the United Nations Scientific Committee on the Effects of Atomic Radiation [*unscear.org*].

Medical radiation and industrial activity exposure limits are set by each country for the population-at-large and the workers concerned, such as medical staff or nuclear power plant staff. For example:

- The limit of any additional exposure for the population-at-large due to industrial activities is set at 1 mSv/yr in a number of countries.
- The exposure limit for an individual working in the nuclear industry is typically set at 20 mSv/yr.

Safety and Its Implications

The specific risks associated with radioactive materials and the levels of energy involved have led to:

- Nuclear power plants engineering designs with several successive barriers to prevent any radioactivity from reaching the environment. In addition, a number of systems are redundant, diversified and independent to limit accident outcomes. Finally, to reduce the consequences of human errors, the safety systems are designed such that several successive errors have to be made to lead to an incident.

- Individual and group training for power plant operators and third-party staff to instill a "safety culture".
- The creation, by the relevant public authorities in each concerned country, of national safety agencies, which are independent of the power plant manufacturers as well as of their owners.

At their onset, nuclear industries and the associated regulations were largely national. The awareness that nuclear accidents can have transnational consequences, especially following the three grave accidents further described below, has led to the need for sharing experiences and the exchange of "best practices" worldwide. This led to the reinforcement of:

- The role of the IAEA, which reviews the operation of national safety agencies of relevant countries.
- The actions by the World Association of Nuclear Operators (WANO) [*wano.info*], which organizes peer reviews by experts and operators from other countries of nuclear power plants from technical and organizational points of view.

The issues related to the nuclear industry have led to the creation of non-government organizations (NGOs) and/or to activities by such organizations. It is also important to note that beyond scientific, technical, organizational and institutional safety analyses, the safety perception by the public-at-large, which may well be different, also matters; in turn, it influences all other relevant parties.

Incidents and Accidents

International Nuclear and Radiological Event Scale (INES)[17]

It was developed in 1990 by the IAEA and the Nuclear Energy Agency of the Organization for Economic Co-operation and Development (OECD/NEA). Similarly to the Richter scale for earthquakes, it applies to all nuclear events and accidents, including the use, storage and transport of

---

[17] In an effort of accuracy, the description below of INES is largely based on text from the IAEA [*iaea.org*].

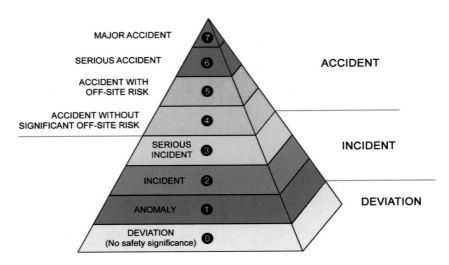

Figure 8.   International nuclear events scale (INES) / © *IAEA*

radioactive material. Figure 8 illustrates the seven levels of the INES scale; levels 1–3 are referred to as "incidents" and levels 4–7 as "accidents".[18] As in any industrial facility, if incidents occur without any safety impact, they are said to be below scale or at level 0. Nuclear and radiological incidents and accidents are categorized, taking three areas of impact into consideration:

- Population and the environment, i.e., the radiation doses received by persons close to the event as well as radiological releases further away from the installation.
- Radiological barriers and control, i.e., events without any direct impact on the population or the environment and which only involve the interior of installations.

---

[18]The INES scale is not intended to compare nuclear safety performances or levels between facilities, entities or countries. The reporting methods and standards differ from one country to the next, making such comparisons cumbersome.

- In-depth protection, which also considers events without any direct impact on the population or the environment but for which existing event prevention measures did not operate as intended.

Three Major Nuclear Power Plant Accidents

The three often referred to major nuclear power plant accidents occurred in three different continents, during three different decades and involved three different technologies. They all involved a major human factor. They are briefly presented below; more detailed presentations can be found under *"Major Nuclear Power Plant Accidents"* in the Companion Document.

- *The Three Mile Island accident — US, March 28, 1979.* A series of incidents and human errors led to the partial meltdown of the reactor core. The containment building resisted, thus protecting the plant operators and population. Outdoor releases were limited to low-toxicity radioactive elements. The event was categorized as level 5 on the INES scale.
- *The Chernobyl accident — Ukraine, Soviet Union, April 26, 1986.* It is the most severe accident in the nuclear power industry's history. It was the first accident classified at level 7 of the INES scale. The accident occurred as a result of a test where several safety measures were deliberately inhibited. A series of erroneous actions led to several severe outcomes, including the fusion of the core, hydrogen explosions, open-air exposure of the reactor and a fire of the graphite moderator. These chemical reactions resulted in radioactive clouds which spread over Europe. More than one hundred emergency personnel who intervened during the crisis died more or less rapidly afterward. The late evacuation of the local population had severe health consequences, including cancers. Severe damage was also done to the local fauna and flora. The access to a zone around the plant remains prohibited.
- *The Fukushima accident — Japan, March 11, 2011.* This accident, which was classified at level 7 on the INES scale, resulted from a tsunami, which was caused by an earthquake in the Pacific Ocean

that registered level 9 on the Richter scale. The protection dike was insufficiently high to protect the plant (which is located on the ocean shore) from the high waves. This resulted in the flooding of the back-up diesel generators, which were also not installed high enough above the ocean. The electricity emergency supply was lost as a consequence. The residual heat in three of the six on-site units resulted in the fusion of their cores and hydrogen explosions. However, the reactor vessels did not rupture. Reactive releases occurred in the air, ground and sea. The evacuation of the local population was fast enough such that major health consequences could be avoided.

Lessons learned

Revisiting the controversies following these serious accidents is not the purpose of this book. However, it is important to note that they have led to profound operating procedure reformulations and engineering modifications of existing nuclear power plants, as well as in the design of plants either under construction or in the planning phases [*iaea*.org].

Nuclear power plant operational and impact factors can be found under "Power Plant Indicators" in the Companion Document.

Nuclear power plants do not create any CO_2 emissions when in operation. They can produce large amounts of baseload electricity and their production can be operator-adjusted within certain limits. Given the overarching need for decarbonization and the ensuing need to largely integrate variable renewable energy sources, nuclear energy may still have a role within electric system operations despite the associated risks and waste management issues.

Solar Energy

Using the Sun

The energy radiated by the Sun and received on Earth each year is several thousand times larger than the energy consumed by Mankind. The challenge is to harvest and use this abundant energy, which is inexhaustible at Mankind's time scale. Solar energy has been used for a long time to warm oneself, dry food (rice, corn, etc.) and recuperate salt from seawater. It is increasingly used for hot water production. Solar residential hot water heaters, with or without electric or natural gas supplement, are used by hundreds of millions of people; solar water heating technologies are presented in the "Housing" chapter.

In this chapter, the focus is on the use of solar energy to produce electricity. Two types of technologies are increasingly used worldwide:

- *Photovoltaic (PV)* panels, often simply designated as solar panels, directly convert some of the Sun's light spectrum into electricity. They contain materials, referred to as semiconductors, inside of which electrons can be put into motion when excited by light, as further explained later.
- *Concentrating solar power (CSP)* plants use radiated energy to heat a fluid. The heat recuperated by the fluid is subsequently converted into mechanical energy and then into electricity. As for classical thermal power plants, the coolant goes through a thermodynamic cycle.

Statistics

Solar Hot Water Production[1]

The installed thermal capacity of solar hot water heaters worldwide increased from 62 GW_{th} in 2000 to 472 GW_{th} in 2017; the energy produced increased from 51 TWh_{th} to 388 TWh_{th} [*iea-shc.org*], i.e., a multiplication of over seven times. The total solar hot water panel area was estimated at 675 Mm^2 in 2017; as a result, the average yearly production was 574 $kWh_{th}/m^2 \cdot yr$. Most of these installations were residential hot water heaters; the remaining ones were primarily used to heat water in offices, hotels, shopping centers or swimming pools. The two leading countries were China and the United States (US).

Solar Electricity Production — [*iea.org*], [*irena.org*], [*ren21.net*]

As seen in the "Introductory Remarks" chapter under "How Electricity Is Produced", the overall solar electricity production in 2018 was 562 TWh. It represented 2.1% of the worldwide electricity production; hydro production represented 15.8%, wind, 4.7%, and geothermal, 0.3%.

Solar photovoltaic electricity production

Also in 2018, the worldwide PV installed capacity was 481 GW, more than a 100-fold increase from 4.5 GW in 2005. The installed capacities are still increasing dramatically: almost 100 GW more in 2019, reaching 579 GW. This increase between 2018 and 2019 was the strongest among all electricity-producing technologies worldwide.

In 2018, with a production of 550 TWh for an installed capacity of 481 GW by the year's end, the capacity factor[2] was 13.1%, i.e., 1 305 h/yr at

---

[1] While, as noted above, the chapter's focus is on solar energy electricity production, the statistics related to hot water production are included here to facilitate comparisons, rather than in the "Housing" and "Tertiary and Services" chapters.

[2] Capacity factor; ref. "Introductory Remarks" chapter under "How Electricity Is Produced". In this case: 550 000 GWh / (8 760 h · 481 GW) = 13.1% or 8 760 h/yr · 13.1% = 1 305 h/yr at full power.

full power.[3] The worldwide PV capacity factor reached 10 %, i.e., 876 h/yr at full power, for the first time in 2013.

Concentrating solar plants (CSP)

The first CSP installations date back to the 1980s, primarily in Europe and the United States (US). The 2018 installed capacity of the CSPs in operation was some 5.7 GW producing roughly 12 TWh; the resulting capacity factor thus was 24.0%, i.e., 2 105 h/yr at full power. CSP power plants were primarily located in Spain (2.3 GW) and the US (1.7 GW); new ones are under construction in a number of countries. The emergence of CSP technologies is recent and their deployment is growing. The competition from PV plants is strong, primarily from investment and operational costs points of view and also since fluids and mechanical elements are not required. In addition, photovoltaics can be deployed at all scales, from a watch up to power plants with installed capacities reaching several hundreds of MW.

Solar Radiation Reaching the Ground

The solar radiation reaching the Earth's surface, referred to as irradiation, depends on the thickness and the characteristics of the atmosphere traversed, i.e., on the:

- Time of day and the season.
- Physical location — latitude and altitude; it increases by 7% with each 1 000 m of altitude elevation.
- Climate — nebulosity, fog and humidity.
- Air pollution.

When traversing the atmosphere, solar radiation is partially absorbed, diffused in all directions, including by clouds, and also reflected back into space. One distinguishes:

---

[3] Given the rapid increase in the installed capacity, the average installed capacity during 2018 was lower than 481 GW; the capacity factor was thus slightly higher. The same is also true for CSPs.

- The direct irradiation, i.e., the rays which directly reach the Earth's surface,
- The diffused irradiation, primarily resulting from the diffusion due to clouds,
- The irradiation reflected by the ground when considering the irradiation on a surface.

A thick cloud can temporarily suppress the direct radiation such that only the diffuse radiation remains; the solar power can then be reduced by a factor of 2 to 5 in a few seconds. The definitions above lead to indicators commonly used to quantify the solar energy reaching a surface:

- The Direct Normal Irradiation (DNI) quantifies the direct irradiation reaching a surface that is perpendicular to the solar flux.
- The Global Horizontal Irradiation (GHI) quantifies the total irradiation reaching a horizontal surface: direct, diffuse and reflected.[4]

These indicators are generally measured in $kWh/m^2 \cdot yr$, i.e., per square meter and per year. The differences between the GHI and DNI are not only due to cloud coverages, but also to the radiation reflected by the ground. In regions with moderate climates, the DNI is generally between 60 to 80% of the GHI. These indicators can be obtained from local measurements or using satellites. More or less precise models provide for measurement interpolations between sites; others make it possible to obtain instantaneous values using average values over a certain time period.

Maps are available at selected scales, which provide the various indicators (see Fig. 1 for an example). Data for all countries can be found on [*irena.org*]. Information for a particular location, identified by its latitude and longitude, can be found on the website of the US National Aeronautics

---

[4]As quantified by the albedo factor, which is the ratio between the light energy reflected and the incident light energy, the Earth's surface reflects more or less light. The albedo factor of fresh snow is above 0.9, and that of asphalt is roughly 0.1. The average albedo factor for the Earth's surface is 0.3. Houses in warm countries, such as around the Mediterranean Sea, are often white (ref. Fig. 2 in the "Housing" chapter); they have a high albedo factor and thus do not absorb much solar energy.

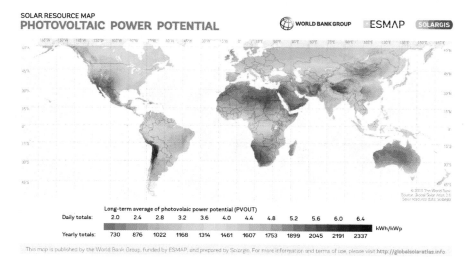

Figure 1.   World map of solar energy global irradiation / © *World Bank group, Solargis*

and Space Administration (NASA) [*eosweb.larc.nasa.gov*]. Regions located between latitude 15° and 40° North and South are the most favored ones from a solar energy point of view. Regions close to the equator are often quite humid with significant cloud cover. Depending on the regions, the daily values can change more or less widely between summer and winter due to the duration of daylight and to the Sun's inclination. For example, in moderate climate European countries, three-quarter of the solar energy is received between April 15 and October 15. The inter-season variations are generally small for regions close to the equator.

Broadly speaking, depending on whether the location is more or less favorable from latitude and/or altitude as well as climate points of view, the GHI on the Earth's surface ranges from 900 to 2 700 $kWh/m^2 \cdot yr$. The corresponding solar radiation energy ranges from 2.5 to 7.5 $kWh/m^2 \cdot day$. In regions with moderate climates, the maximum power received under dry and fair weather conditions on a surface perpendicular to the Sun's rays is close to 1 kW/m^2. This corresponds to energy of 1 kWh per hour at "full sun" on a 1 m^2 surface.

The 1 kW/m^2 power is also used to compare the efficiencies of solar PV panels. The Standard Test Conditions (STC) stipulates the test conditions for solar PV panels: solar radiation of 1 kW/m^2, temperature at 25°C, solar spectrum at AM1.5, which corresponds to the attenuation due to the atmosphere multiplied by 1.5. This attenuation is an average value for the solar rays reaching latitudes with moderate climates such as in Europe, the US, China and Japan.

Solar Photovoltaics: From Principle to Panel

In 1839, **Alexandre Edmond Becquerel**[5] discovered the PV effect. The first PV panels supplied electricity to satellites. Their large scale development dates back to the 1990s.

The Principle

A PV cell absorbs a portion of the solar radiation energy it receives to extract some electrons from the material used to build it. In turn, these electrons create an electric current as a load is connected across the PV cell's terminals. The current is said to be direct current (DC); the voltage across the cell is also DC.[6] To enable this phenomenon, a material is used such that, thanks to the energy received from light photons,[7] some electrons can leave their orbits around an atom and then move inside of the material itself. Such materials are referred to as semiconductors.[8]

Solar Cells and Panel Manufacturing

As of 2019, over 90% of all new PV panels installed worldwide used silicon; it is the most abundant material in the Earth's crust, aside from oxygen. This low cost and non-toxic semiconductor is found in sand and quartz, primarily in the form of silica. It has been used in microelectronics

---

[5] French physicist. He was the father of Henri Becquerel who discovered radioactivity.

[6] This is the reverse phenomenon of that used in Light Emitting Diodes (LED) (ref. "Housing" chapter), where a DC current results in the emission of light.

[7] A photon is a corpuscle associated with light. **Albert Einstein** introduced this concept in 1905. Its energy increases with frequency. It is higher for ultraviolet light than for red light; this is why skin protection is needed against ultraviolet light using sunscreens. The Sun's spectrum contains photons at all energy levels.

[8] These materials have properties between isolating materials, which have almost no free electrons and conductors, which have free electrons associated with all atoms.

industries — transistors and microprocessors — long before the massive development of photovoltaics.

The main silicon panel manufacturing stages are:

1. Extraction of the silicon from the silica using an arc furnace followed by purification to obtain solar grade purity, which is lower than the purity required for microelectronics.
2. Crystallization of the purified silicon followed by cutting of the silicon ingots into thin wafers.
3. Treatment of the wafers to facilitate the current flow through them.
4. Assembly of the PV cells by deposition of a thin metallic electric contact sheet on the back face and a thin metallic contact grid on the front face, to reduce light losses, along with the connectors and an anti-reflection sheet to enhance the light absorption and to protect the cell.
5. Assembly of the cells into rectangular panels, generally by connecting them in series to obtain the desired output voltage. The entire panel is mounted into a frame for added rigidity and connections to neighboring panels.

Depending on the crystallization method used, a cell may contain a single silicon cell or be polycrystalline. The two types of cells — monocrystalline and polycrystalline — are easy to recognize (Fig. 2):

• By the shape of the cells: Circular or rounded squares (to limit the materials losses) for the monocrystalline cells since they are made

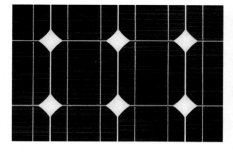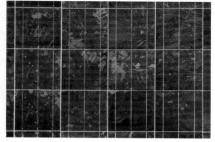

Figure 2. (a) Monocrystalline cell PV panel / © *Martin D. Vonka, Shutterstock,* (b) Polycrystalline cell PV panel / © *Studio 2013, Shutterstock*

from cylindrical wafers, or square for the polycrystalline cells since they are made from square wafers.

- By the visual aspect of the front face: Uniform for monocrystalline and with visible crystal lines for polycrystalline.

Solar Energy Conversion Efficiency

It is the ratio between the electric energy produced and the solar radiation energy received. The efficiency decreases with the temperature, typically 0.5%/°C, as well as with age, typically 0.5%/yr due to the aging of the materials involved. As a result, the cells' efficiency typically decreases some 10% after 20 years. The efficiency of an entire PV panel is lower than that of its cells. While the efficiency of monocrystalline cells had reached 26.1% under laboratory conditions in 2018 (double what it was in 1978!), the efficiency of commercially available cells in 2019 was typically between 18 and 20%, the best ones reaching 23%.

Peak Power

Rather than the panel's efficiency, its performance may also be indicated by its peak power, P_p, under standard STC test specifications. It is also given for a specific panel size. For a 19% efficient, 1.65-m^2 panel with a peak power of 313 W, the relationship between the two indicators is:

$$P_p = 1\ 000\ \text{W/m}^2 \cdot 19\ \% \cdot 1.65\ \text{m}^2 = 313\ \text{W}$$

The Quest for Improved Efficiencies

The theoretical maximum efficiency of silicon PV cells is 32% due to the characteristics of the material involved and the Sun's radiation spectrum, which, in reality, it only partially absorbs. However, cells with higher efficiencies have been developed by way of several technological paths:

- The first one consists of layering cells of different materials, each catching a different portion of the Sun's radiation spectrum. Such multi-junction cells are more difficult and hence more costly to manufacture.[9]

- Another method consists in concentrating the radiation using optics (lenses and/or mirrors). This decreases the cell area while still increasing its production.

In 2018, under laboratory conditions, a record efficiency of 47.1% was achieved using a combination of the two methods described above, i.e., quadri-junction cells and concentration.

The search for less complex manufacturing and/or less material-intensive and/or less costly and/or more flexible processes has led to the development of several new types of cells such as:

- "Thin film" cells, 2 μm thick rather than 200 μm are:
  ○ Amorphous cells, manufactured starting from gaseous silicon that is deposited under vacuum on a rigid or flexible subtract, thus avoiding the intermediary wafer stage.
  ○ CIGS (Copper, Indium, Gallium, and Selenium), i.e., the constituents of the semiconducting alloy that is deposited on a rigid or flexible subtract.
- Organic cells, for which one of the main materials is a carbon-based semiconductor, i.e., an abundant and low cost material.[10]
- Dye-sensitive cells (DSCs), also designated as Graetzel cells,[11] mimic the plant photosynthesis process[12] by separating the photon absorption from their transfer inside of the material.
- Starting in 2013, a new type of metallic, non-organic cell, perovskite material, opens new development avenues, having already reached efficiencies of 29% in 2020.

---

[9] They were initially developed for space applications.

[10] The material is not made of living cells. The designation "organic" comes from the use of carbon, which is the atomic element basis for life on Earth.

[11] Named after the Swiss-Austrian chemist who invented them.

[12] Plants manufacture the glucose, $C_6H_{12}O_6$, and the sugar that allow them to live and grow from CO_2 and the water vapor in the air, and then releasing part of it by way of the chlorophyll, $C_{55}H_{17}O_5N_4Mg$, a light-absorbing pigment, except for green light (which explains the color of plants).

The National Renewable Energies Laboratory (NREL) [*nrel.gov*] in the US regularly updates a chart showing the initial development dates of various PV technologies and the progress of their efficiencies.

Photovoltaics: From Panel to Deployment

The output voltage of PV panels inherently is a DC. As seen in the "Electric Power Systems" chapter, almost all applications of electricity and almost all electric transmission and distribution systems use AC voltages.[13] An inverter, an apparatus converting DC voltages into AC voltages, is therefore required to connect the panels to the local network or most end-use applications.

A PV installation can be:

- Stationary and connected to the local electric distribution network. These installations can either be directly on the ground or the roofs of individual houses, apartment buildings, shopping centers or on industrial buildings and hangars. Some ground-installed PV plants can reach several hundreds of MW.
- Stationary but not connected to a local electric distribution network, as is becoming more frequently the case for islands or isolated villages as well as remote mining operations with no access to local electrical infrastructures. The proper energy and power management of such systems is often complex and may require the installation of energy storage systems carefully selected taking the local circumstances into account.
- Embarked, such as on ships or vehicles as well as for electronic equipment such as telephones and watches, although at lower scales. Providing additional energy is often required leading to complex autarkic energy management.

For stationary installations, aside from the panel's performance, the level of solar irradiation, i.e., physical location of the installation,

---

[13] At 50 Hz in most countries in the world except North America, parts of Japan and Latin America, for example, where it is 60 Hz.

availability of ground or roof surfaces as well as the proximity and installed capacity of the local electric infrastructures have to be taken in consideration. The proper panel orientation and tilt, given the local solar radiation conditions, have an important impact on the electricity production levels. The installation's conversion and connection losses are also important factors along with shading from nearby buildings or other obstacles.

NREL provides an online performance estimation tool for a residential PV installation connected to a local electric network [*pvwatts.nrel. gov*]. A number of losses along the conversion process, typically around 15%, can be included along with the PV panel efficiencies and tilt angle.

The following three inserts further illustrate various aspects of solar PV panel installations.

The first insert illustrates the importance of the actual location of the PV installation and the impact of the installation's tilt.

Residential installation in Stockholm, Sweden, and Nairobi, Kenya

In both cases, the NREL simulation tool is used to obtain the results. The standard default losses proposed by the model are used. A "rule of thumb" is that for optimum performance over an entire year, the panel tilt angle should be equal to the latitude at the installation location; it is used below.

The input data used is:
- DC peak power: 4 kW DC.
- PV panels at 19% efficiency (premium in the NREL simulation).

As a result, the area needed for the 4 kW DC output is:
$(4 \text{ kW} / 1 \text{ kW/m}^2) / 19\% = 21 \text{ m}^2$

	Stockholm	Nairobi
Latitude:	59.65°N	1.3°S — below the equator
Optimum panel tilt:	59.6°	1.3°
Panel orientation:	Directly south	Directly north
Solar energy received yearly:	1 084 kWh/m²·yr	1 840 kWh/m²·yr
Electricity production:	3 541 kWh/yr	5 820 kWh/yr
Capacity factor:	10.1%	16.6%

As a result, to obtain the same energy, the panel area required in Nairobi is some 40% lower than in Stockholm.

The second insert highlights the total electricity which can be produced from a building's rooftop and relates this amount to the number of occupants of the building.

Solar energy production on an apartment building roof

Each floor of an apartment building has four 70-m^2 units and a 20-m^2 stairwell. One assumes here that the 300-m^2 roof is entirely covered with PV panels, which is optimistic given the need to properly tilt and space them to avoid mutual shading. Referring to the previous example, one assumes a solar energy production between that in Stockholm and Nairobi, i.e., 25 m^2 for a production of 5 MWh/yr. At 300 m^2, the building's yearly solar production would be 60 MWh/yr.

In 2015, the average residential electricity consumption per person in European OECD countries was 1.5 MWh/yr [*iea.org*]. The electricity consumption of up to 40 residents could thus be covered using the solar energy produced on the roof. At 2.5 inhabitants per unit, 16 units could be supplied; therefore, the building could have up to four stories if only solar electricity is to be used to supply its electricity needs. However, the overall electricity demand for the entire building could not be met at all times, even if coupled with proper energy storage.

The example illustrates that the PV electricity can provide for some self-consumption by the building's occupants and for some common services. It also illustrates the limitations of solar PV production on the roofs of high-rise buildings.

The third insert provides information for a shopping center PV installation.

PV power plant on the roofs of the City Mall shopping center, Tagum, Philippines

Completed in 2018 by Solar Pacific, shown in Fig. 3, this power plant covers 1 ha of the shopping center's roofs. Using panels with an average efficiency of

(Continued)

(Continued)

16.4%, its peak power is 1.1 MW. The expected yearly electricity production is 1.52 GWh; i.e., with a capacity factor of 15.7%, or 1 380 h/yr at full power. The plant covers 80 to 85% of the shopping center's consumption under good solar conditions.

The picture also shows the rooftop mounted HVAC installations; ref. rooftop installations under the "Tertiary and Services" chapter.

Figure 3. PV roof from Tagum City Mall (Philippines) / © *Solar Pacific*

Concentrating Solar Power Plant (CSP)

They use mirrors to deflect and concentrate the Sun's rays. Only direct radiation can be taken advantage of since diffuse radiation cannot be directed. The two main types presently used are: solar tower and cylindrical-parabolic.

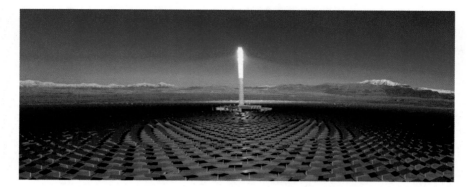

Figure 4.   CSP solar tower, Noor III, Ouarzazate, Morocco / © *Masen*

Solar Tower Power Plants

In these power plants, shown in Fig. 4, mirrors are used to redirect the solar rays toward a receptacle located at the top of a tower to heat a coolant. The mirrors are mounted on supports such that they can rotate around two perpendicular axes to provide for precise focusing of the Sun's rays on the tower receptacle, hence the heliostat designation. They can be from 1 to 100 m^2 large and are generally slightly focusing. When clean, they reflect 95% of the solar radiation received. Depending on the particular installation, several thousand to over 100 000 mirrors may be used. The light concentration on the receptacle[14] at the top of the tower is such that the coolant's temperature may reach 1 000°C. The working transfer fluid can be water, which is turned into steam that is supplied to steam turbines. Molten salts can also be used combined with a heat exchanger where the molten salt's heat is transferred to water to create the steam.[15]

---

[14]The concentration factor is the ratio between the area of the mirror collectors and that of the receptacle area. For a solar tower power plant, it can exceed 1 000 depending on the design. As per the Carnot Principle, higher temperatures, i.e., higher concentrating factors, lead to better energy efficiencies.

[15]Using water as the coolant avoids the need for a heat exchanger, which must be used with molten salts. However, the temperatures that can be reached are higher using molten salts. In addition, it is easier to store the heat directly using molten salts rather than water vapor under pressure (either directly or using an additional water vapor — molten salt heat exchanger). Finally, using molten salts requires that they be maintained in a liquid state, which consumes energy.

Operationally, the main difference from standard steam thermal power plants is the variability of the amount of heat available: a minute to a few tens of minutes due to clouds, some 12 hours between day and night, and the changes from one day to the next along with the weather conditions and the seasonal variations. The fluid's thermal inertia can dampen the shorter variations. To address the slower variations, one can enhance the plant's thermal inertia by augmenting the amount of working transfer fluid and also by providing storage for the fluid. Figure 5 illustrates an installation with two molten salt storage reservoirs. Larger overall mirror areas and receptacle surfaces are required to make more heat available during the day, not only for direct electricity production but also to store it to allow for electricity production during periods with no solar radiation, even during several hours at full power overnight. The plant's capacity factor can thus be increased. A number of solar tower power plants can operate at full output power for several hours after sunset.

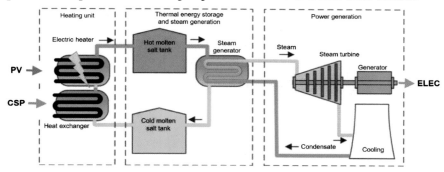

Figure 5.   Layout of hybrid CSP with storage / © *www.dlr.de/tt*

One can also add a fossil fuel-based vapor production to provide for a steadier production; the power plant is then said to be "hybrid"; its operation is then no longer CO_2-free.

The efficiency of this type of power plant can be assessed in several ways:

- The maximum solar-electric efficiency, i.e., the maximum instantaneous ratio between the electric energy produced and the energy received by the mirrors. It depends on the temperature reached by the receptacle and thus on the concentration factor as well as on the particular thermodynamic cycle implemented. It can almost reach 25%.

- The yearly net efficiency, i.e., the ratio between the electricity produced over a year and the total heat received. It can reach 15%. It differs from the gross efficiency, which does not take the consumption of auxiliary system losses into account; they can reach several percent.

Cylindrical-Parabolic Mirror Power Plants

As shown in Fig. 6, the receiving mirrors in such power plants have a cylindrical-parabolic shape, such that the light is focused on a tube, containing the working fluid, which is parallel to the axis of the cylindrical parabola and located at its focus.[16] The mirrors' support allows for their rotation around a single axis; as a result, the tracking of the Sun is less accurate than for tower concentrating power plants. The light concentration is also less intense; the temperatures reached are thus not as high, resulting in lower thermodynamic efficiencies. The coolant's circulation path is also longer since it collects the heat from all cylindrical-parabolic mirrors heliostats. The actual electricity production section of the installation is similar to that used for tower concentrating plants; the same energy storage technologies can also be implemented.

Figure 6.   CSP with cylindrical-parabolic mirrors, Noor I, Ouarzazate, Morocco   / © *Masen*

---

[16] If a ray from the Sun arrives parallel to the symmetry axis of a parabola, the ray will be reflected such that it passes through the point designated as the parabola's focus. Therefore, if the parabola is suitably oriented, the Sun's rays will be focused on the tube carrying the coolant.

The 160 MW cylindrical-parabolic mirror Noor 1 power plant, Morocco

This is the first of several stages of a power plant close to Ouarzazate. The local incoming solar irradiation is 2 635 kWh/m^2·yr. The power plant, shown in Fig. 6, occupies 480 ha. Each of the 19 200 receivers has 28 mirrors to provide the parabolic shape. The total area occupied by the mirrors is 1.45 Mm^2. The coolant is oil; its temperature increases from 293 to 393°C, thanks to the heat received. Using a heat exchanger, the heat received produces steam driving a turbine using an open cooling circuit. In addition, two molten salt reservoirs can store the heat by way of intermediary heat exchangers; this provides for operation at full power for three hours.

Inaugurated in February 2016, the power plant is designed to produce 600 GWh yearly, which would correspond to a net yearly efficiency of 15.7%. The capacity factor is 43%, i.e., 3 750 h/yr at full power.

Hybrid CSPs

The molten salts in the installation shown in Fig. 5 are heated by way of solar mirrors and/or by PV electricity. From the "hot" reservoir, they are used to create steam in a heat exchanger before returning to the "cold" reservoir from where they are reheated. A few molten salts storage installations only featuring PV panel heating are under construction in a few countries.

Environmental Impacts and Risks

Concentration and ground-installed PV power plants require large ground surfaces; this is, of course, not the case for roof-mounted systems. During normal operation, neither concentration nor PV power plants emit any greenhouse gases or polluting particles. However, the manufacturing of their components and their installation does. Globally speaking, the manufacturing of the panels requires electric energy of the same proportion they produce during their first to fourth years of operation. As it is the case for all manufacturing, the amount of CO_2 thus emitted depends on the electricity production mix in the particular region where they are manufactured. The CO_2 content of the electricity produced using silicon PV panels depends on their lifetime. It can be estimated to be between 10 and 40 gCO_2/kWh over a normal 20–25-year life span.

Water is required to periodically clean the mirrors and panels; the quantities heavily depend on the particular site — dust and pollutants — and can range from 10 to 100 l/MWh. Similar to classical steam power plants, concentration power plants also require cooling systems, which could induce additional environmental impacts.

Concentration power plants, which use fluids under pressure and at high temperatures as well as rotating equipment, inherently present risks similar to those of standard thermal power plants in case of the degradation of some components. Other risks are associated with the heliostats and the accuracy and reliability of their tracking systems required to obtain the desired efficiencies, as well as to avoid other equipment degradations, such as above or below the tower receptacle.

The potential impact of large heliostat fields on birds has become a subject of increased studies.

Solar energy power plant operational and impact factors can be found under "Power Plant Indicators" in the Companion Document.

The development of solar energy electricity production technologies is on-going, especially the PV path. While inducing several in-the-field deployment challenges, the contribution of solar electricity production will increase considerably due to its limitless availability on Earth and since it does not produce any CO_2 emissions under operation. PV energy can be deployed at multiple levels, from a few kW for residential applications to several hundred MW for large power plants. Though PV deployment on building roofs has zero land footprint, it competes for rooftop space with hot water and/or HVAC applications.

Wind Energy

Using Wind

Man has been using wind energy for over two thousand years for navigation as well as to grind grains and pump water using wind mills. The first
electricity-producing experiments date back to the late 1800s. In 1941, the
first MW wind machine, rated at 1.25 MW, was connected to the local
utility network in Vermont, United States (US); the two-bladed machine
suffered a fatal blade failure after only 1 100 hours of operation. The rapid
development of wind energy started well after the Second World War;
while far below hydroelectric energy production, as of 2020, wind energy
remains the second-largest renewable electricity source.

Statistics [*gwec.net*], [*irena.org*], [*ren21.net*]

As seen in the "Introductory Remarks" chapter under "How Electricity Is
Produced", the overall wind energy electricity production in 2018 was
1 263 TWh and represented 4.7% of the worldwide electricity production.
China led the way in overall installed capacity and additions, followed by
the US, Germany and India.

Also, in 2018, the on-shore wind electricity production was 1 195
TWh; the installed capacity was 540 GW at the year's end, i.e., more than
20 times higher than the 24 GW of 2001. The capacity factor was 25.3%.[1,2]
In 2018, the capacity addition was some 50 GW.

---

[1] Capacity factor; ref. "Introductory Remarks" chapter under "How Electricity Is
Produced". In this case: 1 195 000 GWh / (8 760 h · 540 GW) = 25.3% or 8 760 h/yr ·
25.3% = 2 213 h/yr at full power.

[2] Given the rapid increase in the installed capacity, the average installed capacity during
2018 was lower than 540 GW; the capacity factor was thus slightly higher. The same is
also true for off-shore installations.

While lagging on-shore installations, off-shore wind energy electricity production is developing rapidly with an installed capacity of 24 GW and a production of 68 TWh in 2018, such that the overall worldwide capacity factor was 32%, i.e., 2 833 h/yr at full power, significantly higher than for on-shore installations. The 2018 capacity addition was some 4 GW.

Wind Creation and Measurement

Varying solar irradiance induces temperature differences on Earth, which, in turn, induce pressure differences. Wind is primarily due to atmospheric pressure differences with wind blowing from high to low-pressure zones; higher-pressure differences induce higher wind speeds. Temperature differences between sea and land induce coastal breezes, from the sea toward land daytime and reverse during nights. Mountain-induced winds also affect local wind patterns.

The Earth's natural rotation around its axis induces two additional wind patterns:

- Planetary wind patterns, which explain, for example, why transatlantic flights from North America to Europe are shorter than in the opposite direction. Their overall directions are in the opposite direction in the northern and southern hemispheres; their strengths change during the year.
- Around high and low-pressure zones. Their circulation directions are in opposite directions in the northern and southern hemispheres.

Incorporating increasingly abundant ground-level wind measurements and a growing number of precise satellite measurements, numerical three-dimensional models lead to improved weather forecasting using some of the most powerful computers in the world; ref. "Tertiary and Services" chapter. While longer-term wind predictions, over several weeks, remain elusive, the accuracy of shorter-term predictions continues to improve.[3]

---

[3] Upon take-off, passengers of a 12-hour flight are informed of its estimated duration with an accuracy of minutes.

Wind speed is measured in m/s or km/h.[4] The horizontal wind speed component, the main one, is generally measured using anemometers with small cups rotating around a vertical axis; a vane is used to determine its direction. Radar-based systems, called lidars, can take measurements along a line as opposed to single point anemometer measurements.

Wind Energy Variability

Similarly to solar energy, wind energy is inherently variable as illustrated in Fig. 1 as the figure shows, the variation amplitudes can be substantial even during short time spans. Wind speeds can vary, sometimes significantly, between even close-by locations. As a result, the cumulative wind power at any specific time is below or far below the cumulative installed capacity of any one region. Potential wind energy resource data is widely available worldwide; Fig. 2 shows the wind energy resources available in the US and Europe. Wind resources are generally higher in the open sea than on land.

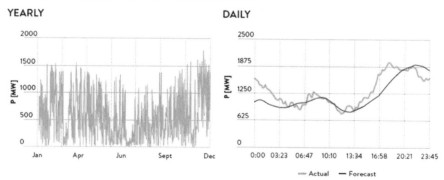

Figure 1. Yearly and daily wind variability in Ireland / © *World Energy Council*

---

[4]The knot, nautical miles per hour, is sometimes used; one nautical mile is 1 852 meters.

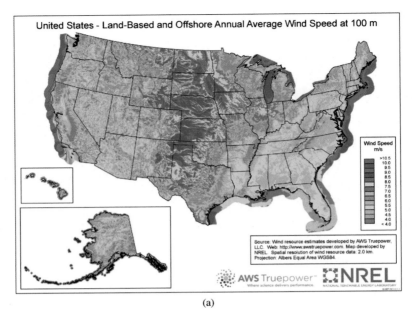

(a)

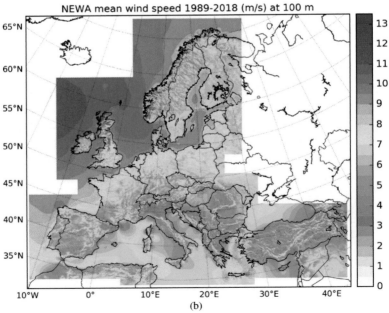

(b)

Figure 2. Wind energy resources maps, (a) in United States / © *US NREL,* and (b) in Europe / © *Björn Witha, Fraunhofer IWES*

Horizontal Axis Wind Turbines

The vast majority of wind turbines have a horizontal axis.

Lift Creation and Torque

The lift creation mechanism on a wind turbine blade is the same as that on an aircraft wing (ref. "Transport and Travel" chapter) except that it is the wind motion on the airfoil that creates lift on the passive airfoil. The airfoil, or blade, can generally pivot around its own axis to adjust its angle of attack and thus the lift. The wind's kinetic energy is converted to blade lift, thus creating torque on the rotor hub.

A typical horizontal axis wind turbine configuration is shown in Fig. 3. Further details as to the configuration and description of the multiple forces to which the turbine blades are subjected can be found under "*Wind Turbine Engineering*" in the Companion Document.

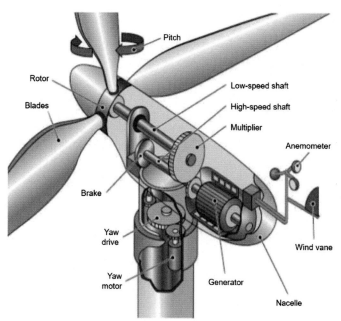

Figure 3. Horizontal axis windmill nacelle / *http://www.lemoniteur.fr/article/ comment-ca-marche-energie-eolienne*

Wind Machine Performance

The wind power transferred to the turbine hub can be expressed as:

$$P = \tfrac{1}{2} \cdot \rho \cdot A \cdot v^3 \cdot C_p$$

where: P is the wind power available, W.

ρ is the density of air = 1.225 kg/m^3 at sea level and 15°C.

A is the area swept by the turbine blades, m^2.

v is the wind velocity, m/s.

C_p is a performance coefficient taking the configuration and losses into account. It is generally between 0.35 and 0.45.

As per the formula above, the available wind power:

- Quadruples if the diameter doubles, hence the trend toward higher rotor diameters.
- Is multiplied by eight if the wind speed doubles, hence the need for careful site selection.

Air density, ρ:

- Decreases when the temperature increases.
- Increases with humidity.
- Increases with pressure and thus decreases with altitude.

As a result, the best sites are at sea level — higher pressure and humidity — and in cold climates, hence the North Sea and Baltic Sea sites in Europe, for example.

Cut-in and Cut-off Speeds

The turbine starts rotating at the "cut-in" wind speed; careful blade design is such that it is as low as possible. The design also ensures that the turbine can safely operate up to a "cut-off" wind speed, which is as high as possible, beyond which the blades are feathered. A turbine with suitable cut-in and cut-off speeds can be selected for the intended location through detailed wind speed measurements over extended periods.

Off-shore Wind Turbines

Off-shore locations address the visual impact and noise concerns of the public-at-large and environmental organizations; they are also attractive due to often higher, steadier and more predictable wind speeds. The investment required are higher due to more complex and costly underwater cables and related infrastructures. Maintenance costs are generally also higher due to faster equipment deteriorations from the marine corrosive environment and due to blade fouling, which can be a concern requiring special, and costly, blade surface treatments. Floating wind turbines are under development to address coastline profiles with great depths, even close to shore.

---

8 MW wind turbine

In 2019, the installed capacity of the largest wind turbine in operation was 8.0 MW. Intended for off-shore installation, its key specifications are:

- Diameter: 164 m.
- Three blades, each 80 m long with a weight of 34 tons.
- Swept area: 21 124 m^2, i.e., larger than three soccer fields.
- Nacelle dimensions: 12 m wide, 7.5 m high and 24 m long.
- Nacelle and hub weight: 390 tons.
- Total weight at the top of the mast: 500 tons, i.e., 15 large trucks.
- Mast height: 105–140 m, depending on location.
- Cut-in wind speed: 4 m/s — below this speed, the turbine does not rotate.
- Rated wind speed: 11 m/s.
- Cut-out wind speed: 25 m/s — above this speed, the blades are feathered and the rotor is locked.
- Rotational speed: 4.8–12.1 RPM.
  At 12 RPM, the peripheral tip speed is 370 km/h.

Such a turbine, located in northern Denmark, set a 24-hour production record of 192 MWh resulting in a capacity factor of 100%, i.e., it ran at full capacity during the entire 24-hour span.

While maintaining the same rotor size, the turbine's rating has been increased to 10 MW, thanks to airflow, gearbox and cooling enhancements.

Wind Farms

For economic and practical reasons, wind farms include up to several hundred turbines, as shown in Fig. 4, for both on and off-shore wind farms. When siting wind turbines, proper compromises need to be reached between proximity to load centers, to limit electric energy transmission and distribution infrastructure requirements, on one hand, and, on the other, availability of significant wind resources, to maximize the yearly electricity production. In addition to visual and environmental considerations, proper wind farm siting and configuration must take wake effects into account; the "rule of thumb" is that the space between turbine rows in the dominating wind direction should be some three to four times wider than the rotor diameter.

Large wind farms

While China is in the process of building very large wind farms with installed capacities exceeding several GWs, as of 2018, the Alta Wind Energy Center in California remains one of the largest in the world. It comprises 11 separate projects, primarily for financial and electricity contractual reasons. Construction on the Alta I project started early 2010, and Alta XI was completed in December 2012, i.e., less than three years later.

The total installed capacity is 1 550 MW and comprises 100 turbines rated at 1.5 MW and 466 turbines rated at 3 MW. The site's area is 12.9 km^2. The 2014–2018 average yearly production was 3.2 TWh such that the capacity factor was 23.6%, i.e., 2 064 h/yr at full power.

The Alta Wind Energy Center confirms how quickly large-scale wind energy farms can be completed despite logistics challenges.

Environmental Impacts and Accidents

The main oppositions to wind turbines are societal and relate to visual impacts on the environment, especially with rotor tip heights reaching over 200 m. Other implementation issues include:

- Noise: It decreases by the square of the distance from the sound source; for example, the noise level at 200 m from a wind turbine is four times lower than at 100 m. To reduce both the visual and noise

(a)

(b)

Figure 4. (a) On-shore windfarm / © *EDF_Photo: Marc DIDIER/PWP* and (b) Off-shore windfarm / © *Blue Planet Studio, Shutterstock*

impacts, minimum installation distances are specified from the nearest residences.

- Radar interference: Multiple tests have shown that radar detection and tracking of aircraft flying at low heights above wind turbines deteriorates, occasionally significantly, under some circumstances. To mitigate these effects, tip height limitations are imposed for wind turbines located close to airports.

- Wildlife impacts: They primarily concern birds and bats. A number of studies have shown these impacts to be small and of no danger to any particular species.

Human death or injuries related to wind turbines primarily involve accidents during installation or repair and maintenance operations. The four main wind turbine failure causes are: blade and mast structural failures, nacelle fires and lightning strikes. In colder climates, blade icing is a concern, not only because of reduced airfoil performance and material performance deterioration but also regarding the potential for injuries to nearby persons as the ice blocks are detached by centrifugal forces; blade heating can be implemented to mitigate the icing effects.

Wind energy operational and impact factors can be found under "Power Plant Indicators" in the Companion Document.

The development potential for both on and off-shore wind energy remains significant worldwide. The enabling technology is well developed with on-going incremental capacity increases, especially for off-shore applications. While wind energy is variable, compared to solar energy, the variability is spread more evenly between day and night as well as between seasons.

Geothermal Energy

Using the Earth's Heat

Over 99% of the Earth's volume is at temperatures exceeding 1 000°C. Geothermal energy[1] is abundant and inexhaustible at Mankind's time scale. As illustrated in Fig. 1, geothermal reservoirs, resources, are hot groundwater tables that are replenished by infiltrations from rivers, oceans or rain at various rates depending on the geological situations. Further information can be found under *"Earth's Structure and Geothermal Sources"* in the Companion Document.

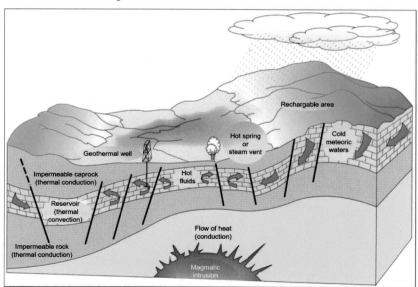

Figure 1. Geothermal resources / © *International Geothermal Association*

[1] In Greek, "geo" for Earth and "thermos" for heat.

Unless used directly for baths, as it has been for centuries, extracting heat from the Earth's sub-surface generally requires more or less sophisticated systems by way of a fluid, generally water and/or vapor. Geothermal energy is renewable.[2] It is also non-intermittent,[3] i.e., usable continuously. It often involves greenhouse gas (GHG) emissions.

One distinguishes two main types of geothermal energy systems:

- *Direct use*:
  - Either using a standard heat exchanger to transfer the heat to water or another fluid when the geothermal fluid's temperature is between 30 and 150°C. The heat exchanger is generally required to avoid corrosion from elements contained in the water extracted, such as sulfur and phosphor. The main applications include: heating of buildings and neighborhoods, hot water, fish farms, as well as drying of agricultural and industrial products. A single resource can have multiple uses.
  - Or using a heat pump when the geothermal fluid's temperature is between 10 and 60°C such that it must frequently be increased. Applications also include individual home heating or industrial uses.
- *Electricity production*, which is the focus of this chapter; one then refers to geothermal power plants. The first one was put into service, in 1904, in Larderello, Italy.

Statistics

Direct Use

The direct heat production installed capacity in 83 countries has increased from 15.1 in 2000 to 70.3 GW in 2015[4] with a heat production of 163 TWh

---

[2] In fact, it is the heat from the Earth that is renewable. To be sustainable, the water extracted must be replenished, either naturally or by way of various processes.

[3] Some famous geysers, such as the Old Faithful in Yellowstone National Park in the United States, are intermittent.

[4] Most recent data when this chapter was written. This is an estimate; the values vary significantly between data sources.

[*geothermal-energy.org*] such that the capacity factor was 26.5%.[5] Also, in 2015, the worldwide installed thermal capacity of all geothermal heat pumps was 49.9 GW, i.e., 71% of the total installed capacity; it continues to grow, which decreases the overall geothermal capacity factor given the seasonal uses of heat pumps.

Electricity Production [*iea.org*], [*irena.org*], [*ren21.net*]

As seen in the "Introductory Remarks" chapter under "How Electricity Is Produced", the 2018 geothermal electricity production was 88 TWh in 27 countries for an installed capacity of 13.2 GW; the capacity factor thus was 76.1%, i.e., 6 667 h/yr at full power. The overall installed capacity had reached 13.9 GW in 2019. Countries with seismic regions had the largest installed capacities: the United States (US), followed by Indonesia, the Philippines and Turkey. The plants typically consist of one or more units with installed capacities from a few hundreds of kW to less than 100 MW.

Geothermal Power Plants

Depending on the characteristics of the geothermal resource (pressure, temperature, geothermal water/vapor flow rate and the fluid's chemical composition), one of three technology avenues can be used.

Dry Steam Geothermal Power Plants

This is the least complex technology; the Larderello plant was dry steam. As of 2015, some 25% of the worldwide geothermal electricity production installed capacity was from dry steam. It is used when the steam is: (a) at temperatures exceeding 150°C, (b) sufficiently dry, and (c) not excessively corrosive such that it can be directly fed to a turbine. The still humid vapor from the turbine can:

---

[5] Capacity factor; ref. "Introductory Remarks" chapter under "How Electricity Is Produced". In this case: 163 000 GWh / (8 760 h · 70.3 GW) = 26.5% or 8 760 h/yr · 26.5% = 2 320 h/yr at full power.

- Either be directly released into the atmosphere, which, while convenient, can induce environmental issues and also results in water wastes.
- Or be fed to a condenser, which enhances the plant's efficiency and provides for the reinjection of some of the water extracted back into the ground to replenish the reservoir and thus reduce the risk of resource depletion. The reinjection wells must be sufficiently far from the extraction wells to avoid cooling the vapor resource. The condensation is generally implemented via evaporation, which consumes some water; ref. "Combustion Thermal Power Plants" chapter.

The Geysers geothermal power plants in California, US

This group of 16 active dry steam power plants is located 120 km north of San Francisco and covers some 100 km^2. Electricity production started in 1960 by way of an 11 MW plant. As of 2019, it was the largest geothermal power plant in the world.

In 2017, the largest among three operators operated 13 units on the site with a total installed capacity of 725 MW; the combined electricity production was 5.63 TWh leading to a capacity factor of 89.3%, i.e., 7 820 hours at full capacity. Two other operators produced one TWh in addition. The overall GHG emissions were 45 gCO_{2e}/kWh [*calepa.ca.gov*].

The heat comes from a magmatic reservoir situated 7 km deep that provides dry steam at some 180°C and an average pressure of 5.2 bar over the 350 active wells. The average depth of all wells is 2 600 m. The wells and production units are interconnected by way of a 130-km-long steam pipeline network to ensure a proper steam distribution. The vapor contains hydrogen sulfite, which is extracted after the turbines to produce sulfur, which, in turn, is used for agriculture.

Starting in 1980, increased production levels combined with low rain levels, which no longer provided enough water to replace the water required for the evaporation cooling, led to decreasing vapor pressures. By 1987, the output power had reached 2 043 MW. To sustain the production, the overall production capacity was reduced and some 60 dry wells are presently used to reinject water from two nearby community water treatment plants by way of a 110-km water pipeline network. On average, some 75 million liters of water are reinjected daily, i.e., some 4 l/kWh. The cold-water injections occasionally induce micro earthquakes, generally of intensities below three on the Richter scale; they can be mitigated by proper injection distributions.

Flash, Condensing, Geothermal Power Plants

Over 60% of the worldwide geothermal power plant installed capacity is implemented using this technology in which the geothermal fluid is a mixture of water and vapor at a pressure of several bars and a temperature of at least 180°C. From the well, the fluid is supplied to a vessel at a pressure below that of the fluid; this results in fast and partial vaporization of the fluid, designated as "flash vaporization". The result is hot and almost dry steam, on one hand, and water at a lower temperature, on the other. The dry steam obtained from the flash vaporization is sent to a steam turbine such that only the heat from the steam is used and not that from the water. Provided the temperature is sufficient after the vessel, the process can be repeated using a second or even third vessel at lower temperatures; the installation is then referred to as double or triple condensation.

Binary Geothermal Power Plants

More recently, this type of geothermal power plant provides for the use of heat below 100°C or the use of geothermal fluids with higher corrosive contents. The geothermal fluid's heat is transferred, using a heat exchanger, to a secondary fluid before it is reinjected. The secondary fluid circulates in a closed loop; it is selected such that, at a suitable pressure, it vaporizes below the geothermal fluid's temperature. The secondary fluid, once vaporized, is sent to a steam turbine and then a condenser before being sent back to the heat exchanger. The secondary fluid thus goes through a thermodynamic cycle with two-phase shifts, hence this plant's designation.

The installed capacity of binary cycle geothermal power plants generally does not exceed 10 MW. By 2015, they accounted for some 15% of the installed geothermal power plant capacity worldwide.

Enhanced Geothermal Systems

The three geothermal power plant technologies described above require that a hot water and/or steam reservoir be available. In the presence of hot but dry roc, a recently developed technology consists of injecting water

Hellisheidi combined heat and power plant, Iceland

Located on the Hengill volcano some 30 km east of Reykjavik, the capital of Iceland, the Hellisheidi combined heat and power plant was constructed in several stages between 2006 and 2011. The maximum depth reached by the 50 wells is 2 200 m. Presently, this flash plant has an installed electrical capacity of 303 MW provided by six turbines, each rated 45 MW and one rated 33 MW. The plant also provides up to 400 MW_{th} of thermal heat for Reykjavik's district heating system. In 2018, it was the largest cogeneration geothermal power plant in the world in terms of total installed capacity.

into existing or created[6] fractures in the roc. As illustrated by Fig. 2, once heated by the roc, part of the water is recuperated before being reinjected; its heat is valorized using one of the technologies described above. Designated as enhanced systems, they provide for broader uses of geothermal systems. The wells are lined down to a certain depth to reduce the risk of perturbation or contamination of layers above the roc heat source, especially in the presence of an aquifer.

Geothermal Power Plant Efficiencies and Emissions

Geothermal power plant efficiencies, i.e., the ratios between the electricity produced and the heat extracted at the well, vary between 6 and 20% for the three types of plants with an average of 12%. These values, lower than for thermal power plants, are due to the lower temperatures involved and the auxiliaries' consumptions such as geothermal fluid purification and reinjections. When technically and economically feasible, electricity and heat cogeneration plants or multi-purpose plants are deployed.

---

[6] Fractures in the roc can be created by injecting hot water, generally containing chemical additives, under pressure into the roc. The technique, similar to what is used for shale gas extraction, is referred to as fracking.

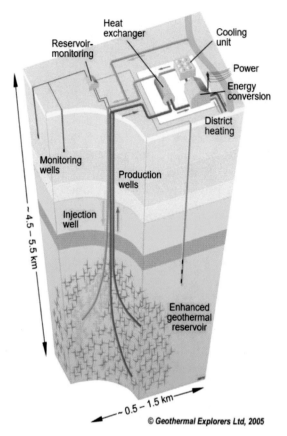

Figure 2. Enhanced geothermal system / © *Geo Explorers Inc.*

As illustrated by the Geysers power plant presented above, geothermal reservoirs gradually provide less energy if the heat extraction is too intensive; geothermal heat is renewable provided it is carefully managed.

During operations, geothermal power plants emissions typically are between 13 and 380 CO_{2e}/kWh [*geothermal-energy.org*]. Binary plants have zero emissions during operation since the fluids are reinjected.

Environmental Impacts and Risks

The environmental impact of geothermal power plants heavily depends on local geological characteristics, particularly on the contents of geothermal fluids: CO_2, NH_3, H_2S and CH_4.

The risks are primarily due to:

- Gas emissions into the atmosphere (greenhouse or acid or toxic).
- Bacterial emissions, which are limited due to higher temperatures.
- Equipment and installation corrosion.
- Micro earthquakes, primarily during preparatory high-pressure injections or during operations.
- Occasional volcanic activity.

Geothermal power plant operational and impact factors can be found under "Power Plant Indicators" in the Companion Document.

Geothermal power plants produce renewable, stable and predictable electricity, provided the underlying geothermal resource is carefully managed to avoid overexploitation. The greenhouse gas content varies considerably between sites. Future drilling and resource enhancement technologies may help additional sites become technically and economically attractive.

An Ever-more Ubiquitous Electricity

Energy Storage

Storage is ubiquitous. The water we use, the food we eat, the fuel we consume, the numerical data on our computers, reach us after one or more intermediate storage, which enhances the continuity of service and the efficiencies of the underlying systems.[1]

This chapter's primary focus is on energy storage systems which can contribute, at multiple timescales and capacities, to the proper operation of the electric system. The main applications are:

- Consumption time shift to store electricity when it is abundant and thus less costly, such as during over-production and/or low demand periods, to restitute it during periods of potential shortage, such as during low production and/or high demand when it would be more valuable.
- Renewable energy support, in particular solar and wind, to mitigate their intermittencies thereby providing for their increased contribution in the overall electricity production mix.
- Deferrals of transmission and/or distribution system upgrades which might otherwise be required to accommodate increased solar and wind energy penetrations.
- Short duration congestion relief of transmission systems.
- Behind the meter, to shift electricity consumption from high to low-tariff periods, such as for electric hot water heaters and especially for commercial and industrial applications.

---

[1] From an energy system perspective, human beings produce heat, mechanical energy, electricity for the nervous system, and waste from all forms of food stored in the stomach during meals.

- System ancillary services, such as frequency and voltage support.
- Microgrid renewable energy integration.
- Power system "black start", i.e., restart of the system from a complete collapse.

Since electricity cannot be stored, as such, the storage is in the form of chemical, mechanical, thermal and electrostatic energy. They can be categorized into two broad types:

- Storage of energy produced using electricity to subsequently restitute it as electricity. This includes:
  - Pumped-storage hydro plants — gravity potential energy.
  - Compressed air energy storage — compression potential energy.
  - Flywheels — kinetic energy.
  - Batteries — chemical energy.
  - Supercapacitors — electrostatic energy and superconducting magnets[2] — magnetic energy.
- Storage of energy produced using electricity to subsequently restitute it in a different energy form. This includes:
  - Residential or tertiary electric hot water heaters with integrated storage capacity; ref. "Housing" and "Tertiary and Services" chapters — thermal energy.
  - Hot or cold water storage for heating or cooling applications; ref. "Housing" and "Tertiary and Services" chapters — thermal energy.
  - Electrolyzers to produce hydrogen; ref. "Industry and Agriculture" chapter — chemical energy, and synthetic gas or fuel production; ref. "Multi-Energy Systems" chapter — chemical energy.

As for all energy systems, the ratio between the energy restituted and the energy provided (i.e., the storage system's efficiency) is always below 100%.

---

[2] Applications of Superconducting Magnetic Energy Storage (SMES) systems remain quite limited; they are not further discussed.

Two situations arise as to the CO_2 content of the electricity restituted:

- The energy storage system itself does not cause any additional emissions, such as for pumped-storage hydro systems or flywheels. However, due to the storage system losses, the CO_2/kWh content of the electricity restituted is higher than for the electricity provided.[3]
- The energy storage system itself causes additional emissions, for example, as will be further explained, for compressed air energy storage systems which require natural gas to be burned. In this case, the emissions of the storage process itself must also be taken into account.

In general, one must compare the CO_2 content of the electricity restituted to the CO_2 content of the electricity that would have had to be produced at the time of its restitution had the storage not taken place.

The main characteristics or specifications of energy storage systems include:

- Energy storage capacity, in Wh or its multiples.
- The maximum powers during storage and restitution, which may not be the same depending on the particular technology. Depending on the technology, they may be expressed in W or Wh/s.
- Efficiency per cycle.
- Lifetime or number of cycles the system is capable of without significant performance degradation.
- Time required to reach full operational power, not only during restitution but also when charging.
- Output voltage, typically for each cell, which is then connected in series and/or parallel to constitute an energy storage system.

---

[3] If, for example, 1 kWh is provided to an energy storage system with 80% efficiency, 0.8 kWh will be restituted. Since the energy storage itself does not cause any emissions, if the energy provided had a CO_2 content of 200 gCO_2/kWh, the restituted energy would have a CO_2 content of 200 gCO_2/0.8 kWh, i.e., 250 gCO_2/kWh.

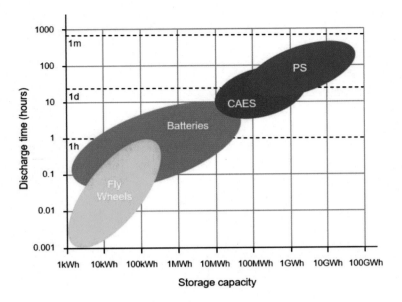

Figure 1. Energy storage systems comparisons / © *HBP and YB*

- The energy storage capacities per unit weight, in Wh/kg, or per unit volume, in Wh/m^3.
- Investment, typically in $ or € per Wh.

As illustrated in Fig. 1, in terms of energy storage capacities and restitution powers, a wide variety of energy storage technologies are used to best suit an increasing number of applications.

Statistics [*hydropower.org*], [*eia.gov*], [*ren21.net*]

By 2018, the total installed capacity of over 1 200 network-connected energy storage systems worldwide was some 165 GW, i.e., 2.5% of the total worldwide generation capacity at the time;[4] it was distributed as follows:

- 158 GW for pump-storage facilities with an overall storage capacity of 9 000 GWh such that the average full power restitution could be

---

[4]The 165 GW does not include smaller batteries used in vehicles or residential applications.

sustained for some 55 hours. China led the way with 30 GW, followed by Japan with 28 GW and the United States (US) with 23 GW.

- 3.2 GW for thermal energy storage facilities integrated within Concentrated Solar Plants (CSP); ref. "Solar Energy" chapter. While they are not directly connected to a network, they play an integral role in CSPs' contributions to overall network operations. Their total storage capacity was some 16.6 GWh such that full power restitution could be sustained during some 5 hours, on average.
- 1.6 GW for electrochemical systems, i.e., batteries.
- 1.6 GW for electromechanical systems, primarily flywheels and compressed air systems.

66 GW of additional pump-storage capacity was added between 2000 and 2018 versus 6 GW for all other technologies. This confirms the dominance of pump-storage.

Pump-Storage Energy Storage Systems

Pump-storage energy storage is presented in the "Hydro Energy" chapter. It remains the energy storage technology that provides for the highest powers at any one location, up to 3 000 MW, as well as the highest energy storage capabilities, up to several hundred GWh. The energy storage capabilities are limited by the effective volumes of both reservoirs as well as by the level variations allowed from a regulatory viewpoint, taking irrigation and tourism, for example, into consideration. The upper and lower reservoirs often have capabilities to provide for a full power operation from 12 to 60 hours, corresponding to day/night transfers or transfers between low demand night and weekends toward high demand weekday periods.

Power electronic variable speed installations provide for optimized operations in both directions and over broad operating ranges.

Compressed Air Energy Storage (CAES)

Their operational principle consists in compressing air using an electric compressor and storing it in a cavity and then restituting the energy by expanding it in a turbine; practically, the expansion takes place by way of a gas turbine. In a gas power plant (ref. "Combustion Thermal Power Plants" chapter), the air is compressed just before entering the combustion chamber; the compression of the air, which also heats it, consumes more than half of the energy delivered. In a CAES, compressed air from the cavity replaces the compression energy provided by the turbine in a gas power plant. The overall efficiency of CAES systems is roughly 40%. The use of gas during the restitution causes CO_2 emissions.

CAES makes it possible to store very large amounts of energy. While less efficient than pump-storage, it provides for such large storage capacities where siting hydro plants is impractical. As of 2019, two such systems were operational at an industrial level, one in Germany and the other in the US, each capable of restituting more than 200 MW during a few hours.

Flywheel Energy Storage

As shown in Fig. 2, in this type of system, an electrical machine is mechanically coupled to a flywheel.[5] When storing energy, the electrical machine operates as a motor driving the flywheel such that its rotational speed increases — electric energy is converted into rotational kinetic energy. When restituting electricity, the flywheel drives the electrical machine, which operates as a generator — kinetic rotational energy of the flywheel is converted into electric energy as its rotational speed decreases. To reduce rotational losses, modern flywheel systems feature magnetic bearings and the entire assembly is under vacuum inside the containment.

Flywheel energy storage systems commercially available as of 2018 have storage capacities from a few kWh up to 120 kWh and are typically capable of operating at full power during some 15 minutes. As highlighted in the next insert, they are primarily deployed for short duration network support. Their installation is not restricted by the availability of suitable sites.

---

[5] Flywheels have been used since the Antiquity to smooth rotational motions, for example a potter's wheel as it is pushed using her/his feet. They also smooth the rotational torque of steam engines or internal combustion engines in spite of the intermittent piston thrusts.

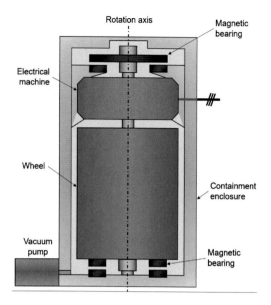

Figure 2. Flywheel energy storage system layout / © *HBP and YB*

Stephentown, US, 5 MWh and 20 MW — Flywheel energy storage system

Composed of 200 flywheels each with an installed capacity of 100 kW and 25 kWh, the system can deliver 20 MW and store 5 MWh. In 2013, when put into service, it was the largest flywheel energy storage plant in the world. It can deliver full power during 15 minutes: 5 MWh / 20 MW = 0.25 hours. It contributes to the local network frequency control by way of its network connection at 115 kV. In this context, it is its power that is important despite its short cycle time. It typically completes from 3 000 to 5 000 cycles per year, i.e., some ten cycles daily.

Batteries

As further discussed, a variety of batteries are presently used for a number of applications. Lead-acid batteries have been used since the early 20th century to start our cars. More recently, nickel-cadmium and nickel metal hydride rechargeable batteries are used in remote controls, which are everywhere in our homes. Following their developments during the 1990s, lithium batteries

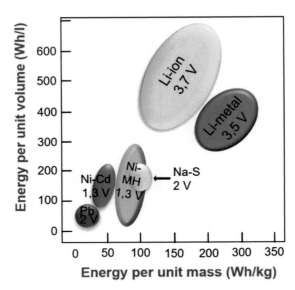

Figure 3.   Battery characteristics comparison / © *Jean-Marie Tarascon (Collège de France)*

have enabled the massive deployment of smartphones and are making electric and hybrid vehicles a reality. Their capabilities are illustrated in Fig. 3.

Principle

The operational principle of a battery, sometimes also referred to as an accumulator, is to store energy by converting electrical energy into chemical energy and vice versa by way of a reversible chemical reaction,[6] as illustrated by Fig. 4.

The main components are:

- Two plates, designated as electrodes, generally composed of two different materials, which provide and receive electrons that "leave" the battery from the negative electrode, designated as the anode, and "return" toward the positive electrode, designated as the cathode, which thus attracts the electrons.

---

[6] This is the major difference with a non-rechargeable battery, in which the chemical reaction is not reversible — once the battery is discharged it can no longer be recharged. The recycling of batteries, which is thus a necessity, is constantly progressing.

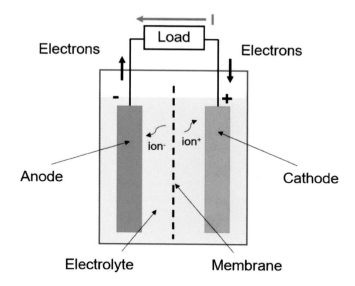

Figure 4.   Discharging battery/ © *HBP and YB*

- A conducting liquid or gel, designated as an electrolyte, which contains mobile ions[7] into which the electrodes are immersed. The ions, which enable the current flow between the electrodes, are produced by the chemical reaction between the electrode and electrolyte materials.
- An isolating membrane that is a material permeable to the ions and which avoids short circuits between the electrodes.

Once the chemical reaction is completed, the battery is fully discharged. To recharge it, the load is replaced by a source, which imposes a current in the opposite direction thus provoking a reverse chemical reaction.

The electrical current provided by the battery is a direct current. The output voltage, V, as well as the discharge rate for any given load, depends on the electrode-electrolyte combination. To obtain a higher voltage,

---

[7]An ion is an atom or a molecule, which, having lost or gained one or more electrons, has a positive, respectively, negative, net charge.

Saltwater, for example, contains positive and negative ions, Na^+ and Cl^- resulting from the decomposition of salt. On the contrary, sugared water does not contain ions since sugar does not decompose as it dissolves. Acids, such as sulfuric acid and alumina salts used in water and aluminum electrolysis, respectively, also are electrolytes.

several cells, elementary batteries, can be connected in series; to obtain higher currents, several cells can be connected in parallel. Series-parallel combinations are used for high-capacity batteries. The voltage to be applied by the source when charging the battery must be adapted to the particular battery — if it is too high, the battery could be damaged. On the contrary, if it is too low, the battery may not charge at all or too slowly.

The energy storage capacity of a battery is the maximum amount of energy, E, it can restitute, measured in Wh or its multiples. A battery's capacity, C, measured in Coulomb, is the product of the maximum time, t, during which the battery can provide a current of intensity, I, such that $C = I \cdot t$; it is also expressed in ampere hours, Ah, especially for lead-acid car batteries. The energy stored then is: $E = V \cdot C$.

The energy density[8] by unit weight, in Wh/kg, or by unit volume, in Wh/l, is one of the battery's key characteristics; continuously improving them is a major stake for vehicle and mobile telephone applications. Progress has recently been achieved to ascertain a battery's state of charge.

Smartphone battery — energy, capacity and heat

For example, a smartphone battery has the following specifications: voltage 3.8 V and capacity 2 500 mAh. When fully charged, it can restitute: $E = 2\,500$ mAh $\cdot$ 3.8 V $= 9\,500$ mWh or 9.5 Wh. Theoretically, it can thus deliver 25 mA during 100 hours or 250 mA during 10 hours. However, as the discharge current increases, the chemical reaction intensifies, thus increasing the electron movements and the associated heat losses, which, in turn, decrease the amount of energy restituted. Temperatures of smartphones increase when running energy-intensive applications and during fast charges.

Main Battery Types

A variety of batteries exist to fulfill a number of stationary and embarked applications; the combinations of materials used for the electrodes and electrolyte differ widely.

---

[8] One refers here to the energy densities on a per cell basis. They are lower for entire battery systems since they also then depend on the cell shapes, the empty space between them when assembled, as well as on the mechanical support structures.

The two main battery types, further detailed later, presently used are:

- *Lead-acid batteries*, which were the first ones developed, remain widely used in vehicles with internal combustion engines since they can deliver the high currents required for the electric starting motor. They are also widely used for a variety of other embarked and stationary applications due to their lower cost and high reliability. The energy storage density per cell typically is 30–40 Wh/kg or 80–90 Wh/l. The storage capacity per cell is from a few tenths of a kWh to a few kWh.
- *Lithium batteries*: The energy density per cell reaches 100–270 Wh/kg or 250–650 Wh/l. The storage capacity per cell ranges from a few Wh to a few kWh.

A number of other types also exist, among them:

- *Sodium-sulfur batteries* whose applications primarily are stationary for electric power network controls. The energy density per cell typically is 110 Wh/kg. In the field, such batteries operate at over 200°C and can reach storage capacities of several MWh.
- *Redox-flow batteries*, which, contrary to the batteries mentioned above, do not have a sealed container. As the name suggests, the electrolyte flows between two containers through the cell itself where the chemical reaction takes place. This configuration provides for very high storage capacities, which can be readily adjusted as storage needs evolve. However, the energy density is only some 20 Wh/kg.

Lead-acid Batteries

When initially invented,[9] they used a combination of lead oxide and lead electrodes along with sulfuric acid diluted in water as an electrolyte, hence the lead-acid designation. The main characteristics are:

- The rated cell voltage is 2.1 V. To reach the 12 V required in a car battery, six cells are connected in series.

---

[9] In 1860 by **Gaston Planté**, a French physicist.

- When discharging, a lead-acid battery can provide high currents exceeding 100 A over short durations of a few seconds without sustaining any damage. This is one advantage of this type of battery.
- The low energy storage density is due to the high specific weight of lead, 11.3 ton/m^3. The storage capacity decreases with lower temperatures.[10]
- The aging depends heavily on the number, duration and types of charging/discharging cycles the battery is subjected to. For example, cycles reaching 80% of the battery's storage capacity will degrade its performance more than cycles reaching only 50%. Typically, the lifetime corresponds to some 1 200 full cycles; for example, two 50% cycles would correspond to one full cycle. In terms of the number of years, the useful lifetime is between 3 and 15 years, depending on the battery's usage.
- The cycle efficiency varies between 50 and 90% — the thermal losses caused by the chemical reactions and the Joule effect losses depend on the charging and discharging rates, i.e., on the currents involved.
- Even if not solicited, the battery typically discharges at a monthly rate of some 2 to 5% at 20°C. When used for back-up applications, a lead-acid battery must thus be continuously and slightly recharged to ensure its full availability when called upon.
- The charging duration of lead-acid batteries is typically from 6 to 12 hours.

Lithium Batteries

Lithium is the lightest metal.[11] Its density is 0.53 ton/m^3, i.e., less than 5% of lead's density, which explains its attractiveness for some applications. The operation of lithium batteries relies on reversible chemical reactions during which positive Li$^+$ ions transit between the two electrodes. The ions are harbored in the electrodes and "leave" them alternatively during the charging or discharging phases.

---

[10]This explains why it is more difficult to start our cars during winters!

[11]Lithium is the third element in Mendeleev's periodic table, just "after" hydrogen and helium.

The two main families of lithium batteries presently used are:

- *Lithium-ion batteries*, Li-ion, which have one graphite electrode, i.e., carbon; the other is in cobalt-lithium dioxide. The electrolyte is a fluor-lithium composite mixed with an organic liquid to limit the degradations of the electrodes and the electrolyte.
- *Lithium polymer batteries*, Lipo, which are a variation of the previous ones where the electrolyte is a polymer, which mitigates the risks associated with liquid electrolytes.

Lithium can ignite spontaneously and burn in the presence of air or if exposed to water or water vapor. It can potentially explode when its temperature increases. This underlines the need for the special precautionary measures to be implemented when using these batteries even though steps are taken to protect against such risks.[12] As of 2018, the main characteristics are:

- The rated cell voltage of 3.6 V.
- As seen above, the energy storage capacity is much higher than for lead-acid batteries.
- The performance degrades when deeply discharged. It is thus preferable to recharge them before reaching an 80 or 90% discharge. Under such conditions, their usable lifetime can reach 1 000 cycles, i.e., one cycle per day over three years, or even longer depending on the particular technology.[13]
- The safe allowable discharge currents for each cell are significantly lower than for lead-acid batteries.
- The charging voltages must be maintained within strict limits to avoid temperature increases that could damage the cells. Fast charging, in half an hour, can be done at 4.1 V ± 0.1 V but only until the battery charge reaches 80% of its capacity.[14] After this point, the charge can continue during three to four hours at gradually slower rates.

---

[12] Special fire extinction systems are installed for large-size containerized batteries.

[13] Cell phone batteries' performances often degrade after two years while space applications are expected to last over 10 years.

[14] This implies that during a trip (to avoid discharging the battery below 20% of its capacity and since fast charges should be limited to 80% of the capacity) the usable capacity during each stage is 60% of the battery's capacity.

- Under normal operating conditions, the charge-discharge cycle efficiency can reach 90 to 95%.
- The no-load discharge rate typically is of the same order as for lead-acid batteries.

A growing number of lithium battery types are reaching the market to enhance specific performances as required by specific applications, such as electric vehicles[15]: higher energy storage densities, higher number of cycles without performance degradations and higher charging currents, etc. For example, in lithium-titanate (Li-Ti) batteries, the graphite electrode is replaced by a lithium-titanate electrode. The charging rate is ten times faster than for Li-ion batteries such that a 90% charge can be reached in 5 to 10 minutes. The lifetime can reach 10 000 cycles; however, the energy storage density is typically only 50 Wh/kg. In addition, the fire hazard related to the graphite electrode is eliminated.

Car batteries

In an internal combustion engine (ICE) car, the battery foremost supplies the electrical starting motor. Subsequently, it is continuously recharged by the car's alternator driven by the ICE. Typically, a 12 V, 70 Ah lead-acid car battery weighs some 20 kg. Fully charged, it can store $E = 12\ V \cdot 70\ Ah = 840\ Wh$, or 0.84 kWh. It can supply over 200 A during a few seconds as required to start an ICE. However, it could only deliver 40 A during 1.75 hours, making it ill-suited for an electric vehicle.

For a full-electric vehicle, the battery provides for all on-board energy requirements, including propulsion. However, it does not need to deliver the high currents required by the starting motor of an ICE. A high-performance electric car may, for example, have a lithium battery rated at 85 kWh and 375 V; it would typically weigh 540 kg.

- Using lead-acid batteries to store the same energy would weigh: $(85/0.84) \cdot 20\ kg = 2\,024\ kg$, which confirms the lithium technology choice.
- The lithium car capacity is 85 000 VAh/375 V = 226.7 Ah. It could thus deliver, for example, 40 A, i.e., 15 kW or 20 HP during more than 5.5 hours, again confirming the lithium battery choice for electric cars.

---

[15] Not only electric cars but also bicycles, scooters and tricycles.

Batteries' Environmental Impacts

As seen above, batteries contain polluting or even dangerous materials such as lead, cadmium, sulfuric acid or molten sodium, to name but a few. The safety and health of the personnel manufacturing and handling batteries is thus a major concern. Strict fire hazard and safety standards and regulations must be implemented and adhered to, including related to potential leaks of hazardous materials. Batteries can also contain rare metals. The recovery and recycling of various types of batteries are gradually being implemented and enforced internationally. The recovery rate of used batteries under proper safety conditions, followed by the recycling of a maximum amount of materials they contain, is, therefore, a major challenge that requires technical, logistical and economical solutions. The recovery and recycling rate is quite different from one country to the next.

Recycling of lead-acid batteries reaches 90% in some countries; the recycling of lithium batteries is progressing.

While the performance of some types of batteries may no longer be sufficient for their original application, it may be quite adequate for other purposes. Thus organizing a "second life" for such batteries reduces the need for recycling. For example, while the performance of electric vehicle batteries may no longer be adequate for the vehicle themselves, their performance for stationary applications, such as houses, apartment buildings or shops, might still be quite adequate.

Super Capacitors

Super capacities can be described as being between batteries and capacitors. Energy is stored in the electrical field between two plates, generally made of porous carbon, which are separated by a solid electrolyte that harbors the ions depending on the electrical field. Since a chemical reaction is not involved, the charging and discharging can be very rapid. As a result, significant powers can be reached while the energy stored remains quite low, at some 10 Wh/kg. One sometimes describes supercapacitors as "power storage". In addition, using an electrostatic physics principle rather than a chemical reaction also makes it possible to safely sustain several tens of thousands of deep discharges during the device's lifetime.

Thermal Daily and Inter-Seasonal Energy Storage

One can seek to store thermal energy:

- In the form of heat or "cold".
- Using water or other materials as support material. Water often is the support of choice in view of its broad availability and high specific heat as well as for environmental reasons. Molten salt, kept in reservoirs, is used in CSPs to store some of the heat from the concentrating mirrors to use it nightly to produce electricity; ref. "Solar Energy" chapter. Recently, molten salts are also used in conjunction with PV power plants.
- Using temperature changes without phase changes, or, on the contrary, using phase changes. The latter allows for higher energy levels to be involved than using a simple temperature variation while also requiring smaller volumes of water or other materials; ref. "Housing" chapter.
- The cycles may be daily, from one day to the next, or between seasons.

The most commonly used thermal energy storage systems are residential hot water systems and those found in office buildings. They store heat in the water that they heat daily.[16] As to daily cold storage cycles, selected air-conditioning systems feature heat pumps that produce and store ice, for example, during nights, to subsequently use the melting ice for cooling the next day.

Broadly speaking, tertiary sector or large residential building heating-cooling systems often rely on several heat and cold sources by way of buffer reservoirs, which contain, depending on the time of day and/or the seasons, hot or cold water. They consume electricity since pumps are required. One must then decide which operating criteria to favor: low cost or low emissions — they may be in conflict!

Heating and cooling requirements for dwellings, offices and shopping centers are seasonal and represent significant portions of their overall energy demands. Therefore, inter season energy storages could result in significant energy savings. Two types of thermal energy storage systems are primarily used that can partially, or even entirely, cover the annual

---

[16]This is also the case for hot water buffer storages used in some heat pump installations. Using them, the heat pumps can be turned off during high electricity demands.

heating requirements of an entire city neighborhood. They both take advantage of the fact that, at some depth, the temperature of the ground is stable and equal to the average annual outside temperature over several years.[17]

Aquifer Thermal Energy Storage (ATES)

As shown in Fig. 5, a bidirectional system is created by way of two wells in an aquifer at a depth of a few hundred meters, for example. Referring to the left hand side of the figure, during summers, when cooling is required, water is pumped from the left side of the aquifer, which is colder than the outside air temperature. The water absorbs the heat from the cooling/air-conditioning system either by way of a heat exchanger or a heat pump to increase the temperature difference. The warmer water is then reinjected into the aquifer at a certain distance from the extraction well where it heats the aquifer. During winters, when heating is required, as illustrated by the right hand side of the figure, the inverse operation takes place. Water is pumped from the right side of the aquifer, which has been heated during the summer. It releases its heat in the heating system before

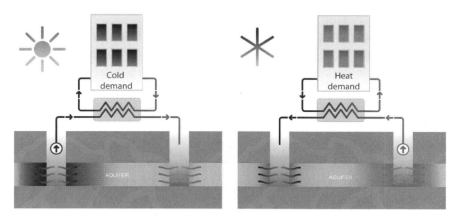

Figure 5.   Aquifer Thermal Energy Storage (ATES) concept / © *IF Technology*

---

[17] This is generally true from a depth of at least 6 m, with the potential exception of volcanic regions with significant geothermal heat.

it is reinjected in the aquifer at a colder temperature, which is close to its left side's natural temperature.

The most favorable configurations are those where the aquifer water is locally "mobile", such that it can readily be pumped out and reinjected, but also where the overall aquifer is stationary or only moving quite slowly, for example, at 1 km per 100 years. Otherwise, the heat exchange would occur directly between the pumping and reinjection well locations. In addition, to ensure that the system operates durably, the aquifer's overall temperature should not drift over the years; an overall balance between the heating and cooling requirements must thus be maintained. If not, additional heat or cold compensation measures must be implemented, which is frequently the case, depending on the local climate and on the requirements of the specific application.

If the temperature difference between the hot and cold water is 25°C, the energy storage capacity is 30 Wh/l,[18] which is of the same order of magnitude as for lead-acid batteries.

As of 2019, several thousand ATES systems are in operation worldwide, some for several decades, primarily in northern Europe.

Borehole Thermal Energy Storage (BTES)

In the absence of an aquifer, one can heat, during summers, or cool, during winters, a cylindrical underground volume by way of a large number — several tens or hundreds — of U-shaped tubes in which water flows. A BTES is more expensive to install than an ATES; the energy storage density is some 15 Wh/l. As of 2020, only a few BTES systems were in operation; one of them is described in the next insert.

ATES and BTES can reach energy storage capacities from several hundred MWh to several GWh. For example, a residential building with a hundred apartments, each requiring an average of 10 MWh of heating yearly, requires 1 GWh of heat yearly that could readily be stored by way of a BTES system.

---

[18] The specific heat of water is 1.16 Wh/l·°C.

The Drake Landing community, Canada — a BTES example

This community is located in the town of Okotoks, at an altitude of 1 000 m with a good solar irradiation of 1 600 kW/m². The temperature can reach −33°C during winters and the HDD is 5 200 degree·days — ref: "Housing" chapter. The community includes 52 detached, one-story and well-insulated family houses built in 2007. More than 90% of the yearly heating requirements is covered using solar energy by way of three piping systems connected to two 120-m³ storage reservoirs, designated as short-term storage.

- The first system recuperates the heat produced during spring, summer and fall from the 44 m² of solar thermal panels located on each of the 52 garages. The thermal power can reach 1.5 MW.
- The second system sends the hot water, between 40 and 50°C, from the short-term storage reservoirs to the inside of the houses when heating is required.
- The third system is an inter-seasonal heat storage in the ground. With an overall diameter of 35 m, it is comprised of 144 tubes, each with a diameter of 15 cm and a height of 35 m. It transfers the heat from the solar panels to the ground when heating is not required and recuperates it when it is. The ground temperature at the center can reach 80°C by the end of the summer.

The overall optimization of the system limits the use of the auxiliary gas furnace to 10%, or even zero, of the yearly heating demand. The overall COP of the entire system, which includes several heat pumps, is of the order of 30. This high value is explained by the rather high temperature of the ground all year round.

The [*dlsc.ca*] website contains further information on this exemplary system.

Energy storage systems are essential in the quest to limit greenhouse gas emissions and to contain the associated investment and operational costs. They are presently enabling the electrification of road transport, the integration of intermittent renewable energy sources and the development of synergies between thermal energy requirements, and more broadly, between energy requirements. The tendency toward integrated and hybrid multi-energy, multi-scale and multi-technologies energy storage systems will accelerate.

Electric Power Systems

From Power Plants to End-users

While we are increasingly aware of how much our daily activities at home or work, or when traveling or ill, have become inconceivable without electricity, we often overlook the complexity of the technologies and systems required to deliver it to wherever and whenever we wish to use it. Though over one billion people on Earth still lack proper access to electricity, modern power systems deliver reliable energy at reasonable costs to well over six billion people worldwide.

This chapter intends to provide an overview of electric power systems' physical layouts as well as their monitoring and control systems. The designation of electric power systems includes:

- The power plants and distributed resources.
- The transmission and distribution networks. Figuratively, they are the freeways and local streets bringing electricity to the consumers.
- The loads, i.e., the equipment and systems that consume electricity; they play an active and key role in the system's overall operations.

The direct current (DC) networks installed by **Thomas Edison** during the 1880s to provide electricity to a few pioneering customers rapidly gave way to alternating current (AC) networks based on the inventions of **Nikola Tesla**[1] and other engineers and entrepreneurs. Often referred to as one of the most complex systems ever build by Man, today's electric

---

[1] A Serbo-American electrical and mechanical engineer (1856–1943). Most of his career was in the United States.

power systems span entire continents and feature thousands of electric power plants supplying electricity to millions of loads by way of meshed and interconnected high voltage transmission networks and lower voltage distribution networks. The increasing deployment of highly decentralized renewable electricity sources, particularly solar photovoltaic, coupled with suitable energy storage at multiple scales, induce new power system challenges, including two-way local distribution network energy flows. Modern power systems, often referred to as smart grids, would simply not be functional without integrated and hierarchical control systems relying on a broad range of information and telecommunication infrastructures. The electric utility industry is one of the most intensive users of powerful computers for planning and operations.

While similar systems exist in North and Latin America and China, and as more are constructed or planned on all continents, the European network, shown in Fig. 1, remains the most complex one.

Transmission system voltages increased from 100 kV in 1910 to 400 kV in the 1950s. A 950-km-long 380 kV transmission line was completed in Sweden in 1952. The first extra high voltage (EHV) 735 kV transmission line was put into service in Québec, Canada, in 1965. The first 1 200 kV ultra-high voltage (UHV) line followed in the Soviet Union in 1982. These increasingly high voltages enable the transport of larger amounts of electricity over longer distances while limiting the losses to a few per cent of the overall demand. In turn, massive electric power plants are built at carefully selected sites, taking proximity to natural resources or close access to port facilities and/or availability of proper cooling resources into account.

With the advent of high-capacity and performance power electronic systems during the second half of the 20th century, DC transmission systems, while more costly, have definite advantages for special applications such as long-distance lines, undersea transmission and connection between systems operating at different frequencies. The 1954 DC connection between the mainland and Gotland Island in Sweden set the stage for many underwater DC connections. In 1970, the Pacific DC Intertie, between the city of The Dalles in northern Oregon State and the City of Los Angeles in the United States (US), set the stage for long-distance and high power DC transmission with its 1 360 km length and 3 100 MW rating.

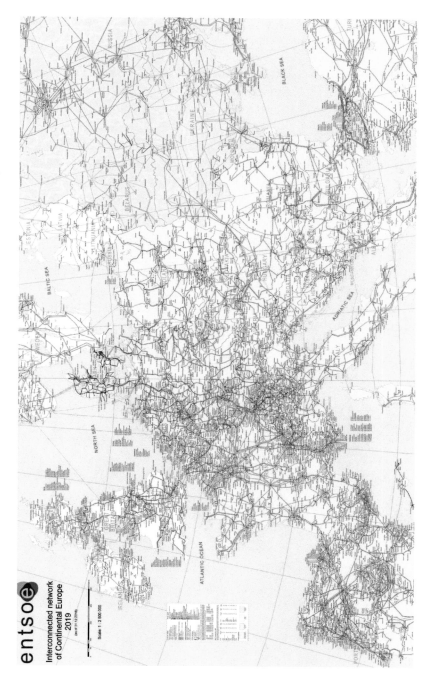

Figure 1. The integrated electric power system — Europa and beyond / © *ENTSO-E*

Electricity: A Carefully Manufactured Product to be Consumed Instantly

As illustrated by Fig. 2, electricity is produced by combustion or nuclear thermal power plants as well as by centralized or distributed renewable sources. Jointly, at any time, they must cover all loads in the system, plus the losses, even though:

- Loads throughout the system change, sometimes rapidly, with the users' needs. Daytime loads differ from nighttime; weekday and weekend loads are different; the seasons affect the requirements while special events such as major sporting events, conferences and concerts affect the load.[2]
- Weather conditions change:
  - Local sunshine and winds change, even drastically within minutes, which affects production.
  - Customers' needs change between hot days or cold nights and from one day to the next.
  - Hot summer days may force power plant production reductions due to a lack of proper cooling.

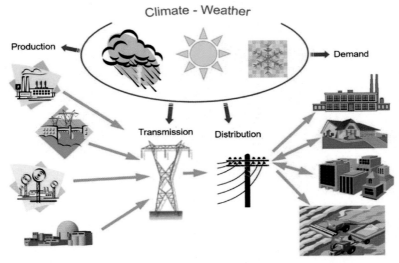

Figure 2. Complexity of modern electric power systems / © *Adapted from US Argonne National Laboratory*

---

[2] De facto, the customers "perturb" the system's operation by changing their consumptions!

- o Storms can create havoc with transmission and distribution networks.
- Power plant operations change due to planned operational changes or following unexpected events. Stringent safety measures are imposed by a number of standards and regulations to protect electricity end-users and personnel working in the electric utility industry.
- In addition to safety precautions, the system must be operated to ensure high reliability of the electricity supply, typically measured in hours of outage per year,[3] and to ensure low tariffs.

Therefore:

- The system frequency must be maintained within a fraction of a percent around 50 or 60 Hz.
- All system voltages must be maintained within a few percent of their respected set values.

Electricity has become a commodity — as are water and natural gas — carefully conditioned within strict norms, which the consumer expects to be available at all times without delay. Water and gas utilities can condition their products well in advance and then store them, centrally or locally, using suitable reservoirs, ready for instant delivery. Since, at this time, there is no way to store electricity, as such, in bulk quantities[4] ready for delivery, it has to be produced instantly when the consumer turns on the switch. It is this latter requirement that sets the electricity commodity apart from water and natural gas commodities.

Synchronous Machines — Electric Power Systems' Work Horses

As seen in the "Introductory Remarks" chapter under "How Electricity Is Produced", in 2018, over 95% of all electricity was produced using rotating

---

[3] Typically, electricity outages are held to no more than a few hours per year in an increasing number of countries worldwide. Yearly outages of three hours, for example, mean that electricity is provided more than 99.96% of the time!

[4] For example, when stored in a battery, electricity must be converted from DC to AC before delivery to the end-user.

synchronous generators driven by steam, gas or hydro turbines.[5] It is thus worthwhile to elaborate on the configuration and operation of synchronous machines. While the next few paragraphs may be a bit arduous, it is important to gain some appreciation of the key technologies that have enabled the development, in just over a hundred years, of so many truly disruptive applications we have come to rely upon.[6] A synchronous machine consists of:

- The *stator*, solidly anchored to the power plant's foundation, where a number of individual windings are implemented in slots around its periphery, as shown in Fig. 3. Each winding occupies two slots. They

Figure 3. Stator windings with front connections / © *Soportography, Shutterstock*

---

[5]Wind turbines are either directly connected to the local network or rely on power electronic interfaces to accommodate the variable frequency from the turbine–generator assembly with the set network frequency.

[6]Readers already familiar with synchronous machines or wishing to delay reading the description may skip ahead to the next main paragraph dealing with Electric Power Systems Controls, page 297.

are connected in series to create three identical but separate stator windings. As also shown in the figure, complex winding layouts are required at each end of the stator slots to make room for the rotor.

The actual spatial distribution of the three separate stator windings in the slots around the stator periphery is such that when they are connected to a local electric network providing what is referred to as balanced three-phase AC voltages, a rotating magnetic field results:

o Its rotational speed, referred to as the electromagnetic rotational speed, is set by the network frequency, i.e., 50 Hz or 50 rev/s or 3 000 RPM.

o Its magnitude depends on the amplitude of the currents in the three stator windings.

The magnitude is time-invariant for any particular stator current.

The three-stator windings, when supplied by three-phase balanced voltages, create a virtual rotating <u>permanent</u> magnet spinning at 3 000 RPM around the common stator-rotor axis. This is the case for all stators connected to the same network operating at 50 Hz,[7] such as all stators throughout the European network shown in Fig. 1.

• The *rotor* is a rotating magnet with a north and a south pole.

The rotor thus creates a rotating magnetic field centered at the common rotor-stator axis. Its magnitude is also time invariant.

Nikola Tesla made the key observation that if two rotating magnetic fields, both with time-invariant magnitudes, irrespective of their magnitudes, rotate around the same axis at the same speeds, a time invariant torque is created between them. The electromagnetic torque, as it is referred to, tries to align the two rotating magnetic fields.

As long as the rotor speed is 3 000 RPM, a constant electromagnetic torque will be created between the stator and the rotor.
The rotors of all synchronous machines with stators connected to the same network, rotate at the same speed, 3 000 RPM; they are all in synchronism. Hence, the designation "synchronous machine".

---

[7] For other networks, such as North America or South Japan, it is 60 Hz.

In the electric power systems context, a synchronous machine can be operated as a generator or motor; it shares the same shaft as a mechanical machine, operated as a turbine or pump.

- When operated as a *generator*, the synchronous machine's stator windings are connected to the local network to which it *supplies* electricity. It is driven by a hydro, steam or gas turbine. The *braking* electromagnetic torque is counter-balanced by the turbine's mechanical *accelerating* torque. As the electrical network load increases, the stator winding currents increase, thus increasing the electromagnetic braking torque. If the turbine's mechanical torque can be further increased, a new steady-state is reached, still at the synchronous speed. If the turbine's capabilities have been reached, the electromagnetic braking torque can no longer be balanced by the turbine's accelerating torque and synchronism is lost. Since the synchronism is lost, the electromagnetic breaking torque disappears and rotating speed of the generator-turbine assembly increases rapidly, requiring a quick suppression of the turbine torque to avoid damages.

- When operated as a *motor*, the synchronous machine's stator windings are also connected to the local network from which it *consumes* electricity. It drives a mechanical load, such as a pump. The *accelerating* electromagnetic torque is counter-balanced by the pump's mechanical *braking* torque. As the mechanical load increases, the stator winding currents increase, thus increasing the electromagnetic accelerating torque. If the stator currents can be further increased, a new steady-state is reached, still at synchronous speed. If the stator current limits have been reached, the electromagnetic accelerating torque can no longer balance the pump's braking torque and synchronism is lost. As synchronism is lost, the rotor, i.e., the pump, stops and the stator currents increase rapidly requiring their interruption by way of adequate protection capabilities.[8]

---

[8]As seen in the "Hydro Energy" chapter, electricity can be produced by tidal dam power plants when the tide is coming in or out. Furthermore, to suit the demand curve, the power plant's head can be artificially increased by pumping water in or out of the reservoir behind the dam, such that the plant becomes a pump-storage plant while the mechanical machine

Rotor Layout and Synchronization

In reality, the rotors of most synchronous machines are not permanent magnets. Rather, the rotor carries a winding supplied by a DC source by way of two slip rings. The winding is such that a north (N) and south (S) pole are created at the rotor periphery similarly to what is described above. Furthermore, the rotor winding can be configured such that pairs of N–S poles are created at the rotor periphery. For example, when two pairs of N–S poles are created and if the rotor's mechanical rotating speed is 1 500 RPM, the electromagnetic rotor field would still have a rotating speed, referred to as electromagnetic speed, of 3 000 RPM as required to create the electromagnetic torque.[9] Indeed, it is the electromagnetic field rotating speed which needs to be in synchronism, whatever the mechanical rotation speed is.

Similarly to the power of an internal combustion engine, which depends on its displacement, i.e., the piston diameter and the length of its travel inside the cylinders, the power developed by a synchronous machine is proportional to its active rotor volume, i.e., its diameter and length.

- Steam and gas turbines have high rotational speeds reaching 3 000 RPM for 50 Hz or 3 600 RPM for 60 Hz; to limit the rotor's peripheral speed and the associated centrifugal forces, as shown at the bottom of Fig. 4, the rotors are long with smaller diameters. The largest generators built for nuclear power plants are rated above 1 500 MW; their weight exceeds 400 tons, and the combined length of the steam turbine and generator can reach 70 m.
- Hydro turbines have much lower rotational speeds. As a result, a large number of pairs of poles is needed to adapt the mechanical rotating speed to the local network frequency. As shown at the top of Fig. 4, their rotor diameters can be larger than 10 m, while the length is smaller.

---

operates as a turbine (or pump), and the synchronous machine operates as a generator (or motor). All the while, the rotor's speed, set by the network frequency, does not change.
[9] For a 50 Hz network, the mechanical rotor speeds, i.e., the turbine speeds, can thus be: 3 000 RPM (2 poles), 1 500 RPM (4 poles), 1 000 RPM (6 poles), 750 RPM (8 poles), etc. If the network frequency is 60, the turbine rotational speeds then need to be 3 600, 1 800, 1 200, 900 RPM, etc.

Both the turbine and generator are therefore specifically designed and constructed for a 50 or 60 Hz operation.

Figure 4.   Top: Synchronous machine salient pole rotor (Canada) / © *Voith Hydro.*
Bottom: Synchronous machine round rotor / © *Siemens*

Similarly to internal combustion engines, synchronous machines are
not self-starting. Indeed, at start-up, the electromagnetic rotor field is at a
standstill while the stator field already rotates at 3 000 or 3 600 RPM as
set by the local network. As a result, before the synchronous machine can
be operated as a generator or motor, its rotor needs to be "synchronized"

with the stator by way of a well-known automated process, the description of which is beyond the scope of this book.

> **Two illustrations to further highlight the operation of synchronous machines**
>
> When a number of cars travel at the same speed on a freeway, i.e., at synchronous speed, their engines may not rotate at identical RPMs, thanks to their transmissions; this is similar to the number of rotor pole pairs. Assuming that all cars are equipped with adaptive cruise controls, if the first one accelerates or decelerates, the other cars will follow as their engine torques are adjusted; this is similar to electromagnetic torque adjustments.
>
> When a car travels on a hilly road using a speed control, the engine torque, i.e., the electromagnetic torque, must increase when going uphill to maintain the speed; when going downhill, the engine torque can be decreased such that the car's gravity actually "pulls" the engine along. Again, it does so to maintain the set, synchronous, speed.

Electric Power Systems Controls

Transmission and distribution networks, which will be presented later in this chapter, are the backbone of any power system. To focus on the basic power systems control mechanisms, it is assumed that they are lossless.[10] In reality, these losses are small, compared to the electricity flowing through them, typically less than 10%.

Under the lossless assumption, the power system is reduced to only power plants and loads:

- *Power plants.* Electricity is produced and injected into transmission and distribution networks in one of two ways:
  - By way of synchronous generators, presented above.
  - By way of direct conversion of the Sun's radiation into electricity. The DC output from photovoltaic panels is converted into an AC voltage using a power electronic converter, referred to as an inverter.

---

[10]This is sometimes referred to as the "copper plate" model.

- *Loads.* The vast majority of electric loads used daily by the public-at-large are:
  - Passive: For example, electric appliances — cooktops, stoves, kettles, lighting elements, heaters, etc.
  - Active: Primarily pumps, fans, compressors, such as those found in heating and cooling systems, and in heat pumps, as well as motors in conveyors and rollers in a number of industrial applications.
  - Power electronic: For example, loads that are controlled by way of a power electronic interface, such as for variable speed drives. A growing number of air-conditioners and heat pumps rely on variable speed controllers to improve their efficiency compared to the more traditional "on-off" controls.

  Overall, the electrical load on a power system will generally decrease with the system frequency, and to a lesser extent, with the supply voltage.

Three main control schemes are intended to keep a complex power system operating safely and as economically as possible. They are:

- Automatic Generation Control (AGC) to ensure that the frequency is kept within very tight bounds and that the energy balances between regions are adhered to. AGC is a real-time and system-wide control scheme.
- Voltage controls, to ensure that the voltages at all system nodes, generally designated as buses, are maintained within preset boundaries. Voltage control is a real-time and local control scheme.
- Economic dispatch, to ensure that once the system is safely operating within set frequency and voltage bounds, it is also operated as economically as possible. Economic dispatch is a "quasi-real-time" and regional control scheme.

Automatic Generation Control (AGC)

Three power systems control mechanisms are generally included under the AGC designation:

- *Primary*: To *stabilize* the system frequency after a perturbation, intended or not, of the overall electricity production and/or consumption as well as following a configuration change of the transmission and/or distribution networks.

- *Secondary*: To *restore* the system frequency to its set value in relief of the primary control and to ensure that inter-regional electricity sales-purchase contracts are adhered to.
- *Tertiary*: To *adjust* the set values of the secondary controls taking into account the most recent adjustments made by the secondary control following major perturbations.

The overall AGC control scheme is illustrated by an analogy in the following insert.

AGC — minivan shuttle service

A 15-seat passenger van is used for a shuttle service between a small city and a regional airport. The trip is almost entirely on a freeway with generally light traffic. It is important that the shuttle service adheres to a rigorous schedule to satisfy the service's passengers at both ends of the trip. This means that it must not only depart on schedule but also arrive at the airport at set times to ensure that the allocated arrival space is physically available. As a result, early arrivals are problematic, not only late ones.

The van has a hybrid propulsion system with a gasoline engine driving the front wheels and an electric motor driving the rear ones; clearly, both the gasoline tank and the battery capacity are of finite sizes. Finally, the van's roof has solar panels to provide some battery recharging during the trip. The number of passengers and their luggage varies from one trip to the next. However, they do not change during each trip; this corresponds to the baseload of an electric power system.

Given these conditions, the average freeway speed, i.e., the power system's frequency, can be determined such that the cruise control can be properly set.

The van's load torque varies with the square of its speed, such that a 10% speed change will induce close to a 20% change in the load torque to be delivered, jointly, by the two engines.

During the trip itself, the following perturbations will occur:

- Increased load when the van climbs a hill and/or when the wind is head-on.
- Decreased load when the van is going downhill and/or when the wind is "pushing" it.

(Continued)

(Continued)

- Unexpected traffic conditions may require unusual acceleration and braking sequences, which affect both the fuel and electricity consumed during the trip.

In all cases, the van's inertia, due to its weight as well as that of the passengers and their luggage, will dampen the speed variations. The same is true for the rotating masses of all synchronous machines.

Speed stabilization — primary control

Following a load variation, the van's cruise control will initially adjust the torque of both engines, according to a preset sharing rule, such that the speed is stabilized to a new value; higher if the load decreased, lower if it increased. The cruise control will also ensure that the torque of each engine stays below the set maximum values.

This is equivalent to the primary control of an electric power system.

Regaining the set speed — secondary control

Once the speed has been stabilized to a new value, the cruise control will seek to return the van's speed to its set value, by increasing the torque of the engines if the speed is too low or decreasing it if it is too high. In doing so, the cruise control will also seek to ensure that the fuel and battery reserves are again set for proper reserves as required for the primary controls.

This is equivalent to the secondary control of a power system.

Adjusting the torque sharing rule — tertiary controls

As the effective load keeps changing and as the cruise control keeps adjusting the torque of both engines, sufficient fuel and battery margins must be maintained for future adjustments. If, for example, the fuel consumption has been unusually high while the PV panels have provided above-expected battery recharging contributions, the sharing rule between both engines should be changed to favor the electric motor while reducing the future role of the gasoline engine.

These additional adjustments correspond to the power system's tertiary controls.

Primary control

The purpose of primary controls is to stabilize the system frequency after a sudden change, intended or not, in the overall system load and/or in the overall generation; a sudden change in the power system's configuration, such as a line outage, may also cause the primary control to engage. Primary controls will result in a lower frequency following an increase in the system load and/or a decrease in the generation; conversely, it will result in a frequency increase following a decrease in the system load and/or an increase in the generation. In both cases, the actual frequency changes. Since, as seen above, the overall system load decreases with frequency, if the frequency settles to a level below the set value, the system in fact "borrows" some kinetic energy from the rotating masses of all turbine-generators connected to the power system until a new equilibrium is reached. Conversely, if the new frequency settles to a higher value, additional kinetic energy is stored in all rotating masses.

The rotating masses of all synchronous turbine-generators thus contribute to the stability of the overall system by effectively reducing the frequency variations after each perturbation.[11] Photovoltaic (PV) systems, connected to the local network by way of power electronic interfaces, do not provide such stabilizing inertia; indeed, they are "static" power plants. This situation presents new power systems control challenges as the electricity production mix becomes increasingly "inertia-less" with the penetration of PV systems.

Primary controls are implemented in real-time, i.e., within seconds or tens of seconds after each perturbation. They ensure that the operation of all power plants remains within safe boundaries.

Secondary control

Once the primary controls have stabilized the system frequency at a value typically different from the original set value, the purpose of the

---

[11] To some extent, the rotating masses of electric motors directly connected to the network also contribute.

secondary controls is to bring the system frequency back to its original set value or a new value set by the operator.

As for the primary controls, a designated subset of power plants takes part in the secondary controls. In turn, the secondary controls ensure that the participating power plants are kept within their respective safe operating limits.

Large interconnected power systems are typically subdivided into several control areas, which may be the service area of a major electric utility or a geographical territory covered by several utilities or a combination of both. The equilibrium between the total production and the overall demand must be maintained within each control area.[12]

A number of contracts will generally have been entered into between:

- Electricity producers within a control area and customers within the same area. This represents the vast majority of situations where the local utility supplies electricity to the local residential, commercial, small industrial loads, and municipal utilities.
- Customers within a control area and electricity producers outside of the particular control area.
- Power producers within a particular control area and customers outside the particular control area.

In the latter two situations, the consumers may be large commercial and/or industrial facilities and/or municipal utilities seeking to secure lower-cost supplies from outside their service area.

As these individual contracts to sell or purchase electricity are summed up within each control area, overall, each one will either be a net electricity exporter or importer. This net control area export or import is referred to as the control area consignment, $P_{area\ cons}$; it is negative if the control area is a net importer and positive if it is a net exporter. Since the system losses are neglected, the sum of all control area consignments is zero. Within each control area, an operator is assigned the task of adjusting the electricity productions such that:

$$Production_{area} = Demand_{area} + P_{area\ cons}$$

---

[12] The system is assumed to be lossless.

This equilibrium is perturbed each time primary controls are exercised since the participating power plants are typically dispersed among all control areas. As a result, the control area consignments may need to be adjusted.

The secondary controls:

- Adjust the control set points for all power plants participating in the primary and secondary controls.
- Adjust the operating set points for all power plants participating in the primary and secondary controls to satisfy the control area consignments.
- If needed, send out new set points to restore the overall system frequency to its original value.

The secondary controls are set to avoid interfering with primary controls to avoid any instabilities. As a result, they are slower and typically act within minutes.

Tertiary controls

If the control reserves are not deemed to be sufficient to safely face further system perturbations after the primary and secondary controls have brought the system frequency back to a set point, the tertiary controls adjust the set points of the power plants participating in the primary and secondary controls. Additional plants may also be called upon for tertiary control purposes.

The tertiary controls are set not to interfere with the secondary controls; as a result, they are slower and typically only act beyond 10–15 minutes after a disturbance.

Spinning and stand-by reserves

As the primary, secondary and tertiary control actions are implemented, additional power plants may be required to restore sufficient control margins at the entire system level.

One distinguishes between two types of reserves:

- *Spinning reserve*: These power plants are synchronized to the network but operated quite far below their installed capacities. As a result, their reserve margins are quite high such that they can rapidly be called upon like any other synchronized power plant. When available, hydro storage power plants are well suited for spinning reserve purposes.
- *Stand-by reserves*: These power plants are not yet synchronized to the network but could rapidly be. Typical candidates are gas turbine power plants and diesel generator sets.

Economic Dispatch

Once the power system has reached steady state such that all loads are satisfied within the set frequency and voltage limits, the issue of achieving the lowest overall system operational cost remains; this is the objective of economic dispatch under the following constraints:

- The overall system demand must be met such that each load is supplied.
- The output of each power plant must not exceed its installed capacity.

The main components of the operational cost of a power plant are:

- A fixed operational cost, in \$/hr or €/hr, incurred whether the plant is running or not. It includes the taxes and fees incurred by the plant as well as the financial costs. Some maintenance operations are also required at regular intervals whether the plant is running or not.
- An hourly operating cost, in \$/MWh or €/MWh, which is directly related to the actual energy produced such as some of the fuel costs as well as fees paid to local authorities. Some maintenance operations are scheduled based on the number of operating hours.
- For fossil power plants, the fuel costs generally increase more rapidly than proportionally to the output power, typically as the square of the output power. Indeed, the plant efficiencies change as the output power changes; this is especially true for older plants.

Once the operational cost is introduced, it is useful to also introduce the notion of incremental, or marginal, cost of a particular power plant. It is the additional operating cost to produce one more MWh, i.e., an increase of one MW for one hour, from the present operating point. It depends on the present operating point.

Economic dispatch — neglecting transmission losses

Assuming that the system is lossless, the lowest overall generation cost is achieved if the marginal costs at all power plants are the same,[13] such that the additional cost of producing one more MWh is the same irrespective of the particular power plant solicited. The operating point resulting from the economic dispatch changes with time, particularly as the demand configuration and levels change. As a result, economic dispatch continuously affects the output levels of all power plants; however, slower than the primary, secondary and tertiary controls to avoid interference with the latter.

Economic dispatch — taking transmission losses into account

In reality, system losses are unavoidable, mainly in the transmission and distribution networks. When the output power at a particular plant increases, the output powers at other plants also have to change to accommodate the overall system demand and the losses. These production adjustments affect the system's overall losses. When increasing the output at some plants, the result might be that the system losses decrease; these plants are said to be "electrically close"[14] to the main demand centers, for

---

[13]To justify this outcome intuitively, it is important to remember that: (a) the incremental cost at each power plant increases as its output increases, and (b) the sum of all power plant outputs remains the same, i.e., equal to the total demand (the system is assumed to be lossless). Thus, if the incremental cost at one power plant is lower than those at the others, it would mean that a higher output at that plant would have to be compensated by a lower output at one or more other plants with higher incremental costs. It would lead to a decrease in the global cost, which would mean that the original operating point was not optimal! It follows that the lowest overall cost is achieved when all plants operate with the same incremental cost.

[14]The physical configuration of the power system in general, as well as the topology of the lines, play a dominant role, not only the geographical proximity.

example, close to a major city. On the contrary, increasing the output power at other plants might increase the overall system losses, for example, at a power plant located far away from major demand concentrations. If transmission network losses are to be considered, it seems sensible to favor power plants "electrically closer" to the main demand centers while also considering their operational costs. This leads to a merit order ranking of the available plants, which is used to set the output powers of all plants to achieve the lowest overall operational cost.

It is important to note that economic dispatch seeks to minimize the overall system operational cost at a particular time and not over longer durations, such as a year. Furthermore, it does not explicitly seek to minimize emission levels.

Electric Power Systems Infrastructures

Electric power systems were first developed at local, primarily urban, levels and later at increasingly larger regional levels. Before the Second World War, interconnections between electric power infrastructures were embryonic and primarily served as mutual support between networks. After the war, they developed quickly to reach the present situation; they have become indispensable in a global market context. In 1958, the 380-kV interconnection between the French, German and Swiss networks at Laufenburg, in northern Switzerland, marked the beginning of transnational interconnections.

Aside from the European interconnected electric power system, shown in Fig. 1, which covers over 30 countries, major power systems also exist, for example, in the US (Fig. 5), China, (Fig. 6), and southern Africa under the auspices of the southern African Power Pool (Fig. 7). India's five regional networks were recently synchronously interconnected. Major interconnection projects are either under construction or being planned in other parts of the world.

Main Electric Power Subsystems

As illustrated in Fig. 8, the main subsystems and components of power systems are:

- *Power plants*: Presented in the six "How Electricity Is Produced" chapters.

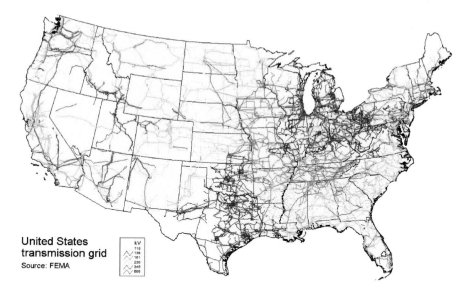

Figure 5. United States transmission grid / © *US Federal Emergency Management Authority*

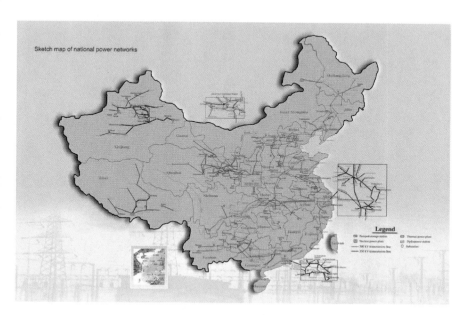

Figure 6. China transmission grid / © *Global Energy Network Institute*

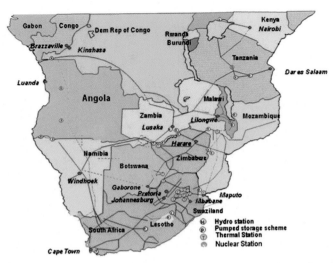

Figure 7.   Southern Africa transmission grid / © *Southern African Power Pool*

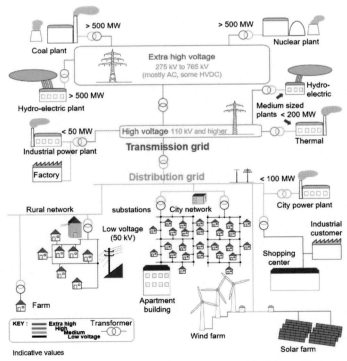

Figure 8.   Electric transmission and distribution systems / © *Adapted from Stefan Riepl, Wikimedia.org, Electricity Grid Schema*

- *Step-up and step-down transformers*: Further detailed in the next insert, they are essential to reduce the transmission network losses by increasing the generator voltages, typically at 25 kV or less, up to the transmission network voltages and then to reduce them down to user-level voltages in successive steps.

The basic physics principle on which transformers are based only works if the voltage and currents involved are AC. This observation alone explains why modern electric power systems use 50 or 60 Hz voltages and currents. Since AC voltages and currents are required, Nikola Tesla's inventions related to AC machines, including synchronous machines, lifted a major road-block toward the ubiquitous development of electric power systems.

Transformers — essential electric power systems building blocks

Transformers are static devices. Their concept is simple — two coils/windings, with N_1 and N_2 turns respectively, wound around a magnetic core. Energy is supplied to the transformer's primary winding; the secondary winding is connected to the load. Neglecting the transformer losses, which are of the order of one percent at rated power for large transformers, the ratio between the primary and secondary voltages, V_1 and V_2, is N_1/N_2, referred to as the turn's ratio. Since the input and output powers, $P = V \cdot I$, are the same since the transformer is assumed to be lossless: $V_1 \cdot I_1 = V_2 \cdot I_2$ such that the ratio between the primary and secondary currents, I_1 and I_2, is: $I_1/I_2 = N_2/N_1$.

Assuming that the primary voltage is 5 and that the input power is 100, the primary current is 20. If the turn's ratio is 10, then the secondary voltage is 50 and the secondary current is 2. Since the losses in a transmission line is proportional to the square of the current, using a step-up and a step-down transformer with turn's ratios of 10 at either end of a transmission line reduces the losses by a factor of 100 in the example above.[15]

- *Transmission networks*: They are meshed for enhanced overall system reliability by providing several paths for electricity flows between production and demand centers. Major urban areas, as well as selected

---

[15] Note for readers with power systems engineering knowledge: to simplify the exposé, the power factor is assumed to be 1 in the example.

Figure 9.   High voltage towers / © *ssguy, Shutterstock*

industrial and commercial loads, are supplied at the transmission system level.

Figure 9 shows a few typical network transmission system tower configurations. Some of the towers support one three-phase line (three conductors) while others support two, as shown to the left forefront. As seen to the right of the picture, double conductors (bundle) are sometimes used to increase the current capabilities. The two conductors above the main conductors are lightning shield/arrester wires to prevent lightning from striking the transmission lines; in turn, they are solidly connected to ground.

In addition to maintaining sufficient separation between conductors to avoid short circuits, sufficient conductor clearances above ground must be maintained at all times, taking into account the lines' elongations as their temperatures increase with the ambient temperature and with the current carried. Finally, the isolation provided between the conductors and their anchoring to the towers must be sufficient, as evidenced by the longer isolation strings as the rated voltage increases.

- *Transmission substations*: Used to interconnect several transmission lines; they provide for transmission system reconfigurations as may be required to enhance reliability margins or economic dispatch.

- *Distribution networks*: The vast majority of urban, industrial, tertiary, commercial and residential loads are connected to distribution networks; their voltages are typically below 40 kV. They are generally configured as radial systems in what is referred to as "feeders", each connected to a distribution substation, which, in turn, is connected to the transmission systems using transformers. While typically operated as radial systems, distribution network switching capabilities may allow for automatic reconfigurations to ensure the continuation of service in case of incidents, especially in urban regions.
- *Distribution loads* can either be:
  - Three-phase as in continental Europe, for example. Heavier loads such as cooktops, ovens and washing machines are directly connected to the three-phase, 380 V supply, whereas the remaining outlets are single-phase at 220 V and distributed among the three phases to maintain a three-phase balance as closely as possible.
  - Single-phase, as in the US, for example.

SCADA and Central Dispatching

As mentioned earlier, the operation of thousands of power plants along with several hundred thousand transmission and distribution lines and transformers that make up modern electric power systems must be carefully coordinated to ensure a high-reliability electricity supply to all customers. This is only possible if reliable data is available at central and local command centers, designated as dispatching centers. This is the purpose of Supervisory Control and Data Acquisition (SCADA) systems. The resulting coherent data is displayed at dispatching centers where system operators can issue commands to ensure appropriate operational margins following unpredicted events or in preparation of planned ones.

AC Versus DC Transmission Lines

Higher performance and lower cost power electronic converters are making the deployment of DC transmission systems more attractive. The investments in the AC-DC and DC-AC converters at either end of the DC lines and their additional losses need to be considered in view of the main advantages of high voltage DC, HVDC, transmission lines, among them:

- Only two conductors are typically required, rather than three for three-phase AC systems, a clear advantage for underwater and underground applications as well as for above-ground applications in view of the narrower corridors required.
- Transmission line voltage drops due to the line inductances, which are inherent with AC lines, are eliminated with the DC, i.e., non-time varying, voltages and currents. This is a considerable advantage for longer transmission distances.
- AC systems operating at the same frequency but not in synchronism, such as in North America, or at different frequencies, such as between North and South Japan, can be connected using DC interties.

The breakeven line length for which DC systems can be economically justified continues to decrease.

Overhead Versus Underground Transmissions Lines

The most commonly used overhead transmission conductor is Aluminum Conductor Steel Reinforced (ACSR). As the name suggests, the core is made of steel with aluminum strands surrounding the steel core. Aluminum is lighter and a better electricity conductor than steel but does not have the tensile strength of steel, which provides the mechanical tension required for the long distances between the suspension towers. Overhead conductors are generally not insulated; air provides the necessary insulation as well as cooling.

While recent urban distribution networks are almost entirely constructed underground using conduits often shared with other telecommunication, water and gas infrastructures, older ones are also being reconstructed underground. Rural distribution networks are often overhead due to their lower installation and repair costs. Increasing public resistance toward overhead transmission lines is primarily due to visual impact and perceived health hazards associated with increased magnetic fields around the conductors. As a result, constructing new transmission lines is often only possible underground using sophisticated cable systems.

Compared to overhead transmission networks, underground networks often are an order of magnitude more expensive; among the primary reasons are:

- Underground or underwater cables are significantly more expensive than overhead ones due to the complex insulation system required around each conductor in contrast with exposed overhead conductors.
- Cables, be they overhead or underground, are delivered on reels. Depending on the cable's diameter, the length on each reel may be quite short. As a result, for longer distances, several cable lengths must be "spliced" together, which is a simple clamp for overhead lines but complex joints to be installed by highly trained technicians for underground or underwater cables.
- The most convenient way to install an underground cable is in a trench dug for this purpose and then refilled upon the cable's installation. The trench must remain accessible for maintenance and repair purposes, which restricts possible routes. In other situations, this may not be possible and a tunnel might have to be drilled under a mountain to install a cable.
- Laying underwater cables are complex operations, especially if the seabed has severe configurations or if done in regions susceptible to earthquakes.

Power System Protections

A number of power system protections are routinely implemented. The most common ones are:

- Over and under-voltage to protect both electricity production and end-use equipment against voltages that could endanger their insulation, on one hand, and against excess currents caused by insufficient voltages on the other hand, for example, in electric motors.

- Over-currents, to protect equipment against high currents caused by short circuits or by lightning.
- Under-frequency situations that occur when, as a result of a disturbance, the overall system load exceeds the available electricity production capabilities, thus causing the entire system to "slow down" as it "borrows" kinetic energy from the system inertia. If the frequency cannot be stabilized and then restored using primary and secondary controls, load shedding is initiated to disconnect loads based on preset scenarios, starting with customers with contracts stipulating that they can be disconnected under certain circumstances, with lower rates in compensation. Loads deemed to be non-critical are shed next. If at all possible, shedding critical loads, such as hospitals, data centers, central monitoring and control centers, should be avoided. Industrial processes for which interruptions may cause major damage, such as glass or aluminum manufacturing, are often protected using standby auxiliary supplies.

In each case, the disturbance is detected by special purpose relays, which then issue control signals to circuit breakers, which, in turn, interrupt the supply.

Power Plant Scheduling — Electricity Production Mix

Year-long, the overall electricity demand is supplied using several categories of power plants, the production of which may be operator-adjusted or not:

- *Baseload*: Operated over longer periods, like weeks, with little or no change in their output powers. Their cumulative production is less than the minimum demand during the time period considered. Nuclear and coal power plants are typically base-loaded, as are some geothermal and biomass power plants.
- *Intermediate*: Their production can be readily adjusted to accommodate changing load levels. Gas power plants are well suited for this purpose since their production levels can be adjusted over broad ranges within minutes or even started and stopped within a few tens

of minutes. Modern coal plants can also be operated as intermediate plants even though their production adjustments are slower. Some nuclear power plants in France and Germany, for example, are also operated as intermediate plants with more limited adjustment ranges.
- *Peaking*: Accommodate, as the name suggests, peak loads as well as rapidly changing loads. Gas power plants are well suited for this purpose.

The following values are indicative; they may change from one country to the next and/or from one year to the next:

- Base-loaded units are typically used more than 45% of the time, or more than 4 000 hours per year.
- Intermediate units are typically used less than 45% of the time, or less than 4 000 hours per year.
- Peaking units are typically only used some 10% of the time, or some 900 hours per year.

Hydro power plants inherently are:

- Base-loaded for run-of-the-river plants; their production levels are generally set during weeks at a time. However, they typically vary with the seasons. Since "spilling" energy is to be avoided, the production power is thus not dispatchable.
- Peaking units for storage facilities. In fact, given their ease of output control (as seen earlier, it is sufficient to open and close the "faucet"), they are the most suited type of peaking units; they are the ultimate dispatchable units. Occasionally, they can also be operated as intermediate units.

The scheduling of solar and wind power plants requires special care due to their variability, including potential production peaks, and their non-dispatchability, except when coupled with energy storage to better leverage their contributions.

Taken together, all available power plants are referred to as the production fleet. The contributions of each source to the overall production, given in

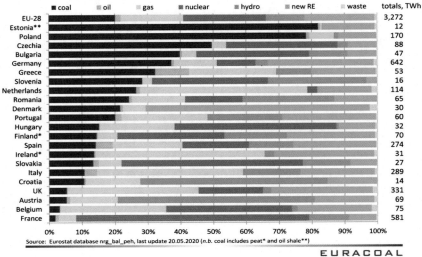

Figure 10.   2018 Electricity generation — EU / © *Euracoal*

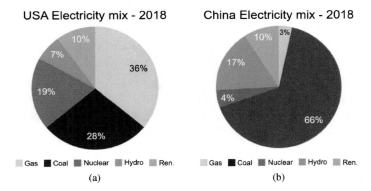

Figure 11   2018 electricity mixes: (a) US and (b) China / *Data from BP Statistics*

percent, is referred to as the production mix. As can be seen from the figures above, the electricity mix is quite different from country to country:

- Figure 10 for the European Union (EU) countries.
- Figure 11 for the US and China.

Electricity Demand Curves

The daily electricity demand changes:

- *During the day*: It is typically higher during the day than at night; the load patterns also depend on the load type.
- *With the day of the week*: It is typically higher on working days than on weekends.
- *With the seasons*: In some countries, such as in northern Europe and North America, the demand is higher during winter than in summer due to heating loads. In other regions with warmer climates, it is often higher in the summer than winter due to air-conditioning loads. Weather conditions change from one season to the next, as illustrated in Fig. 12. The demand pattern may also significantly change between winter and summer at the same location.

Load Forecasting

The proper scheduling at all duration levels of all available power plants requires the proper forecasting of the overall load to be served. Increasingly sophisticated load forecasting techniques are used taking historical load data, weather forecasting and local population evolution models into

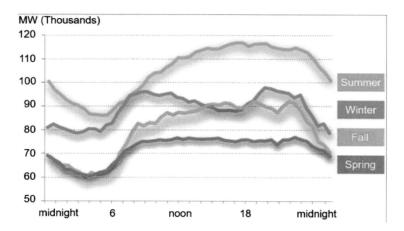

Figure 12.   Example of daily loads seasonal variations / © *PJM*

account. Additional planning information is incorporated such as neighborhood developments, implementation of new industries and shopping centers or future closures, major infrastructure changes, etc. Decisions related to the construction of new power plants as well as the removal from service of older ones, along with longer duration maintenance and upgrades are included in the forecasting process. Existing and anticipated regulatory and public policy impacts are also important in longer-term load forecasting.

Longer-term load forecasts, reaching several years ahead, are refined as the forecasting horizon becomes shorter all the way down to "day-ahead" forecasts, which are used for the precise scheduling of all available plants.

Yearly Scheduling of Hydro Energy Resources

The yearly scheduling of hydro storage power plants requires special consideration. In most parts of the world, hydro energy resources vary with the seasons, as illustrated by Fig. 13, which shows the cumulative hydro dam storage levels in Switzerland[16] over 40 years.[17] The figure highlights the significant differences from one year to the next.

Given their typically low operational costs, simply including hydro storage power plants in an economic dispatch process would lead to the use of all available hydro resources first and then of the more costly plants. However, using this approach, starting in January, for example, during an entire year would probably result in fully emptying the storage dams well before the end of the year, after which the production mix would almost entirely include more costly plants. Improved distribution of

---

[16] In the following presentation, some figures rely on data from Switzerland. The main reason is that the Swiss Federal Energy Office [*bfe.admin.ch*] publishes a set of coherent graphs and data yearly, thus facilitating a homogeneous presentation and not that one of the authors is Swiss! Switzerland's electric production mix, with a significant hydro energy contribution along with its geographical location at the heart of Europe, illustrates an extensive use of renewable and dispatchable energy and the need for inter-regions exchanges.

[17] The hydrological year, as used in the figure, lasts from September until August the next year.

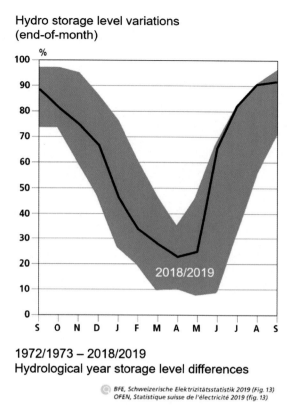

Figure 13. Hydro dam storage levels in Switzerland from 1972 to 2019 / © *Swiss Federal Office of Energy*

the hydro resources over an entire year could decrease the need to call upon the most costly plants.[18] The optimization must, therefore, be done over an entire year using well-known mathematical techniques.

Upon completion of the yearly scheduling of all hydro resources, the other resources can be scheduled. The monthly 2019 electricity load and production for Switzerland, again as an illustrative example, is shown in Fig. 14. As can be observed from the figure:

---

[18]The most costly plants often also have the highest emissions.

Monthly electricity production shares
and consumptions - 2019

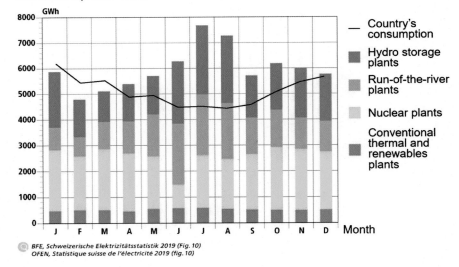

BFE, *Schweizerische Elektrizitätsstatistik 2019 (Fig. 10)*
OFEN, *Statistique suisse de l'électricité 2019 (fig. 10)*

Figure 14.   Monthly electricity production and consumption in Switzerland in 2019 / ©
Swiss Federal Office of Energy

- The Swiss production mix is almost entirely made of:
  - Run-of-the-river and storage hydro plants.
  - Nuclear plants.
  The small thermal power plant production is almost entirely from
  waste and biomass incineration.
  The country does not have any fossil fuel power plants of signifi-
  cance, which is unusual.
- The peak production occurs during the summer when the winter snow
  melts since roughly 60% of the production is from hydro power
  plants. The peak consumption occurs during the winter due to the
  shorter daylight hours and the heating requirements.
- Switzerland imported electricity during three months and exported
  it during eight months; equilibrium was reached in December. The
  1984 addition of a nuclear power plant was the last major extension to
  the production fleet; as a consequence, the country has become more
  dependent on electricity imports from its neighbors.

The winter–summer production-consumption imbalances are not unusual in the northern hemisphere.

Daily Scheduling of All Power Plants

Once the daily load curve has been forecasted — generally at least one day ahead — and that the yearly hydro resources have been scheduled, the actual daily scheduling of all available power plants can proceed based on their merit order rankings. The actual loads will slightly deviate from the forecasted ones for a number of reasons, including unforeseen changes in weather conditions, power plants schedules and network configurations. The primary, secondary and tertiary controls thus ensure that the power system's operation remains within all set constraints, occasionally calling upon rotating or stand-by reserves, and, under extreme conditions, implementing load shedding.

Again, for illustrative purposes, a detailed presentation based on data for Switzerland can be found under *"Power Plant Scheduling — Swiss Case Study"* in the Companion Document. The overall outcome in 2019 is summarized in the table below.

TWh	France	Germany	Austria	Italy	Net
Winter	−2.0	−10.5	−3.4	+11.3	−4.6
Summer	−1.7	+ 2.7	−0.3	+10.2	+10.9
Year	**−3.7**	**− 7.8**	**−3.7**	**+21.5**	**+6.3**

Electricity exchanges between Switzerland and its neighbors — 2019.
Import: negative   Export: positive

The country was a net exporter of electricity, 6.3 TWh[19]; it imported 4.6 TWh during the six winter months and exported 10.9 TWh during the summer months. This situation changes from year to year; for example, Switzerland was a net importer in 2017. The data highlights the need for electricity transfers between and transits through countries; this is further highlighted by the German situation presented next.

---

[19] The country's final electricity consumption was 57.2 TWh in 2019.

Germany — Energiewende: The Impact of Deliberate and Sustained Public Policies

During the early 2000s, successive German government coalitions decided to reduce the contribution of nuclear energy to its electricity mix. Following the Fukushima nuclear reactor accident, policies were enacted to accelerate the penetration of renewable energies, primarily wind and solar PV, as the shutdown of all remaining nuclear power plants was set for 2022.

As of 2017, the outcome of German public policies and incentives is one of the most spectacular energy transitions — actually electricity mix transitions — in the world. The evolution of the German electricity production mix clearly illustrates that:

- Deliberate and sustained public policies can have significant impacts on a country's electricity production mix.
- The integration of a significant proportion of renewable energies, primarily solar PV and wind, does not systematically result in a low emissions electricity mix. Indeed, the German electricity mix still resulted in emissions of 424 gCO_2/kWh as of 2017, one of the highest in Europe.
- The 2017 overall operation with a significantly increased solar PV and wind installed capacities would not be possible without the inter-ties the country has with its nine neighboring countries.

Germany's 2017 final electricity consumption was 530 TWh;[20] it declined slightly from 540.6 TWh in 2010. This is a less than 3% decrease in eight years, despite deliberate public policies and incentives, which confirms that significant electric demand decreases are unlikely to occur in the near future; quite the opposite will occur as electrification of a growing number of end-uses, including private cars, take place.

The following table [*ag-energiebilanzen.de*] shows the installed capacities and gross energies produced in 2017 as well as in 2010, the year before the 2011 Fukushima accident.

---

[20] Germany had 82.3 million inhabitants. Thus, the yearly consumption per capita was 6.4 MWh.

Source	2017			2010		
	Inst. Cap. GW	Gross Production TWh	Gross Production %	Inst. Cap. GW	Gross Production TWh	Gross Production %
Fossil	80.2[21]	333.6	51.0	79.4[21]	360.9	57.0
Nuclear	10.8	76.3	11.7	20.4	140.6	22.2
Hydro	5.6	20.2	3.1	5.4	21.0	3.3
Wind On and Off-shore	56.2	105.7	16.2	26.9	38.6	6.1
Solar PV	43.0	39.4	6.0	17.9	11.7	1.9
Biomass	7.4	45.0	6.9	5.7	29.2	4.6
Total	**203.2**	**620.2**		**155.7**	**602.0**	
Waste and Others		33.5	5.1		31.1	4.9
Total		**653.7**	**100.0**		**633.1**	**100.0**
Net export balance		-52.5			-15.0	
Miscellaneous[22]		-71.2			-77.5	
Domestic cons.		**530.0**			**540.6**	

A number of observations can be derived from the table:

- Power plant installed capacities
  - Overall, the power plant installed production capacity increased from 155.7 GW to 203.2 GW.
  - The net decrease in the combined fossil and nuclear installed capacity was 8.8 GW, consisting of a decrease of 9.6 GW nuclear and an increase of 0.8 GW fossil.
  - The combined installed capacity of wind and solar PV increased by 54.4 GW.
- The 2017 capacity factors were:
  - Wind: 21.5%,[23] i.e., 1 880 h/yr at full power — it was 38.5% for off-shore and 19.9% for on-shore.

---

[21] In 2017: coal 46.3; oil 4.4; gas 29.5 — In 2010: coal 49.7; oil 5.9; gas 23.8.

[22] Miscellaneous: power plant auxiliaries, network losses and pump-storage consumptions.

[23] Capacity factor; ref. "Introductory Remarks" chapter under "How Electricity Is Produced".
In this case: 105 700 GWh / (8 760 h · 56.2 GW) = 21.5% or 8 760 h/yr · 21.5% = 1 880 h/yr.

o   Solar PV: 10.5%, i.e., 920 h/yr at full power.

To be compared with 80.1%, i.e., 7 020 h/yr at full power, for the nuclear power plant fleet. This explains the need for significantly higher solar and wind installed capacities to produce the same amount of energy.

- CO_2 emissions

  In 2017, Germany's generation mix emissions were 424 gCO_2/kWh, still among the highest in Europe but slightly down from 483 gCO_2/kWh in 2010 [*eea.europa.eu*].[24]

- Yearly electricity export-import

  Due to its higher overall installed capacity for an almost unchanged demand and its geographical location with nine neighboring countries, the German electric system in 2017 was occasionally exporting electricity to its neighbors (80.3 TWh), mostly during summers when solar PV production is high; on other occasions, electricity was imported (27.8 TWh), especially during high demand winters when solar PV production is low. The 2017 net electricity export was 52.5 TWh; it was 15 TWh in 2010.

- Network losses

  The overall network losses provided in the [*bundesnetzagentur.de/monitoringbericht*] website were 27.5 TWh in 2017 and 21 TWh in 2010. While still low relative to the overall total domestic demand, compared to most other industrialized countries, they increased by 30% even though the total domestic demand decreased. The reasons for this evolution include: (a) the intensified north-south electricity transit (the off-shore wind production is primarily increasing in the north) and (b) the intensified exchanges with neighboring countries due to increased variable solar and wind productions.

- Negative electricity prices

  Due to its high wind and PV energy penetration, Germany occasionally is in a position when the output of the base-loaded power plants (nuclear and coal) cannot be further curtailed to make room for the non-dispatchable wind and PV production without incurring significant

---

[24] When applying the values used elsewhere in the book for similar computations, i.e., 950 gCO_2/kWh for coal plants, 650 gCO_2/kWh for oil plants and 475 gCO_2/kWh for gas plants, one finds 440 gCO_2/kWh for 2017 and 500 gCO_2/kWh for 2010.

additional operational costs.[25] The result is what is referred to as negative "day ahead"[26] electricity costs, which is the price the German power producers are willing to pay their neighbors to "take" its excess production by making room for it within their systems. This situation occurred during 144 hours in 2017.

Power System Outages

Even though modern power systems are meshed at the transmission level in an effort to reduce both the number and duration of outages, they do occur. The two primary causes are:

- Sudden changes in the available electricity production fleet generally caused by one or more power plants disconnecting from the system either due to overloads or internal failures. In turn, such failures may cause overloads at other power plants and/or for one or more transmission lines. More recently, sudden changes in PV and/or wind production may cause overloads at other plants.
- Sudden changes in the transmission network configurations, primarily due to overloads but also due to voltage and/or over/under frequency protection.

Typically, wide-area power system outages, often referred to as "blackouts", originate from one specific incident, which then "cascades" until the area of the outage can be confined to a territory, as small as possible, before the system is restored to normal operations.

Using numerical simulation tools, system operators seek to predict the system's behavior at any time should a set of predetermined outages, intended or not, occur: outage of a subset of power plants with particularly high production levels or/and outage of a set of particularly heavily loaded transmission lines. This set of events is referred to as contingencies. Every single potential outage is first simulated one at the time and then taken as

---

[25] Primarily power plants shut down and start-up costs.

[26] "Day ahead" prices describe the price published by utilities on various electricity exchanges 24 hours ahead of the actual contractual transfers.

combinations of two or even three events.[27] Modern power systems are generally operated such that the so-called n–1 rule is fulfilled, i.e., such that an outage of a major infrastructure — power plant or key transmission line — does not cause a blackout.

Four major blackouts are described under *"Power System Blackouts"* in the Companion Document. They relate to:

- The northeast US 2003 blackout, which happened when an overloaded power plant disconnected from the system, thus causing a system-wide voltage collapse.
- The northern Italy 2003 blackout, which was caused by the outage of two major intertie lines after flashovers with nearby trees. A wide area system collapsed due to under frequency.
- The European 2006 blackout was a consequence of incomplete simulation preparation and a lack of full knowledge of the wind productions at the dispatching centers. A planned opening of a transmission line in northern Germany led to the separation of the European interconnected system into three separate zones.
- The India 2012 blackout was a consequence of a broad-based overload of a weakly-interconnected power system covering a very large geographical territory.

Electric Power Systems Operational Organizations

Vertically Integrated Electric Utilities

Until new regulatory frameworks were gradually put in place during the last decades of the 20th century in North America and Europe, and more recently in other parts of the world, electric power systems were structured vertically from an operational standpoint. National or regional entities, often designated as electric utilities or electric power companies, owned and operated the power plants as well as the transmission and distribution systems, such that they delivered electricity directly to all end-users in their assigned service territories and directly billed them for it. The service

---

[27] In technical terms, this is referred to as n–1, n–2 and n–3 contingencies.

territories assigned to the utility might be entire countries, states or smaller geographical entities such as agglomerations or cities. Under a vertical structure, electric utilities operate as local monopolies under the authority of local regulatory bodies which set and update safety and operational regulations; they also approve the rates the utilities can apply to various categories of customers.

Vertically integrated electric utilities might be owned by a public authority and/or by investors, either privately or as publicly traded shares, or in the form of cooperatives.

Aside from North America and Europe, the vast majority of electric power systems are still operated as vertically integrated entities. Alternative structures, as discussed next, are either being gradually implemented or being considered in other parts of the world.

Horizontal Organizational Structures

With the advent of smaller-sized power plants having lower capital and operational costs, such as modular gas, solar and wind power plants, new regulatory dispositions gradually opened electricity production markets to a number of new operators. While the outcomes of these regulatory evolutions have yet to fully settle down, the operational structures of electric power systems have been layered into fully or partially independent "horizontal" entities owning or/and operating:

(a) Electric power plants, referred to as "genco".
(b) Transmissions systems, referred to as "transco".
(c) Distribution systems, referred to as "disco".

A growing number of electricity trading entities are marketing and selling electricity to their commercial, industrial or municipal customers. They contract with "gencos" to procure the electricity and then resell it to their customers while paying a service fee which is regulated by the relevant authorities, to "transcos" and/or "discos".

How these layers are structured varies greatly from one country, group of countries (the EU, for example) or region to the next as primarily driven by local regulatory stipulations. In this type of operational structure, transmission and distribution system operators must treat all electricity producers and consumers in a non-discriminatory manner. While a detailed presentation of the various organizational possibilities is beyond the scope of this book, a few clarifications regarding transmission system operators and independent system operators are provided next.

Transmission system operators

A Transmission System Operator (TSO) owns, maintains and operates a transmission system within an assigned territory; it is responsible for the non-discriminatory bulk transport of electricity between producers and distributors connected to its network. If necessary and in agreement with the relevant public authorities,[28] a TSO can construct or deconstruct transmission lines within its service territory. TSOs operate independently from other entities active in the electricity market, such as gencos, discos and traders.

TSOs primarily exist in Europe, where they were enabled by a number of EU directives and Network Codes, which stipulate how they are to be operated; most of them are part of the European Transmission System Operators — Electricity (ENTSO-E)[29] [*entsoe.eu*].

Independent system operator

Independent System Operators (ISOs) are primarily implemented in the US. They are voluntary organizations between several Regional Transmission Operators (RTOs). They independently coordinate and monitor network operations in a particular region in a non-discriminatory fashion between all relevant parties, as well as regarding data and information sharing. They typically do not own the facilities for which they assume the monitoring. In the US, an ISO is authorized by the Federal Regulatory Commission (FERC) [*ferc.gov*] to monitor and operate a regional transmission system or group of regional transmission systems.

---

[28] Which has become quite complex in most industrialized countries due to public resistance to overhead lines or high investment costs if the line is to be underground or underwater.

[29] Most natural gas transmission system operators are part of the ENTSO-G.

Multi-Energy Systems

As seen throughout the book, energy developments took place in several broad stages:

- Until the 19th century, direct use of a growing number of raw energy resources: first wood, then generally coal and later oil. Access to new resources broadened the applications and hence the use of energy.

- The advent of intermediary forms of energy, designated as energy vectors, such as electricity and natural gas,[1] which, by way of dedicated network infrastructures, facilitate the remote and flexible use of the energy sources. They generally require a first conversion from raw resources into energy vectors transported by network infrastructures, followed, most often, by a second one in conjunction with specific end-use applications. Electricity has made it possible to use solar energy not only directly for heating but also to convert it into an energy vector; it has also enabled the civilian use of nuclear energy. As a consequence of the flexibility of these energy vectors, the direct use of some other resources, such as wood and coal, is declining.

- More recently, conversions between energy vectors appeared, such as the recuperation of waste heat from industrial processes using electricity or natural gas. The production of synthetic gas, especially methane, using hydrogen couples the electricity and natural gas energy vectors. As further discussed later, as well as in the two "Where We Should Be Heading" chapters, these inter-energy vector conversions will play increasingly important roles in future energy systems.

---

[1] Natural gas is an energy source, as well as an energy vector, once purified.

Growing Couplings between Energy Vectors

While not pretended as a detailed historical rendition, the following presentation highlights the global energy system's main development stages since 1950, recognizing that it was not uniform worldwide.

As of the 1950s, as illustrated in Fig. 1, electric and gas network infrastructures were supplying residential, commercial and industrial end-users, particularly so for electric networks in OECD countries. Electricity was primarily produced using hydro energy or coal and, less so, using oil. The early pump-storage power plants were constructed primarily to better adjust the available production capacities to the load profiles. The gas supplied was increasingly becoming natural gas and less so town gas obtained from coal. Large capacity above-ground or underground gas storage capacities were built in support of gas networks; where gas network infrastructures were not available, gas cylinders were used. The electricity and gas network infrastructures were essentially independent of each other even though they were sometimes built and operated by a single company

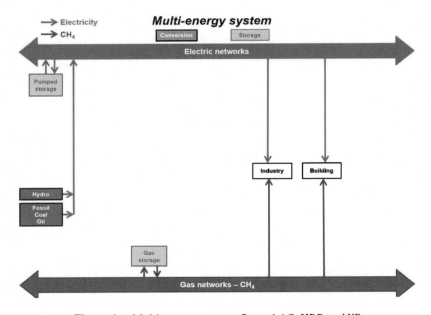

Figure 1.   Multi-energy systems Stage 1 / © *HBP and YB*

selling both products primarily to residential and commercial customers. As seen in the "Electric Power Systems" chapter, electric utilities were generally vertically integrated such that the electricity production adjustments required to follow demand changes were primarily carried out using the production control capabilities at each utility's own power plants. The transportation sector primarily used oil and still coal with the notable exceptions of electric subways in large cities and a few railroad systems.

Developments as of the 1960s

The number of couplings between electricity and natural gas increased. Referring to Fig. 2:

- The advent of the first natural gas electric power plants (bottom left, Fig. 2).
- The first nuclear power plants were added to the electricity networks (bottom left, Fig. 2).

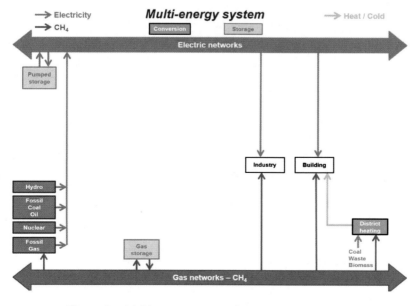

Figure 2. Multi-energy systems Stage 2 / © *HBP and YB*

- District heating systems became operational primarily burning fossil fuels, including natural gas, and also valorizing heat from thermal power plants, mainly coal-fired and more rarely nuclear, as well as using waste incineration (right, Fig. 2).

The use of electricity to produce heat and cold for a variety of applications increased broadly, including for residential applications, such as refrigerators and cooktops, which are further examples of inter energy vector transitions. Electric hot water heaters with integrated heat storage capabilities and electric space heating systems with integrated refractory materials for heat storage enabled, in some countries, the smoothing of electric demand curves and more rational uses of electricity production facilities. The electrification of railroad systems in Japan and Europe intensified.

Development of Renewable Energies

The beginning of the 21st century saw a rapid and massive penetration of alternative renewable energies other than hydro, primarily solar and wind, but also geothermal, marine and biomass (left, Fig. 3). As illustrated by the figure, electricity is the main energy vector used to valorize these new sources. New stakeholders attracted by these electricity production technologies arenas in view of their smaller modular sizes (thus requiring smaller investments) and by public subsidies, contribute to these developments. The variability of these sources call for new electric system controls increasingly relying on additional electricity storage capabilities (top, Fig. 3). The use of electric heat pumps to produce both heat and cold intensified, supported by thermal storage systems (right, Fig. 3).

Increased Inter Energy Vector Couplings

Illustrated in Fig. 4, more recent developments include:

- Electrification of public and private transportation systems is intensifying worldwide. Public transport buses powered by natural gas are increasingly deployed in larger cities.

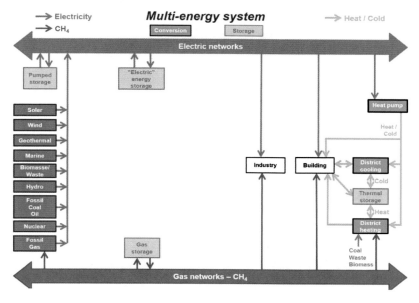

Figure 3. Multi-energy systems Stage 3 / © *HBP and YB*

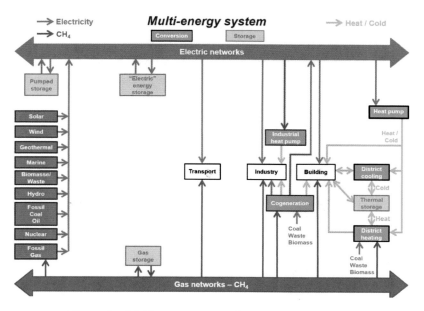

Figure 4. Multi-energy systems Stage 4 / © *HBP and YB*

- The development of heat pumps featuring installed capacities compatible with industrial applications operating at lower temperatures increases the possibilities of valorizing more heat.
- New cogeneration technologies to jointly produce electricity and heat using natural gas and biomass are presently applied for both residential and industrial applications in a growing number of regions.

As will be further discussed, several new multi-energy interactions are made possible using hydrogen as an energy vector.

Biomass

Biomass, the first energy source used by Man, consists of wood and other vegetal organic materials. It is renewable as long as overexploitation of forests and loss of soil fertility are avoided.[2] The biodegradable portion of wastes resulting from human activities, especially urban waste, is increasingly being valorized. As of 2018, including waste, biomass represented 10.2% of the total final energy consumption, along with 2.7% of the electricity production worldwide, i.e., 720 TWh.[3] In 2018, 3.5% of the road transport fuels came from biomass [*iea.org/renewables2019*].

Combustion

The main use of biomass remains the production of heat by direct combustion, particularly for cooking purposes as broadly used in some countries. The use of wood pellets is another combustion application; obtained by wood desiccation, with a humidity below 10%, and compression of sawdust, they can be continuously used in automatically fed furnaces. Decomposition can precede the combustion to obtain a more suitable fuel. When using wood pyrolysis (i.e., the decomposition at high temperature without oxygen), dry charcoal having a high carbon content is obtained.

---

[2] The ethical and economic issues related to the competition between the use of land and vegetation for food production, on one hand, and energy, on the other, are beyond the scope of the book.

[3] Ref. "Energy and Emissions — Where We Are" and "Introductory Remarks" chapters under "How Electricity Is Produced".

Biogas

Biogas is a second utilization of biomass. It is implemented by fermentation, or anaerobic digestion, in a closed container, the digester, in the absence of oxygen to avoid combustion. The biogas obtained contains a varying proportion of methane, CH_4, and CO_2. By purifying the biogas, biomethane is obtained with various degrees of purity depending on the conditions. This process is not to be confused with power to gas processes, further presented below, for which hydrogen is the basis.

The methane obtained can be directly used or injected into a local gas network, or to produce electricity using a gas turbine. A number of such biomass to electricity multi-energy installations exist worldwide with installed capacities between a few hundred kW to one or two MW.

Biofuels

A third biomass utilization is the production of liquid fuels, designated as biofuels. The production process and fuel obtained depend on the plants used, such as:

- Rapeseed or corn can be pressed to obtain an oil, which, following a chemical process called esterification, can be either directly used as fuel or mixed with diesel fuel.
- Starting from beet or sugar cane, sugars can be obtained, which, after fermentation, yield ethanol that can either be directly used as fuel or mixed with gasoline. Brazil is a major producer of this type of biofuel.
- Starting from wheat, corn or potatoes, starch and then sugar can be obtained that can then be used to produce ethanol.

First-generation biofuels only use part of the plants; as a consequence, aside from special circumstances, they have become economically and ecologically unacceptable. Second-generation biofuels using either the entire plant or almost all of it and/or using plants growing on soils not suitable for other crops are under development. For further details, the websites of entities active in the domain may be consulted as well as [*fao.org*] or [*iea.org*].

A multi-energy valorization of wood — Enerbois, Switzerland

This combined electricity and wood pellet production facility valorizes waste products from a large-scale sawmill. Figure 5 shows the layout of the facility. Located in Rueyres, the plant burns bark and wood remnants from the woodcutting. It produces 14 tons of steam per hour. In turn, the steam is used, on one hand, to generate 4 MW of electricity by way of a steam turbine, and, on the other, to dry the sawdust before it is compacted to manufacture pellets. The heat is also used for the plant's offices, living quarters and buildings.

Yearly, some 28 GWh of electricity is delivered to the local utility and 19 500 tons of pellets are sold to local distributors. The efficiency of the electricity production itself is 28%. When including the heat produced, the overall efficiency is 49.5%.

Before the construction of the Enerbois facility, all waste products were trucked to Germany.

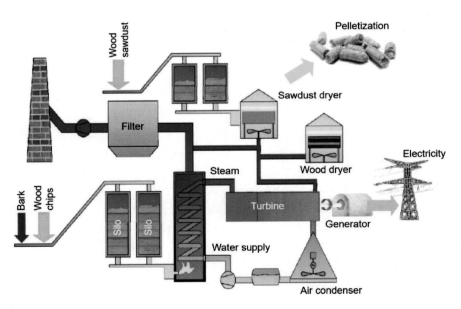

Figure 5. Layout of Enerbois multi-energy facility (Switzerland) / © *Romande Energie*

Hydrogen

In 2018, some 875 million m^3 of hydrogen[4] were used worldwide; compared to the 3 800 billion m^3 of natural gas, this represents some 0.02%. Roughly 50% of the hydrogen was used to manufacture ammonia, a key component of a variety of fertilizers; the refining industry is also a major hydrogen user. For cost considerations, hydrogen is still overwhelmingly obtained (over 95%) using steam reformation of methane (CH_4), as discussed in the "Industry and Agriculture" chapter. Using reformers to produce hydrogen generates CO_2 as a byproduct.[5] The production generally takes place directly at end-use locations or centralized production sites. Hydrogen is transported using dedicated pipelines, primarily in North America and Europe (over 3 000 km) and/or by rail or trucks in pressurized vessels or cylinders. The remaining production is done by electrolysis, which makes it easier to produce hydrogen with higher degrees of purity.

The mass per unit volume of hydrogen at atmospheric pressure and 20°C is quite low at 0.09 g/l, i.e., 90 g/m^3. Under these conditions, the volume occupied by one kg of hydrogen is 11 m^3.

The recombination of hydrogen and oxygen yields water: $H_2 + ½ O_2 = H_2O$.

Since the reaction involves oxygen, it can be referred to as combustion, similarly to the combustion of coal. It produces 34 kWh/kg, i.e., three times more than the combustion of gasoline (12 kWh/kg).[6] The water molecule is quite[7] stable; a temperature above 850°C is required to "crack" it. At room temperatures or at temperatures of only a few hundred degrees, energy as well as catalysts is required to decompose the water molecule, i.e., to separate hydrogen and oxygen.

---

[4] At the per unit volume mass given below, this represents 75 million tons of hydrogen consumed yearly.

[5] Indeed, $CH_4 + O_2 \rightarrow CO_2 + 2 H_2$ produces 11 kg of CO_2 for 1 kg of hydrogen.

[6] However, due to its low mass per unit volume, the energy content of hydrogen per unit volume at atmospheric pressure and 20°C is low: 3 kWh/m^3.

[7] Since at least 34 kWh/kg is required for its decomposition at room temperature!

Hydrogen Production using Electrolyzers

Alkaline electrolysis

Presented in the "Industry and Agriculture" chapter, it is the most commonly used process to decompose water using electricity. The principal steps are:

- The electricity supply is DC (direct current) by way of a rectifier connected to the local utility.
- Water is supplied using a purifier to avoid any system fouling or degradation. After purification, an electrolyte is added to facilitate the current circulation.
- The water decomposition takes place in the "stack" consisting of several electrolyzer cells assembled in series.
- The stack produces hydrogen and water on one side and oxygen and water on the other. Two separators are used to separate the hydrogen from water and oxygen from water. The water is reused.
- Since the process is exothermic, cooling of the stack is typically required to avoid any damage. Heat can be recuperated from the oxygen and hydrogen produced as they are cooled down in dedicated condensers.

A number of electrolytes have been experimented with; potassium hydroxide is most commonly used even though it is corrosive. Similarly, several materials and catalysts, including nickel, are used for the electrodes. Alkaline electrolyzers are a mature and still progressing technology; units with rated capacities up to several MW[8] have been available for several decades.

---

[8]Electrolyzers are characterized by their installed electrical capacity and by the electricity consumed per m^3 produced at room temperature and atmospheric pressure, designated as Nm^3; 4 to 6 kWh/Nm^3 for alkaline electrolyzers. The volume produced per hour is another useful characteristic, for example 220 Nm^3/h for an alkaline electrolyzer with an installed capacity of 1.25 MW.

The overall efficiency[9] of alkaline electrolyzers reaches 65–70% when operated between room temperature and 90°C.[10] They generally operate at pressures of a few bars, which implies, as further discussed regarding hydrogen storage, the use of a compressor thus requiring more energy. To reduce or even eliminate the required compressor, electrolyzers operating at pressures of up to 50 bars have been developed. Compression at the liquid stage is less energy intensive than the compression of hydrogen at the gaseous stage.

Proton exchange membrane (PEM) electrolyzers

This technology was introduced in the 1960s as an alternative to alkaline electrolysis. Rather than using a liquid electrolyte, it uses a special membrane which allows hydrogen protons, H^+, to pass through but not atoms, including oxygen.[11] Rare metals, such as platinum and iridium, are still required for the electrodes. However, the handling of corrosive liquid electrolytes is avoided since only water and electricity are required. In addition, operation under pressure is not a problem. PEM electrolyzers are not only simpler and more compact than the alkaline alternatives; they are also more suitable for applications with rapidly changing electricity supply conditions such as from photovoltaic (PV) or wind. As of 2017, the overall efficiency of PEM electrolyzers is similar to that of alkaline electrolyzers while their cost is higher. Further developments should improve

---

[9] Several efficiencies or efficiency ranges are provided in this chapter. When comparing efficiencies, it is important to be precise: does the efficiency only relate to the main process (the electrolyzer itself in this case), or is the energy consumption of the auxiliary systems (pumps, purifiers, etc.) included? In the efficiency computation, one could also include the heat that could be recuperated from the steam resulting from the combustion of hydrogen. In this chapter, overall efficiencies include the auxiliaries, which affects them negatively, as well as the heat that could be recuperated from the steam, which affects them positively.

[10] The combustion of 1 kilogram of hydrogen produces 34 kWh. As a result, with an efficiency of 68%, which is typical, 50 kWh, i.e., 34 kWh / 68%, is required to produce one kg of hydrogen.

[11] As further discussed, these membranes are also developed for fuel cells, which should decrease their cost, which remains high.

their efficiency and lower their cost. Installations rated at the MWe level are commercially available.

Solid oxide electrolyzer cell (SOEC)

Still under development, they operate at high temperatures, between 600 and 800°C. The energy required is partially provided as electricity and partially as heat. Part of the heat can be recuperated from the hydrogen and oxygen produced. The higher temperatures require the use of ceramic electrodes and membranes. A particularly interesting feature of SOEC electrolyzers is that they are reversible such that they can also be operated as fuel cells, presented below. Overall efficiencies reaching 85% are envisioned. Developments are underway to remedy the fragility and shorter lifetimes of the materials used; the possibility of operating at pressures above 30 bars is also investigated.

Hydrogen Storage

Hydrogen is usually stored in two forms: compressed or liquid. Research efforts are underway toward solid storage by "trapping" hydrogen molecules inside an atomic network or mixed in with liquids.[12]

Compressed hydrogen storage

As seen above, the per unit volume energy content of hydrogen at room temperature and atmospheric pressure is low. Thus, to store sufficient amounts of energy in a reasonable volume, one approach is to compress it before storing it in reservoirs designed to sustain higher pressures and which are also impermeable to hydrogen. For industrial applications, hydrogen is routinely stored at a pressure of 200 bars in cylinders. A 50-liter cylinder can contain 0.9 kg at that pressure[13]; it weighs 70 kg when empty, i.e., 70 times more than the hydrogen it can contain!

---

[12] The advantage of liquid or solid storage is that hydrogen could then be readily transported over long distances.

[13] As seen above, at room temperature and atmospheric pressure, 0.09 g of hydrogen occupies a volume of one liter. Thus, at 200 bars, the same volume can hold 18 g. Finally, a 50-liter reservoir at 200 bars can hold 0.9 kg of hydrogen.

More recently, to accommodate range requirements of hydrogen-fueled cars, hydrogen reservoirs with higher energy contents yet of smaller sizes and weight are required. A pressure of 700 bars is possible using specially designed carbon-reinforced reservoirs made of composite materials. They typically weigh 30 kg and can hold 5 kg of hydrogen; as a result, the ratio between the reservoir's empty weight and its content has been reduced to 6; however, its complexity has increased to ensure safe fill-ups and withdrawals under changing pressure and temperature conditions.

Compression requires energy whether done fast adiabatically, which causes a temperature increase, or slowly at a constant temperature. The required compression energy increases with the desired pressure. It is therefore desirable that the hydrogen be produced at some pressure directly from the electrolyzer, 20 bars, for example, before its compression. Typically, compressing hydrogen from 20 to 200 bars requires some 7% of its calorific energy content; a compression of 20 to 700 bars requires 10% or more,[14] depending on the compression technique used. Given the properties of hydrogen and pressures mentioned above, special transfer piping and compressors are required to ensure appropriate security and safety levels at industrial hydrogen end-use sites.

Given the electrolysis efficiency (some 70%) and that of the compression to 700 bars (some 85%), the overall efficiency is some 60%. From 1 kWh of electricity required for electrolysis and compression, one thus obtains compressed hydrogen at 700 bars, which could produce 0.6 kWh when combusted.

Liquid hydrogen storage

It makes sense to investigate liquid storage possibilities to enhance the amount of energy stored per unit volume. The liquefaction temperature of hydrogen is −253°C or 20 K. At this temperature, hydrogen can be stored as a liquid at atmospheric pressure; its density then is 71 kg/m^3. The energy density of liquid hydrogen is 2.4 kWh/l, i.e., roughly 25% of that of gasoline and somewhat less than 50% of that of liquid natural gas (LNG) at −162°C.

---

[14] The compression energy is compared to the potential combustion energy of the compressed hydrogen.

Liquid hydrogen is stored in properly insulated, cryogenic, reservoirs with capacities from one liter to several thousand liters. In all cases, a hydrogen evaporator is required since, despite the insulation, some heat penetrates the reservoir and causes hydrogen evaporation. A 1% daily evaporation is common; such losses do not occur using pressurized storage.

Liquefying hydrogen requires up to one-third of the stored energy. Taking the electrolysis and the liquefaction efficiencies, 70% and 67%, respectively, into account means that the liquid hydrogen obtained from one kWh of electricity could produce 0.47 kWh when combusted, i.e., less than using compressed hydrogen; however, the storage volume is far less.

Hydrogen as an Energy Vector

Once the hydrogen has been produced using electrolysis, it can either be stored in compressed or liquid form or be directly injected into the local gas distribution network. As long as the electricity used for the electrolysis is renewable or has a low carbon content, the emissions due to the gas are reduced in proportion when used. Direct injection is generally limited to 10–15% to avoid the need to modify the gas distribution network[15] and also to limit the risks in case of gas leakages.

The following insert provides an example of the use of hydrogen as an energy vector.

> **The hydrogen production facility in Mainz, Germany**
>
> The factory produces hydrogen using the PEM technology. Worldwide, it was the largest of this type when put into service in 2015. Three electrolyzer stacks, each rated at 1.3 MW, can support a 2.1 MW_e temporary load, i.e., a 6.3 MW_e peak performance. The load increase rate can reach 10%/s over a broad range of operating conditions. The factory is connected to the local electric distribution system at 20 kV using a dedicated line. An 8 MW wind power farm can either supply the factory or the local network depending on the wind energy available and the load requirements.

(Continued)

---

[15] Since the hydrogen atoms are quite small, hydrogen can potentially permeate through a number of materials. In addition, hydrogen is quite flammable, making the avoidance of any leaks paramount.

(Continued)

The hydrogen is stored as a gas at a maximum pressure of 35 bars. Two reservoirs can store a total of 1 000 kg of hydrogen, which corresponds to an energy of 34 MWh. Some 200 tons are produced annually.

Some of the hydrogen produced is injected into the local gas network at a pressure of 7 to 9 bars. This represents some 7% of the annual production; the demand is low during summers. The remaining production is either transferred to industry or used in hydrogen filling stations at pressures which can reach 22.5 bars. The overall efficiency, i.e., the ratio between the combustion energy of the hydrogen produced and the electric energy consumed, varies with the power drawn. The average overall efficiency is 60%.

Power-to-Gas

The power-to-gas process described below, sometimes also referred to as methanation, should not be confused with the biomass fermentation process presented earlier. Starting from hydrogen, one can produce synthetic methane by one or the other reaction below:

$$CO + 3\ H_2 = CH_4 + H_2O \quad \text{or} \quad CO_2 + 4\ H_2 = CH_4 + 2\ H_2O$$

These two exothermic reactions, named after **Paul Sabatier**,[16] have an efficiency (i.e., the combustion energy of the methane obtained divided by that of the amount of hydrogen required) of some 55% using present technology installations. If the heat produced is valorized, the overall efficiency increases by 12–15%. Using new catalysts could potentially increase the efficiency.

The main components of a power-to-gas installation are:

- A source of electricity.
- An electrolyzer to produce the hydrogen. Intermediate hydrogen storage might be required.
- A CO_2 source, for example, from a combustion thermal power plant, or an industrial site or from a biogas fermentation facility.

---

[16] French chemist who received the Chemistry Nobel Prize in 1912.

- The methanation reactor itself.
- A connection to the local gas network and/or a gas storage facility.

As long as renewable excess and/or low-carbon electricity is used to supply the electrolyzer, the process valorizes CO_2 by converting it into synthetic methane. Once it is available, one can either:

- Inject it into the local gas network, which reduces the emissions pro-portionally when it is combusted, or
- Store it for future use in the industry or to produce electricity in a natural gas power plant.

Given the efficiency of industrial electrolyzers (70%) and that of the methanation reaction (some 55%), the overall efficiency of the process, from electricity to methane, is of the order of 40%, i.e., lower than for a direct injection of hydrogen into the local natural gas network. However, the volume of gas one can inject is not limited since methane is injected as opposed to hydrogen.

The power-to-gas process, illustrated in Fig. 6, is a new coupling between the electricity and gas energy vectors. Pilot projects are under-way in several countries with the aim to valorize CO_2, reduce overall emissions and/or to valorize surplus intermittent electricity productions as presented in the insert.

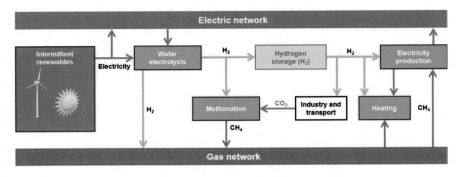

Figure 6.   Power-to-gas functional layout / © *HBP and YB*

The Werlte pilot plant synthetic methane cycle

Audi, which manufactures natural gas internal combustion engine cars, wished to demonstrate the technical feasibility of a complete production — utilization synthetic methane cycle. At the core of the installation is a pilot electrolyzer that was put into service in 2013; its peak capacity is 6.3 MW, which can be reached in five minutes from start-up. The electricity supplied to the electrolyzers comes from excess production from a local wind farm, which occurs, on average, some 4 000 hours yearly.

After it is dried, the hydrogen is combined with CO_2 from a local biomass fermentation plant, which was put into service in 2002. Its existence led to the decision to construct the pilot plant next to it. In turn, the biogas facility is within one km of a pre-existing natural gas network. The heat produced by the methanation reaction is used to enhance the fermentation process. Yearly, 2 800 tons of CO_2 are transformed into 1 000 tons of synthetic gas.[17] Knowing that the combustion of methane yields 14 kWh/kg and assuming that the electrolyzers operate at full power to produce the 1 000 tons of methane, the overall efficiency would be 14 000 MWh / (6.3 MW · 4 000 h) = 55% — this is a low estimate since the electrolyzers are probably not operating at full power during the 4 000 hours. The methane can be directly used in the manufacturer's natural gas cars. It could also be injected into the local gas network, or used to produce electricity by way of a gas turbine.

Fuel Cells

While electrolyzers provide for a conversion from the electricity vector to the hydrogen vector, fuel cells provide for the reverse conversion. A fuel cell is thus the "inverse" of an electrolyzer.

They can be used for both stationary and embarked applications, such as for cars and buses.

The scientific principle of fuel cells has been known since 1839; a 6 kW fuel cell was already operational in 1953. The complexity and cost of some of the elements required explain why the industrial development of

---

[17] The Sabatier reaction shows that one CO_2 molecule yields one methane (CH_4) molecule. Since the mass proportions are 16/44, one gets 1 kg of methane for 2.8 kg of CO_2.

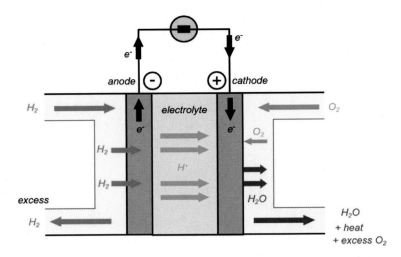

Figure 7.   Layout of a fuel cell / © *HBP and YB*

fuel cells did not occur before the space industry adopted them for on-board satellite and manned space missions.

As illustrated by Fig. 7, fuel cells convert chemical energy into electricity by way of the exothermic combustion reaction: $2\ H_2 + O_2 = 2\ H_2O$.

- The fuel generally is hydrogen; it is supplied continuously. It could also be natural gas. The oxidant is generally oxygen; it could also be air.
- The electrodes contain catalysts, such as platinum, to facilitate the reaction.[18]
- As shown in the figure, the electrolyte usually serves as a membrane that allows the H^+ protons to pass but not the electrons, which are "forced" through the external electric circuit and the load.

The output voltage of an individual cell is some 0.7 V. As for regular batteries, higher voltages can be obtained by connecting several cells in series. The output current depends on the electrode surfaces. The energy

---

[18] Research toward the replacement of platinum by more abundantly available and less costly materials or using lesser amounts of the materials required is on-going. The same is also true for the membranes.

density is generally between 160 and 200 Wh/kg, while the power density is generally between 1 000 and 2 000 W/kg. The efficiency between the energy contained in the hydrogen provided and the electricity produced is some 60%. Hydrogen combustion being exothermic, the efficiency can reach 80% if the heat can be valorized.

The installed capacities range from a few mW, for portable applications, to several hundred kW, for vehicle propulsion applications, and up to a few MW for high-temperature devices operating at between 600 and 1 000°C.

Similarly to electrolyzers, the three main fuel cell technologies are alkaline, Proton Exchange Membrane (PEM) and Solid Oxide Fuel Cell (SOFC), which are operated at higher temperatures. The PEM technology is most commonly used, especially for embarked applications; it operates at low temperatures and can readily follow load changes. The hydrogen provided for the technologies mentioned above must have a high degree of purity, which complicates their operation.

For mobile embarked applications, fuel cells have the advantage of not being locally polluting since the only "emission" is water. However, high pressure embarked reservoirs are required. As long as the electricity used to produce the hydrogen is low-carbon, the overall CO_2 emissions are reduced compared to fossil-fueled vehicles.

From Electricity to Electricity by Way of Hydrogen

To temporarily store surplus electricity, for example from solar or wind production, one solution can be to use hydrogen as the storage vector — electrolysis from electricity to hydrogen, which is then stored before its conversion back to electricity, using a fuel cell, when needed. Intermediate hydrogen storage is thus required.

The electricity-hydrogen-electricity "round-trip" efficiency is the product of efficiencies of electrolysis (70–80%), hydrogen storage (80–85%), and fuel cell (60–65%), i.e., between[19]:

$$70\% \cdot 80\% \cdot 60\% \approx 35\% \quad \text{and} \quad 80\% \cdot 85\% \cdot 65\% \approx 45\%$$

---

[19] The efficiencies mentioned are average to high within the ranges found in most scientific publications.

This evaluation assumes that the entire "round-trip" is executed at the same location. Otherwise, the hydrogen transport efficiency must also be taken into account. This round-trip efficiency is clearly lower than that for pumped storage or chemical batteries. However, the high-pressure compressed hydrogen approach makes it possible to store higher amounts of energy with reduced volume or weight, compared with chemical batteries.

Coal to Liquid (CTL) and Gas to Liquid (GTL) Using Electricity

Invented in 1923, the chemical Fischer–Tropsch[20] process makes it possible to produce liquid synthetic fuels from coal (CTL) or natural gas (GTL). High temperature gasification of coal results in a mixture, referred to as syngas, containing various proportions of CO and hydrogen, and often some CO_2. From the syngas, various liquid or gas carbonated fuels can be obtained depending on the type of process used.

It was deployed from the 1930s.[21] Several variations were developed. Later, its cost compared with that of oil and gas resources from the Middle East led to its demise. However, the prospect of reusing CO_2 and having access to large amounts of hydrogen, in turn, produced using abundantly available and low-carbon electricity could lead to a revival of the technology. A large-scale installation based on the Fischer–Tropsch technology would require the availability of large amounts of CO_2 captured from a power plant or an industrial plant. It would constitute yet another multi-energy interaction. As of 2018, the few installations in the world relying on this technology are tied to strategic fuel availability requirements.

Heat and Cold Networks

Using one or several sources, heating networks provide heat and/or hot water to residential buildings and individual houses as well as to industrial, commercial and tertiary and services facilities. By the same token, even though they are less developed and more recent, cold networks

---

[20] So-named after the two German engineers, **Franz G. Fischer** and **Hans Tropsch**, who invented it.

[21] This process made it possible for Germany, with considerable coal resources but devoid of oil resources, to produce enough fuel during the Second World War.

provide for the cooling of same types of facilities. These networks are also referred to as district or distance heating/cooling networks.

While heating networks have been used at least back to the Roman Empire, it is only since the second half of the 20th century that they have been broadly deployed.

They frequently have the following structure:

- A source provides[22] heat generally to water to simply heat or vaporize it. The heat can be provided from:
  - A cogeneration thermal power plant — coal, natural gas or nuclear.
  - A geothermal source.
  - More frequently, a boiler specifically dedicated to the heat network and using biomass and/or urban waste, at least for part of its requirements. Occasionally, electricity can also be produced when heat requirements are reduced.
  - Occasionally, waste heat from industrial sites.

  The resulting emissions depend on the source(s) used, hence the CO_2 content of the heat provided to the end-users.

- A network of conduits provides for the transport of the heat to intermediate stations that control the pressure and flows toward distribution networks. The end-users are supplied from the distribution networks. Second level distributions may be installed at the entrance of an apartment complex, for example, to provide end-use metering.

Cold networks have similar structures and are primarily supplied using industrial-sized heat pumps.

The City of Göteborg's district heating network, Sweden

Launched in 1953 using a cogeneration plant burning oil, the city's district heating network had reached over 1 200 km of conduits and pipes by 2019 when it supplied 90% of the residential buildings and a growing number of individual houses, presently over 20%. Overall, it provides for the heating requirements of 150 000 families.

(Continued)

---

[22] A combination of several sources may be used for optimization purposes or to improve the system's supply reliability.

(*Continued*)

Also as of 2019, 89% of the energy mix was renewable or waste energies. Its 19 production sites use urban waste, biomass, surplus heat from local refineries, heat from wastewater heat pumps, as well as cogeneration using natural gas. This considerable mix evolution during the last 60 years took place while the annual heat provided reached more than 4 TWh since 2010. The customers can follow their individual use of heat by the hour.

Local ferries are also connected to the district heating system to heat them overnight thus avoiding the use of the ships' on-board fuel; a similar system was used during the 1920s in Paris to heat passenger train carriages overnight using the local urban heating network.

Multi-Energy Systems — The Path Forward

Figure 8 illustrates the potential integration of the various multi-energy interactions discussed in this chapter. As mentioned earlier, the availability of abundant and decarbonated electricity is and will remain the

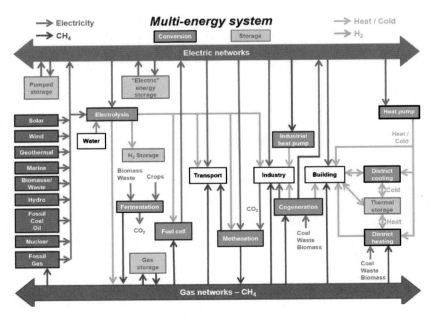

Figure 8.   Multi-energy systems Stage 5 / © *HBP and YB*

foundation toward multi-energy interactions. Compared to the 2018 "state-of-art", illustrated by Fig. 4, a number of multi-energy interactions described in this chapter involve the use of hydrogen as an energy vector. How these technologies are implemented in any one region of the world will depend on local geographical, resource availability, societal, economic, environment and climate considerations.

A number of R&D and demonstration efforts are underway to further improve the efficiencies of the related technologies — especially electrolyzers, hydrogen storage and fuel cells — to better align their investment and operational costs with those of "standard" technologies.

The interactions between the electricity vector and other vectors like heat, natural gas and hydrogen, as well as between these other vectors, will be at the core of the integration of renewable energy sources, coupled with multi-vector and multi-scale energy storage.

Where We Should Be Heading

Public Policy and Carbon Fee

Four Observations Leading to Differentiated Energy Mutations at all Levels

First Observation: The Rapid Reduction of Greenhouse Gas Emissions is a Compelling Necessity

To grasp the magnitude of the overall decarbonization challenge, it is useful to keep in mind the present emissions levels, presented in the "Energy and Emissions — Where We Are" chapter, along with the October 2018 IPCC Special Report, which observes that the overall remaining greenhouse gas emissions budget is 770 $GtCO_2$ to still have a 50% chance of limiting global warming to 1.5°C above pre-industrial levels. It is thus clear that maintaining the emissions at their present level of 45–50 $GtCO_2$ per year, i.e., "business as usual", would be irresponsible since we would reach the IPCC's upper bound in less than 20 years.[1]

As also observed in the same chapter, greenhouse gases (GHGs) accumulate in the atmosphere over decades. It is thus important to reduce their emissions as rapidly as possible.

---

[1] Since the world population was roughly 7.7 billion in 2018, the overall remaining budget per capita is 100 tCO_2/cap, whereas the yearly average has recently been at roughly 5.8 tCO_2/cap, which confirms the less than 20 years mentioned.

Second Observation: Energy Options Must Include Engineering, Societal, Economic and Political Considerations

The impact of energy production and end-use systems on climate change as well as on the local and global environment, broadly speaking, is such that solutions to energy and environmental challenges are often coupled. These challenges are not only technical but also societal, economic and political, in line with the 17 United Nations Sustainable Development Goals, SDGs [*sdgs.un.org*]. Hence the need for balance between:

- The needs of the present generations, on one hand, versus those of future ones, on the other. For example, the use of raw materials and their impact on climate change.
- Solutions at the individuals' level (solar panels at home, cars, etc.), on one hand, versus community solutions (power plants, public transport, etc.), on the other.
- More broadly speaking, individual well-being, on one hand, versus community and general well-being, on the other, such as those related to the implementation of a power plant or a transmission line.
- Financial subsidies in support of housing insulation or to replace appliances and cars with high emission levels, on one hand, versus support of the deployment of photovoltaic (PV) panels, on the other, which have different societal and industrial impacts.

Public officials play a central role in orienting societal options.[2] Their implication in the energy arena has always been felt given its strategic importance. Overall, both the local and broader regulatory contexts have evolved during the last few decades; rather than overall deregulation, the evolution is better described as a re-regulation given the number of stipulations and documents that resulted.

---

[2] Sometimes also at supranational levels such as the European Union.

Third Observation: We Will Continue to Have Enough Energy Resources; the Challenge is to Reduce Their Impact by Way of a More Rational Energy End-Use

The overriding priority is broad decarbonization until 2035 and then pursued beyond. The main energy challenges thus are:

- A rational energy end-use, which is essential since energy utilization always has an impact. In turn, this implies, whenever possible, that:
  - ○ Energy be wisely used and waste be eliminated, for example, by way of waste heat and cold valorization along with waste reduction and valorization and taking advantage of synergies between local end-uses.
  - ○ Recycling be encouraged, facilitated and deployed at all levels. Regulatory and/or fee structures also need to be implemented.[3]
  - ○ High efficiency equipment and systems be deployed at affordable costs. Efficiency, itself, is not the objective, but a means for a rational energy end-use. For example, energy storage is not, in itself, efficient since a storage-restitution cycle results in an energy loss. However, it could turn out to be rational, depending on the energy availabilities and/or the prevailing tariffs. Once the decision to store energy is reached, it is important to do so with the highest efficiency possible. In this case, storing energy is rational; doing so efficiently reduces the losses incurred.
- That priority be given to sources producing no or only a few GHGs when deployed, i.e., renewable energies, and if necessary, nuclear energy. This implies avoiding resources relying on combustion such as coal, oil and gas; this is difficult in view of both their abundance and relatively low cost of use.
- That if the use of fossil fuels cannot be avoided, natural gas be favored over oil and coal given its lower CO_2 emissions for the same amount of heat, i.e., electricity, produced.

---

[3] In Switzerland, for example, curb refuse collection is restricted to garbage in specially intended bags sold at a number of locations, such as grocery stores. Garbage collection decreased by over 30% upon implementation of the regulation as citizens increased glass, paper and aluminum recycling to save on garbage bags.

Fourth Observation: Energy Mutations Must Be Adapted At All Levels

The previous three observations lead to the fourth one.

- It is now more imperative than ever to listen to and heed the aspirations of populations with the lowest living conditions toward a better quality of life, broadly speaking. Satisfying these aspirations requires, among other challenges, improved access to energy at costs they can afford.[4] This implies, de facto, providing for increased energy consumption, particularly of electricity, while limiting environmental impacts and especially the increase of greenhouse gases. This broad objective can only be reached by way of technologies and systems that are well suited to the prevailing local conditions, in particular, to the climate.[5]
- It is unreasonable to build energy scenarios assuming a decrease in the quality of life for future generations. The future does not so much require a reduction of energy consumption but a more rational end-use. To this end, decisions must be reached in terms of:
  - End-use technologies,
  - Electricity and, where relevant, heat production mixes,
  - Incentives and regulations, which are coherent with the decarbonization priority. The best available technologies so provide even though some of them are still more complex to deploy and more expensive in terms of investments and/or operational costs.

Setting GHG targets is and will remain subject to negotiations and compromises at the supranational, national and regional levels. Large-scale experiments, demonstrations and pilot projects confirm that energy mutations must involve stakeholders at all levels, especially in cities where an increasing proportion of the world population concentrates. Indeed, energy synergies can be first decided and implemented at local levels such as in neighborhoods as well as industrial or commercial parks. Local authorities, corporations and citizens are all stakeholders.

---

[4] "Ensuring access to affordable, reliable, sustainable and modern energy for all" is one of the United Nations SDGs.

[5] Indeed, a number of existing technologies and systems were developed in northern hemisphere countries, by northern hemisphere corporations, for northern hemisphere applications.

Four Guiding Principles

Neither a standard nor a unique solution exists for all energy mutations. Decisions as to the outcomes to be reached and their paths depend on a number of factors such as the present energy and climate situations as well as climate change impacts, natural resources available and the desired or feasible energy independence level to be maintained or reached. Additional factors, such the local economic situation, environmental issues and life-styles, also must be considered.

Elaboration of a Systemic Energy Vision

An all-encompassing energy vision is necessary, not only concerning electricity. Indeed, electricity represents less than 25% of the total final energy consumption in most countries.

The production, transport, distribution, storage and end-use of all energy vectors will increasingly interact with each other. A holistic energy vision must give due consideration to these interactions. Promoting or regulating the deployment of a particular technology impacts well beyond the technology considered. For example, public policies to encourage the deployment of PV and/or wind energy have resulted in outcomes that were sometimes not fully envisioned in terms of electric network operations, additional production capacities, CO_2 emissions, or energy costs. The electrification of automobiles provides another example of complex interactions between the reduced reliance on liquid fuels and the increased need for electricity supplies, which, in turn, induce complex arbitrations at the taxation level.

Holistic energy visions should be developed not only at national levels but also at larger geographic scales to fully encompass natural resource availabilities, end-use synergies and impacts. Local energy plans must be integrated within such larger-scale plans. The elaboration of these visions must be based on models that integrate all aspects, from production to end-use. Such models provide for the assessment of the impacts of any potential decision in terms of greenhouse gas emissions and energy resources as well as on residential, commercial, industrial and transportation requirements.

Deployment, Without Further Delay, of the Best Available Technologies

Since fast emissions reduction is urgent, it is important to deploy the best available technologies as quickly as possible without waiting for the "ideal" technology to become available. This is especially important when replacing existing equipment or installing new ones. While infrastructures at the energy production, transport, distribution, and storage levels are generally designed to operate for 50 years and beyond, energy consumption equipment is generally replaced far more often, even several times before 2035, such as for automobiles and home appliances. The notion of "best technology" depends on local circumstances. Between equipment with similar performances at a price one is willing to pay, priority should be given to the one with the lowest emissions when operated.[6] If the local electricity production mix is low-carbon, electric end-use equipment should be preferred; if, on the contrary, its carbon content is high, gas end-use equipment should be favored if the local infrastructure so allows.

Creation of a Universal and Redistributed Carbon Fee

To fulfill the decarbonization requirement, a cost of CO_2 emissions must be set and applied worldwide. A periodically adjusted, universal and redistributed carbon fee appears to be an attractive solution that could set all stakeholders in motion — public authorities, businesses and citizens. The motivation and specifics of this concept are further discussed in a dedicated paragraph.

---

[6] Equipment manufacturing and power plant construction consume energy and thus produce emissions. For equipment and power plants not involving combustions during their regular operation, the emissions during their manufacturing or construction, as well as during their recycling or deconstruction, play a dominant role over their entire life cycle. On the contrary, if combustion is involved during the regular operation of equipment or power plants, their emissions during operation are generally dominant over their life cycle.

Inclusion of End-user Behaviors

Above all other considerations, the behavior of end-users — individuals or corporations — must be taken into account. End-users generally welcome automatisms which preserve, or even improve, the overall performance of their equipment or their comfort while also reducing their expenses. However, past experience in numerous countries has shown that residential end-users gradually lose interest in following their daily consumption routines, for example using a smartphone or a terminal in the kitchen. Indeed, except for rare circumstances, the resulting cost savings are not sufficient to sustain the consumer's interest; in addition, she or he generally has more interesting things to do!

Overall, end-users generally accept that their consumptions be automatically or remotely controlled as long as they can reassume local control whenever desired.[7] This implies the need for information and communication technologies to provide for more rational energy end-uses. Improving the controls of existing equipment often provides for significant consumption reductions at a lower cost than replacement. On a broader scale, the ubiquitous use of inexpensive sensors connected to the Internet will provide significant cost reductions and optimizations at a system level as opposed to only at the individual apparatus level.

However, cost reductions could, in turn, induce energy consumption increases by what is referred to as the "rebound effect". For example, improved home insulation may induce the resident to increase the home's temperature to improve its comfort at a cost that has become acceptable to her or him; thus, the expected consumption and emission reductions do not materialize. This explains many differences between the expected and actual outcomes!

---

[7] For example, a hot water heater that normally only operates overnight must provide for daytime heating when guests are hosted. Automatic process control in a factory must provide for occasional adjustments during unusual production levels.

Rules in the Electric Utility Industry

Even though electricity markets are opening to an increasing number and diversified set of stakeholders, the inherent nature of the electric utility industry still necessitates that it be "re-regulated" under the auspices of local, regional, national and/or supranational regulating bodies entrusted with setting and enforcing rules. Whenever feasible, the rules related to the production, transport, distribution, storage, and end-uses of electricity should focus on the decarbonization priority while also taking personnel safety, supply security and the costs for citizens or other stakeholders into account. The following insert illustrates the importance of basing CO_2 emissions comparisons between alternative solutions — for heating in the next insert — on final energy consumptions and not on primary energy.

As previously noted, the behavior of all residential, commercial and industrial end-users is crucial for a sustained trend toward decarbonization; it is thus essential that, whenever possible, the rules be prepared and published while keeping a broad understanding by the public-at-large in mind.

Rules are set related to electricity pricing, on one hand, and to regulate the safety and performance of electric equipment and systems, on the other.

A primary energy reduction may not result in an emissions reduction

Two identical houses each consume 1 MWh for heating during the winter. The first one is heated using electricity and the second one using natural gas.

The electricity production mix in the country is:

- Hydro: 25%
- Nuclear: 30%
- Natural gas: 30% with an efficiency of 50%
- Wind and PV: 15%

The IEA conversion coefficients used to compute the primary energy based on the final consumption are (ref. "Energy and Emissions — Where We Are" chapter): 3 for nuclear and 1 for the other sources. The emissions resulting from the combustion of natural gas is 200 gCO_2/kWh; ref. "Units and Reference Values" annex.

(Continued)

(Continued)

House heated using electricity

To compute the primary energy and emissions, the average efficiency of the natural gas power plants must be taken into account; 50% in this case.

A 1 MWh of final consumption corresponds to:

$(25\% \cdot 1) + (30\% \cdot 3) + ((30\% \cdot 1) / 50\ \%) + (15\% \cdot 1) = 190\%$, i.e., 1.9 MWh of primary energy.

As a result, a 1 MWh final heating energy consumption corresponds to 1.9 MWh of primary energy.

The emissions due to the final consumption of 1 MWh are:

$30\% \cdot ((200\ gCO_2/kWh)/50\%) = 120\ gCO_2/kWh = 120\ kgCO_2$ for 1 MWh.

House heated using natural gas

In this case, the primary energy is equal to the final energy, i.e., 1 MWh. The emissions are: $200\ kgCO_2$ for 1 MWh.

In summary

- The house heated using electricity consumes 1.9 MWh of primary energy and has emissions of $120\ kgCO_2$.
- The house heated using natural gas consumes 1 MWh of primary energy and has emissions of $200\ kgCO_2$.

In this particular country, replacing a natural gas heating system with an electric one increases the primary energy consumption but decreases the CO_2 emissions. Clearly, the conversion factor between primary energy and final energy consumption for nuclear power plays a major role in the result above.

Electricity Pricing

Electricity invoices are structured along three main components:

- Electricity consumption, which, in turn, is based on tariffs agreed to or set by a regulating entity, for:
  - The actual consumption, in kWh or MWh.
  - The capacity of the customer's connection to the local network, measured in kW or MW.
  - Network access, typically based on consumption.
- Taxes, levied by local and/or national governments.
- Renewable energy subsidies to enact government policies.

Regulations

It is important that:

- Regulations applicable to equipment and the operation of energy systems are mandatory, strict and clear. This will become even more important as the number of both electricity end-users and electric appliances continues to grow. The access of a growing number of actors to open markets also reinforces the need for transparent regulatory contexts. Regulatory harmonization between countries is desirable, for example, to reduce manufacturing costs for larger markets.
- Regulatory bodies ensure that product performance tests are done under real-life operating conditions using standard products and not under laboratory conditions using products specially prepared for the test.
- The results of the performance tests are published based on indicators allowing end-users to rely upon them when comparing products fulfilling identical functionalities.
- Penalties imposed by regulatory authorities in conjunction with equipment having high-energy consumption levels are directly included in the purchase price of the said equipment, for example, at the purchase of an appliance or car.[8]
- The sale of a particular equipment or type of equipment is prohibited only after equipment having identical performances or fulfilling identical services at equivalent costs have been broadly introduced on the market. The typical example of such transitions is the prohibition of selected incandescent light bulbs as they were gradually replaced by LEDs.

Regulatory Setting Process Durability and Transparency

Cycles in the energy domain often last several decades. Indeed, the infrastructures are designed, constructed and maintained to reach very long

---

[8] In some countries, the registration fee, as well as the yearly renewal fees, are more expensive for high fuel consumption cars, which thus have higher emissions levels.

lifetimes. Many power plants still in operation were put in service more than fifty years ago and their decommissioning is still not foreseen.

Cycles in the political arena, i.e., between elections or votes, typically only last a few years in most democratic countries.

> *Political decisions in the energy domain should result in regulations and policy incentives implemented over long durations, for example, 15 years. Revisions must certainly be possible if only to incorporate technology advances; however, they should only be implemented at regular intervals and over long durations. In addition, such revisions should not be retroactively applied, or only exceptionally, to installations already in operation or under construction.*

> *Models and data should be used to support energy planning. More importantly, those used in support of regulatory and public policy decisions should be publicly accessible not only by entities reaching such decisions but also by research and public/private educational institutions, and generally to the public-at-large to encourage open and constructive debates based on quantifiable and shared information.*

Factual Communications Toward the Public-at-Large

To sustain active participation of the public-at-large in energy mutations, it is important to avoid potentially misleading information. For example:

- Some information campaigns suggest that a consumer may purchase "green energy" produced from hydro, PV, wind or a mixture, depending on the situation. Except for rare circumstances, such tariffs do not reflect physical reality. Indeed, electrons coming from all power plants of any sort "co-inhabit" once combined within transmission and/or distribution infrastructures. Thereafter, it is no longer possible to "separate" them as per their production origin. All occupants (i.e., all consumers) in an apartment building consume the same electricity "assemblage"; this is also the case, for example, for tap water, natural gas and heat in the case of district heating.

  It is, therefore, potentially misleading to sell "green energy" to any one consumer. It would be more accurate, hence preferable, to inform a consumer that she or he may elect to encourage the deployment of one or more renewable electricity production technologies by way of a specific contribution added to each bill. The support of renewable energies would be the same without leaving an incorrect impression with the consumer.

- For illustration purposes, one often compares the production of a PV and/or wind facility with the consumption of a number of households — "the facility produces xxx GWh, which is sufficient to supply the needs to yyy thousand households". The impression could be that the particular production facility is sufficient, by itself, to fulfill the requirements of these households. As illustrated by the following insert, such comparisons overlook the infrastructure requirements, especially in terms of supply variabilities.

Case study: 200 zero energy houses in Geneva, Switzerland

The average yearly electricity consumption per household of 3 to 4 persons is 4.2 MWh/yr in Switzerland.

- Two-third during winter, between October 15 and April 15, i.e., 2.8 MWh.
- One-third during summer, between April 15 and October 15, i.e., 1.4 MWh.

For 200 households, the consumption thus is: $200 \cdot 4.2$ MWh = 840 MWh/yr, i.e.:

- 560 MWh during winter.
- 280 MWh during summer.

Using the NREL simulation tool [*pvwatts.nrel.gov*] already used in the "Residential installation in Stockholm and Nairobi" insert in the "Solar Energy" chapter, for a 4 kW DC installation with PV panels of 19% efficiency (i.e., 21 m^2), tilted at 46.2° (i.e., Geneva's latitude) and standard losses, for Geneva, one finds:

- Solar energy received per year: 1.3 MWh/m^2·yr.
- Yearly electricity production: 4.16 MWh/yr.

To produce the 840 MWh consumed by the 200 households, one would need: $(840/4.16) \cdot 21$ m^2 = 4 240 m^2 of panels.

(Continued)

(Continued)

In Switzerland, 75% of the solar energy is received during the summer and the remaining 25% during winter:

- 210 MWh during winter.
- 630 MWh during summer.

Assuming that the 200 households have collectively installed 4 240 m² of PV panels (thus making them "zero energy" over an entire year), they would need to:

- Be supplied with 560 – 210 = 350 MWh from the local network during winter.
- Find one or more buyer(s) for 350 MWh during summer.

It is thus important not to confuse "zero energy" with autarky and to not cut the umbilical cord with the local utility! This is even more important considering that unless they have installed energy storage capabilities, the residential customers will need access to the local network during the night.

Renewable Electricity Subsidies

Direct and indirect subsidies have existed under many auspices for a long time. Among a few examples, one can mention:

- Support of the coal mining industry all over the world, often to reduce unemployment and mitigate the desertification of mining regions.
- Transportation infrastructure construction and maintenance support (i.e., for canals, roads, rail beds, stations, harbors and airports), financed by public authorities to transport coal, oil and natural gas as well as raw materials and manufactured goods.
- Civilian nuclear industry public support to enhance local energy independence.

Renewable energy incentives typically address technologies designated as "new" or "alternative" — hydro energy is often not included. PV, wind and geothermal heat are generally eligible for subsidies; bioenergies are less often eligible. The subsidies are generally formulated in one of two ways:

- *Investment subsidies*: They are generally indexed to the installed capacities of new production facilities and depend on the particular technology. The subsidy levels for new installations are generally adjusted yearly to reflect cost evolutions.
- *Guaranteed energy purchase tariffs, feed-in tariff (FIT)*: They are based on the purchase of the electricity produced and are applicable over a period corresponding to the typical lifetime of the particular technology. This is the most common subsidy form worldwide; further information can be found under *"Feed-In Tariffs"* in the Companion Document.

In general:

- The decision to subsidize a particular renewable energy production facility is left to a public authority based on public policies. The two contracting entities, i.e., the local producer and local energy purchaser, generally have no say in the subsidy decision.
- The tariff and the duration of the contracts are also set by public policy considerations. Again, the two contracting parties generally do not have a say.
- However, in the vast majority of situations, the subsidies paid to the local producers are not financed by a public authority but rather by way of a surcharge included in the electricity invoices paid by residential customers and other customers, in some instances.

Renewable energy FITs, in fact, increase the cost of electricity, particularly for residential customers. While their objective is to encourage renewable energy productions, i.e., electric ones, FITs actually increase the tariffs and thus the price of electricity, thus making it paradoxically less attractive!

They often include contractual frameworks that are overly favorable to renewable energy producers leading to a more expensive access to electricity, especially for less favored populations. Furthermore, it is unfair to have only electricity consumers bear the cost of renewable energy subsidies.

Provisions to encourage energy storage are generally missing in the subsidy mechanisms even though it is an essential element to ensure the proper integration of future renewable and intermittent sources.

Overall, for the reasons outlined above, FITs are not a sustainable solution for the development of renewable energies, i.e., toward low-carbon energy systems.

To ensure that renewable energy public subsidies actually fulfill their low-carbon technologies encouragements, it is essential to:

- *Only provide a one-time investment subsidy for the construction of any renewable energy facility based on its installed capacity and adjusted to the particular technology.*
- *Provide the investment subsidy only if the installation also incorporates an energy storage capacity. The energy storage facility should then also be eligible for an investment subsidy.*
- *Also consider subsidies for stand-alone energy storage installations.*
- *Decrease the subsidies as the investment and operational costs, computed over a reasonable time period, become closer to those of already mature technologies, and remove them when parity is reached.*
- *Connect intermittent renewable electricity production plants at the local distribution network level whenever possible.*
- *Only subsidize renewable electricity production plants directly connected at the transmission level if they integrate sufficient on-site energy storage to enable them to inject electricity into the system with a minimum capacity factor,[9] decided upon by a regulator, which, in turn, would allow them to contribute to the proper overall system operation.*

Micro-grid Subsidies

Subsidies should be allocated for micro-grids in regions without connection possibilities to a local distribution network. This is particularly important in support of the less favored populations or isolated ones. In such situations, the subsidies must not only support the construction of micro-grids but also their evolutions as the energy demand increases along with the standard of living. Issues related to operations and maintenance are also crucial in such situations, including education and training.

---

[9] Capacity factor; ref. "Introductory Remarks" chapter under "How Electricity Is Produced".

Carbon Pricing — Universal and Redistributed Carbon Fee

The fast decarbonization overall objective can only be reached if:

- It is pursued worldwide. Indeed, all inhabitants on Earth share the same atmosphere.
- The use of energies having higher greenhouse gas emissions is made more expensive than those with lower emissions.

As further discussed under *"Carbon Tax and Carbon Cap-and-Trade"* in the Companion Document, past experience with both approaches have largely failed. Indeed, they are either perceived as yet another opaque government-imposed tax, ill-received by the population-at-large, and/or they distort open markets since they are not universally applied world-wide. In addition, the issue is not to lower energy consumptions per se but to lower carbon emissions of energy consumption and production mixes.

In that context, we believe that a carbon fee that is universally collected and integrally redistributed towards households, i.e., to the public-at-large, is a mechanism that warrants serious consideration.

The carbon fee would be collected by a national or regional entity, which, in turn, would be charged with its full redistribution to households based on the number of members in each one.[10] The proposed carbon fee would not be a tax since it would be integrally redistributed to the public-at-large. Hence the designation Universal and Redistributed Carbon Fee (URCF).

In such a system, the carbon fee would be paid based on the amount of CO_2 emissions due to the combustion of a fossil fuel. It would be paid by any entity introducing the fossil fuel into the energy system of the consuming country.[11]

---

[10] The carbon fee redistribution could be assigned, for example, to a local income tax collection agency that already has access to accurate and regularly updated information related to each household. In addition, such an assignment would soften its image with the public-at-large!

[11] For the sake of simplicity, in the following discussion, the country designation includes countries in their political definition or a group of countries, such as the European Union.

It would thus be paid as described here[12]:

- For coal, by:
  - Importers, via ship or train, selling coal to domestic users,
  - Local mining companies, when selling coal from domestic mines to resellers or domestic users.
- For oil, which needs to be refined before it is used as a combustion fuel (i.e., gasoline, diesel, kerosene or heavy oil), by:
  - Importers, via ship, train or pipeline, selling oil to domestic refiners,
  - Refined product importers, via ship, train or pipeline, selling the refined product to domestic users,
  - Domestic petroleum companies, when selling raw oil extracted from domestic wells to domestic refiners.
- For natural gas, which requires some processing or conditioning before it is used as a combustion fuel, by:
  - Importers of processed natural gas, via liquefied natural gas (LNG) ship or pipeline, selling the processed product to domestic users,
  - Domestic natural gas companies when selling processed gas extracted from domestic wells to local gas distributors.

Under the proposed system, the carbon fee would inherently be included in electricity produced using fossil fuels as well as in fossil fuels used for transport or directly for heating purposes; it would inherently also be included in any manufactured product calling upon industrial processes consuming fossil fuels. The production cost of electricity would increase, more so as the local electricity production mix emissions increases, i.e., more heavily relying on fossil fuels; however, the overall carbon fee collected would also increase such that the redistribution would increase as well.

Using Switzerland, which imports all of the fossil fuels it consumes, as an example, a case study included in the chapter's conclusion illustrates the orders of magnitude that could be involved.

---

[12] Fossil fuel exports are addressed in the following dedicated paragraph.

Fossil Fuel Exports

Any domestic coal mining, oil or gas company, would pay the URCF in conjunction with its internal energy consumption, for example, related to resource extraction, oil refining or gas processing. In addition, their sales to domestic customers would also be subject to the carbon fee. However, for fossil fuel exports, the related carbon fee would be paid in the importing country based on the amounts imported.

A Universal Carbon Fee

It would be naive to believe that all countries worldwide would readily apply the carbon fee described above, yet its universal implementation is essential to avoid market distortions, especially if the fee is to be large enough to encourage the industry and end-users to seek economically feasible alternatives to fossil fuels.

Imports into a country applying the URCF from a country not applying it

To avoid market distortions, countries applying the URCF would apply a carbon fee when importing raw or processed fossil fuels as well as manufactured goods — appliances, electronics, industrial equipment, textiles and vehicles — from countries not applying the URCF. As further discussed, the fee for manufactured goods would be determined based on the carbon content of the product or class of products.

In this situation, the URCF would be paid in the importing country and thus be redistributed to its population as opposed to the exporting country's population.

Imports into a country applying the URCF from a country also applying it

In this case, the URCF would have been paid and redistributed in the exporting country and redistributed toward its population. The cost of manufactured goods in countries applying the URCF would inherently include the carbon fee. The importing country would, therefore, not be allowed to apply any additional carbon fee.

Costs of manufactured products

The cost of manufactured goods in countries applying the URCF would inherently include the carbon fee, thus making the said products more expensive both on the domestic market and for export compared to the situation before the URCF was implemented. At the domestic market level, the end-user cost increase would be compensated, on average, by the redistribution; this is further illustrated in the case study below. As to exports, the advantage for manufacturers in the countries applying the URCF is that they would have control of the cost of their products while avoiding carbon-related import fees set by importing countries over which an exporting manufacturer has little control or influence.

A serious challenge to the proposed URCF would be the determination of the carbon fee when manufactured goods from countries not applying the URCF are imported into countries that apply it. These fees would have to be determined by broad groups of products or materials such as steel, textiles, appliances, cars or computers, starting with the classes of products with the largest export-import market shares. These fees would have to be universally agreed upon when the URCF is decided, as further discussed.

Special Exemptions for Emerging Countries

It would be important and prudent to make special exemptions for emerging countries by considering them as applying the URCF even though they would initially be exempted. This would avoid making domestically manufactured goods more expensive on their domestic markets or penalizing their goods when exported. However, such countries would be required to apply and subsequently redistribute, to their populations, the carbon fees applied to goods imported from countries not applying the URCF; otherwise, market distortions would arise between countries applying and not applying the URCF when exporting to countries exempted from the URCF system.

Launch of the URCF System

To ensure that the URCF has a proper impact at its launch, it will be important that a sufficient number of countries adhere to it initially. A reasonable launch criterion could be that 55 countries whose emissions represent 55% of the worldwide emissions agree to adhere to it, which is similar to the criterion used in conjunction with the Kyoto and Paris Agreements.

Once the URCF is launched, countries that have decided *not* to adhere to the URCF system would be encouraged to do so for several reasons, including:

- Their exports toward countries applying the URCF, by then at least 55 countries certainly representing well above 50% of the world-wide gross domestic product (GDP), would be subject to the carbon fee described above.
- The populations of countries not adhering to the URCF system would pay higher costs for goods imported from URCF adhering countries while not benefitting from the redistribution; as a result, they would encourage their governments to adopt the URCF.

Furthermore, given the hopefully growing number of countries adhering to the URCF system, it is unlikely that a group of non-adhering countries could organize themselves to create a coalition living in a quasi-autarky with respect to URCF adhering countries.

Setting the URCF

It would be crucial that the URCF applied to the various forms of fossil fuels be set worldwide and for set periods. The URCFs could be set to coincide with the progress reports due by the parties that have signed the 2015 Paris Agreement, i.e., five years. At the beginning of each five-year-period the following decisions would be reached worldwide:

- The URCF for the coming five years, i.e., its yearly increase during the said five years.

- An upper and lower bound for the URCF for the subsequent five-year period.
- Updates of the carbon fees to be applied for broad categories of manufactured goods by importing countries applying the URCF from countries not doing so.

The provisions mentioned above are essential to encourage all stakeholders, in particular industry, to engage in longer-term planning leading to lower carbon processes while remaining competitive on the worldwide market.

URCF — Conclusion

The URCF proposal outlined above is close to what has been proposed by the Climate Leadership [*clcouncil.org*] (whose objectives we do not entirely share), which was endorsed by the Carbon Tax Center [*carbontax. org*]. In our opinion, our proposal has several advantages over other carbon taxes or cap and trade systems:

- It would be entirely transparent to the public-at-large, which would also be its main beneficiary.
- The redistributions would be toward households and depend on the number of persons living in each one and not on the energy consumed; indeed, the emissions reduction challenge is on a per capita basis. Households with high-energy consumptions and/or significant purchases of manufactured goods would pay more than those with lesser needs or means, while families with lower consumptions would proportionally benefit more from the URCF.
- The allocation of the amounts paid under the URCF auspices would not be subject to political decisions since the amounts collected would be entirely redistributed.
- Being universally applied, it would avoid worldwide market distortions. Provisions could be implemented to accommodate the special needs of emerging countries. Countries having initially decided not to adhere to the URCF system would be encouraged to join to have better control over the prices of their exported products at the entry points of importing countries.

Case study — the URCF applied to Switzerland

Switzerland imports all of its fossil fuels, thus simplifying the computations.[13] The 2019 fossil fuel imports were [*bfe.admin.ch*]:

- Coal: 139 kton; at 9 kWh/kg, this corresponds to 1 251 GWh.
  At 360 gCO$_2$/kWh, the emissions from coal were 450.4 ktCO$_2$.
- Refined oil products: 10 904 kton; at 12 kWh/kg, this corresponds to 130 848 GWh.
  At 250 g/kWh, the emissions from refined oil products were 32 712 ktCO$_2$.
- Natural gas: 34 060 GWh.
  At 200 gCO$_2$/kWh, the emissions from natural gas were 6 812 ktCO$_2$.

Assuming that all fossil fuels were used for combustion purposes during the year, the total related emissions in 2019 were 39 974 ktCO$_2$. The actual emissions due to combustions were slightly lower since some fossil fuels were imported for non-energy use, such as by the chemical industry, and some petroleum products were re-exported.

The 2019 cost of imported fossil fuels was 6.85 billion CHF or 6.37 billion €.

Using a URCF value of 100 €/tonCO$_2$, i.e., 0.10 €/kgCO$_2$, the total URCF collected would be 4.0 billion €, i.e., a 62.8% increase over the 2019 import costs.

Switzerland's population was 8.6 million in 2019.

As a result, the per capita redistribution would be 465 €/cap.

Increased energy end-user prices

- Since the combustion of one liter of gasoline causes emissions of 2.25 kgCO$_2$ (9 kWh/l · 250 gCO$_2$/kWh), the corresponding URCF per liter would be 0.10 €/kgCO$_2$ · 2.25 kgCO$_2$/l = 0.22 €/l. The average price was 1.49 €/l in 2019.
- Since the combustion of natural gas causes emissions of 200 gCO$_2$/kWh, the corresponding URCF would be 0.10 €/kgCO$_2$ · 0.2 kgCO$_2$/kWh = 0.02 €/kWh, or 20 €/MWh. The average price was 95.8 €/MWh in 2019.
- Switzerland's 2019 electricity mix was very low carbon — over 95% hydro, other renewables and nuclear. As a result, the URCF impact on the electricity production cost would be quite low.

(Continued)

---

[13] Reference "Units and Reference Values" annex for the energy content and emissions values.

(Continued)

Household energy consumption cost impacts

For a household of three persons living in an apartment, consuming 10 MWh/yr for heating and 5 MWh/yr for hot water, both using natural gas, and using a car consuming 6 l/100 km for 15 000 km/yr (i.e., consuming 900 l/yr of gasoline), the additional yearly costs due to the application of the URCF would be:

$$(10 + 5) \text{ MWh/yr} \cdot 20 \text{ €/MWh} + 900 \text{ l/yr} \cdot 0.22 \text{ €/l} = 498 \text{ €/yr}.$$

The yearly amount of the URCF redistribution would be: $3 \cdot 465$ €/yr = 1 395 €/yr.

For this family, the yearly URCF redistribution would significantly exceed the additional energy consumption costs.

The situation would be different for a household of only two persons (receiving a $2 \cdot 465$ €/yr redistribution) living in an individual house — thus facing higher heating and hot water costs — and driving two higher-consumption cars for longer distances each year. This confirms the inclusive nature of the proposed URCF.

Both families would, in fact, face higher costs when purchasing other household equipment and calling upon outside services as the URCF is included in the related costs. The assessment of such costs is beyond the scope of this book.

The electricity production cost would also increase in countries with higher CO_2 contents in their production mix.

- Populations in non-adhering countries would encourage their governments to join so that they can benefit from the carbon fee redistributions.
- Setting the URCF over predefined periods with set variation boundaries would allow all stakeholders, in any particular industry, to make long-term plans toward the deployment of lesser emitting processes, both at the energy usage and production levels.

We are fully aware that the adherence to and deployment of the proposed carbon fee system will require a number of complex negotiations. However, should these negotiations find positive outcomes, as it is our hope, the aspirations raised at the conclusion of the 2015 Paris Agreement could be fulfilled!

Energia's Energy Mutation

Principles and Applications

Elaboration of an Energy Mix — A Systemic Approach

While promoting rational energy end-uses, the decarbonization urgency requires that they be decarbonated while relying on low-carbon energy vectors and/or vectors that can be further decarbonated, such as electricity and heat. Deployment options of *direct* solar energy and biomass uses are often limited. The electrification of end-uses combined with low-carbon electricity is thus to be favored along with low-carbon heat whenever possible.

The emission outcomes from energy end-use options are not only coupled between themselves but also with energy production and storage options. For example, public policies seeking to decrease fossil fuels uses for heating by promoting heat pumps or for cars by promoting electric vehicles, will lead to refining activity reductions, which, in turn, will lead to energy consumption reductions in the transformation industry. At the same time, electricity consumption will increase in the earlier examples; as a result, the decarbonization goal will only be reached if the electricity mix is also decarbonated.

Only by way of a systemic approach toward energy mutations (not only electricity mutations) — end-use, production and storage — can a complete analysis be carried out of the impacts that potential public policy and/or regulatory proposals, such as described in the previous chapter, can have on decarbonization objectives.

The chapter is structured into four parts laying out what we believe are options leading to enhanced decarbonization:

- The first one contains an analysis of the four energy end-use sectors. For each one, following our thoughts, concrete implementation proposals are formulated for Energia in 2035. By summing up the results for all four sectors, the final requirements for each vector can be determined[1]: direct consumptions of fossil fuels, coal, oil and gas, along with electricity and heat.
- Once the electricity consumption is known, the electricity production mix can be determined. The second part lays out the main steps of this process.
- Energia's 2035 electricity production mix is presented in the third part by implementing this process, while taking the requirements determined under the first part and the "triple one" emissions objective into account.
- Energia's overall CO_2 mutation is presented in the fourth part, including the heat production distributed by way of the corresponding networks and the consumptions of the transformation industries, such as refineries.

As Energia's population increases from 50 million in 2015 to 52.5 million in 2035, Energia's energy mutation results in:

The overall emissions reduction from 500 MtCO$_2$ in 2015 to 170 MtCO$_2$ in 2035 is made possible — along with a number of final energy consumption measures — by the reduction of electricity and heat production emissions from 426 gCO$_2$/kWh to 111 gCO$_2$/kWh.

The 330 MtCO$_2$ emissions reductions can be further subdivided as follows:

- 170 MtCO$_2$: From 251 MtCO$_2$ to 81 MtCO$_2$, thanks to the reduction of direct consumptions of fossil fuels.

---

[1] The customer, i.e., end-user, consumes and pays for the actual use of energy vectors: coal, oil (in the form of gasoline, diesel, heating oil, etc.), natural gas, biomass (wood), electricity, heat, etc.

- 126 MtCO$_2$: From 198 MtCO$_2$ to 72 MtCO$_2$, thanks to the decarbonization of the electricity and heat productions.
- 34 MtCO$_2$: From 51 MtCO$_2$ to 17 MtCO$_2$, thanks to activity reductions in the transformation industries, such as oil refining, as a result of the two reductions above.

All numerical results related to Energia's energy mutation can be found under *"Energia Worksheet"* in the Companion Document.

Consumption

Decarbonating the various energy end-use sectors requires that clear indicators, fully understandable by and useful to the public-at-large, be available and that equipment performances be assessed against these indicators by independent and properly trained individuals or entities — gCO$_2$/m^2 and kWh/m^2 for buildings or gCO$_2$/km and l/100 km or kWh/100 km for vehicles.

Housing

A significant proportion of existing buildings in a number of countries will still be used in 2035 and beyond. It is thus important not only to construct low-carbon buildings but to also reduce the emissions from existing buildings, be they individual houses or apartment buildings.

Existing building emissions reductions at an acceptable cost

Aside from a building's outer shell, most energy-consuming equipment, such as furnaces, hot water heaters, cookstoves and refrigerators, are replaced at least once in 20 years. Such replacements are low-cost opportunities to install better performing equipment that cause less emissions, such as replacing an oil furnace with a gas one or a heat pump or replacing a B class refrigerator with an A+ one.[2] Depending on the situation, changing

---

[2] In lesser-developed countries, the option might be to replace open cooking with enclosed stoves consuming less biomass, thus making them less polluting and costly to operate and also safer.

the heating system or, as a first step, improving its controls may lead to significant emissions reductions at lower costs than improving the building's insulation. This being said, improving the attic insulation for an individual house or updating doors and windows for a house or an apartment building are always options to be considered. The installation of solar hot water heaters, in support of an existing fossil fuel system, is often a convenient and low-cost emission reduction option for individual houses.

Installing new or increasing the capacity of rooftop photovoltaic (PV) systems should systematically be considered when renovating existing buildings. Available subsidies are an integral part of any related decision.

The options described above for individual houses are also those for apartment buildings. However, depending on whether the building has one single or multiple owners, renovation decisions may be more or less complex. In some cases, for strictly financial reasons, such as covering the cost of an in-depth energy renovation, a decision may be reached to add one or two floors to the building, should its structure and local regulations allow.

Low-carbon construction

When constructing individual houses and apartment buildings, taking the local climate into account is important for their orientation, their wall and roof characteristics as well as their door and window locations and sizes. Natural shades from trees and/or close-by buildings may negatively affect the production of rooftop solar panels but also provide natural cooling depending on the situation. In addition:

- While remaining respectful of local customs, the orientation and slope of the roof should be selected to facilitate the installation of PV and/ or thermal panels; the equilibrium between the two types depends on local climate, economic and regulatory circumstances.
- Using the ground below the house and/or its yard to implement a geothermal heating and/or cooling system is a low emissions solution. Should this not be sufficient and/or possible, air-water or air-air heat pumps generally are low-emission solutions.[3] Connecting to a local

---

[3] If the heat pump has an average COP of 3 over a year's operation, its emissions are lower than a 90% efficient natural gas furnace, provided the local electricity's carbon content is

district heating or cooling network is generally also a low-carbon solution[4]; it is mandatory in an increasing number of territories.

• If the thermal insulation level is high, the hot water system's required installed capacity may be higher than that required for the heating system. This is also the case in warm weather regions where the focus needs to be on the hot water system.

Proper adaptation to the local site is essential for apartment buildings and individual houses. In warm climates, apartments entirely traversing the building facilitate natural cooling. Proper spacing between close-by buildings in warm climate neighborhoods may provide natural air draft cooling. In addition:

• Depending on the building's height and the number of apartments it has, the roof's area may be sufficient to install a collective solar energy system, thus contributing to the hot water production. Individual hot water systems and/or solar panels may also be installed.
• Collective heating and cooling systems can be more elaborate than for individual houses with higher efficiencies and lower emissions consequences. However, energy metering must then be installed for each apartment to curtail wasteful usages.
• Geothermal systems are often attractive heating and cooling low emissions and cost solutions.

Energy management systems and synergies

Apartment buildings and individual houses will increasingly not only become energy-consuming but also energy-producing — PV, coupled

---

below 660 gCO_2/kWh during heating periods, which is the case in most countries.

[4] The emissions of district heating systems depend on the heat's origin: a few tens of gCO_2/kWh, if provided from waste and/or biomass incineration, or from industrial process waste heat or by way of nuclear power plant cooling systems. On the contrary, it can be much higher — several hundred gCO_2/kWh — if natural gas is burnt or higher, if oil or coal is used. District heating operators are increasingly seeking to reduce the emissions of their networks.

with energy storage — batteries and hot water. Energy management systems can be installed to monitor and optimize the energy consumption, production and storage of an entire building based on a particular criterion, such as consumption and/or emissions and/or peak demand reduction. Depending on the integration level — apartment or individual house, building or an entire neighborhood — the energy management can be carried out by the dwelling's occupants or by an energy services operator[5]; the role of each one must be clearly defined — remote control and monitoring of selected equipment/appliances or only comparison of energy consumptions between similar dwellings, for example.

The next insert illustrates potential synergies between buildings.

Synergies between two buildings

The heating energy consumption patterns of two buildings in close proximity are shown in Fig. 1. The peak demand for each one is 10 kW.

- Building 1: Offices with a high consumption starting at 06:00 before the employees arrive. The heating temperature is then lowered as the sun shines and then further lowered between 18:00 and 06:00.
- Building 2: Small apartment building. The consumption is low during the day between 06:00 and 18:00 when the occupants are at work. The consumption is high between 18:00 and 24:00 and then lowered between 24:00 and 06:00.

As shown in the figure, while two individual 10 kW connections are required for the two buildings separately, only one common 13 kW connection would be sufficient to supply both buildings together.

Zero-energy buildings or neighborhoods — not always a great idea

Zero energy or even positive energy buildings/neighborhoods produce as much energy as they consume (or even more) over an entire year. In the

---

[5]Which could be the building's management, or the local district heating operator, or a municipal entity or the local electric utility.

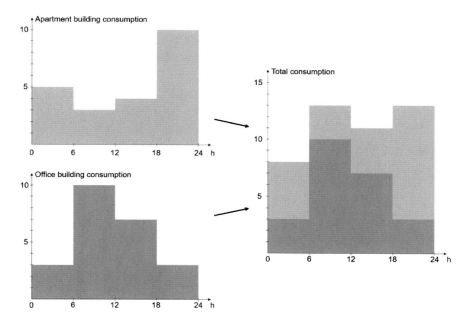

Figure 1. Synergy between two buildings / © *HBP and YB*

north of the northern hemisphere regions, such buildings/neighborhoods featuring PV panels generally are:

- Net consumers during winters when more heating is required and when the PV production is lower.
- Net producers during summers when less heating is required and when the PV production is higher.

Provided the solar systems are suitably dimensioned, a building/ neighborhood could be zero energy or even be a net producer over an entire year. However, unless sufficient energy storage capacities are installed, a self-sufficient operation with respect to the local utility is not possible. A connection to the local distribution system remains necessary to provide some electricity during winters and to absorb excess production during summers. In addition, to overcome situations without any local production, the connection's installed capacity with the local utility can generally not be reduced.[6] Finally, the notion of zero energy does not, itself, provide any assurance as to the building/neighborhood's emissions; indeed, a house equipped with some solar panels but also using coal for heating could be zero or even positive energy!

Thus, the true challenge is the construction of low or even zero-carbon houses or neighborhoods. Recent pilot projects, primarily in Europe, show that this is possible without significant additional costs. There is no need to restrain architectural creativity.

Contributing building — energy storage and resiliency

A building can be referred to as contributory if it can contribute to the overall electric system operation by deferring some of its consumption, such as for washer-dryers, dishwashers or hot water heaters, to later periods, or by way of some local energy storage capability, typically batteries. Such buildings facilitate the integration of intermittent renewable energy sources, primarily PV, either into the building's production capabilities or those of neighboring ones. Suitable business models, properly adapted to local circumstances, can facilitate such community operations.

In view of the evergrowing importance of electricity in daily activities, the integration of some electricity production and/or energy storage capabilities in individual houses and apartment buildings should be considered to mitigate the impacts of potential local or large scale network operational incidents. Their characteristics would have to be adapted to the local context (seismic zone, for example) and the desired level of protection for critical systems — energy capacity and outage duration.

Energia's 2015–2035 Energy Mutation in the Housing Sector

Assuming that the housing density remains at 2.5 inhabitants per dwelling, the number of dwellings will increase from 20 to 21 million between 2015

---

[6] Renouncing a local network connection implies renouncing the associated cost sharing and enhanced supply reliability afforded by the electric system.

and 2035. Four million dwellings are slated to be replaced during the same time period, i.e., a yearly average of 200 000 or 1%. Though 60% of the population lived in individual houses until 2015 and 40% in apartments, the proportion is expected to reverse for new constructions after 2015.

As a result, the 2035 housing stock will consist of:

- 16 million dwellings built before 2015, which should be renovated.
- 5 million new dwellings — 4 million to replace old ones and 1 million to accommodate the population increase.

Along with rational energy end-use, the primary objective of Energia's successive governments remains low-emissions housing at affordable costs. This implies, as priority and whenever feasible, the replacement of fossil combustion systems with alternative low-emission systems as well as improved thermal insulation, at reasonable costs, and a sustained focus on high-efficiency technologies.

Remote control of a variety of electricity and/or heat consumptions that can be shifted time-wise, while not mandated, will be encouraged by suitable energy rates. Solar system installations, thermal and/or PV, will be also encouraged by way of investment subsidies. The types and sizes of the PV production is further detailed in the part dealing with the electricity mix.

The energy required to deconstruct, construct and/or renovate the buildings is not included.

Renovation of the 16 million dwellings constructed before 2015 and assumed to be still existing in 2035

Seven specific measures are intended for the 2015–2035 period — one for thermal insulation, four for heating, one for ventilation and air-conditioning, and one for hot water production.

1. Insulation improvement to reduce the heating requirements by 30% for the following dwellings:
   - All dwellings that were heated using coal or oil.
   - One-third of the dwellings that were heated using natural gas.
   - One-third of the dwellings that were heated using electricity.

2. Replacement of the furnaces for the dwellings mentioned above which were heated using coal, oil or natural gas, i.e., the first two categories mentioned above, with electric heat pumps, preferably ground-water, if not air-water, with an average COP of 3 over a year's operation.

3. Replacement of the electric heating elements in the third category of dwellings mentioned under 1 above with new heating elements to reduce the corresponding consumption by an additional 10%.

4. Replacement of the gas furnaces in dwellings where the insulation was not improved (i.e., for two-third of the dwellings that were heated using natural gas) by condensing furnaces to reduce the corresponding consumption by 10%.

5. Replacement of the electric heating elements in dwellings where the insulation was not improved (i.e., for two-third of the dwellings that were heated using electricity) with new elements to reduce the corresponding consumption by 10%.

6. In the South, increased air-conditioning consumption will be fulfilled using reversible or cooling-specific heat pumps.

7. Replacement of all gas hot water heaters, roughly equal in proportion, by:

   • Solar thermal installations,
   • Electrical heaters,
   • Heat pumps with a COP of 3 over a year's operation.

Construction of five million new dwellings between 2015 and 2035

The construction of the five million new dwellings will consist of two million individual houses, i.e., 40%, and three million apartments, i.e., 60%. As of 2015, improved construction materials and techniques already enabled the reduction of heating requirements to an average of 8 MWh/yr for individual houses rather than 14.5 for houses constructed before 2015; similar reductions will be possible for apartment buildings: 4 MWh/yr rather than 7.25 for those constructed before 2015.

The three main measures related to new constructions are:

1. District heating will be developed for urban regions to which one million new apartments are to be connected. 80% are to be supplied by way of waste heat recuperation, on one hand, and by a combination of

waste incineration and natural gas combustion, on the other. The remaining 20% is to be supplied as cogeneration from new nuclear power plants as per the 2035 electricity mix.
2. The other two million new apartments and two million new individual houses are to be heated using heat pumps with an average COP of 3 over a year's operation.
3. Half of the hot water will be supplied using thermal solar systems. The other half will be supplied by heat pumps with an average COP of 3 over a year's operation.

Cooking, lighting and other consumptions

It is expected that the population increase along with the increased overall use of electricity will balance out with the improved average performance of the devices and systems used, such that the overall related consumption will remain the same in 2035 as it was in 2015. However, the related emissions will be reduced as the electricity production emissions decrease.

Energia's overall 2015–2035 housing energy mutation

TWh	Coal	Oil	Natural Gas	Solar Thermal	Biomass and Waste	Electricity	District Heating	Total	*MtCO$_2$*
2015	10	70	135.0	15.0	5	150.0	15	**400.0**	*118.0*
2035	0	0	62.9	21.7	4	174.7	16	**279.3**	*33.8*

Transport and Travel

Contrary to the housing sector, where replacements may occur only every 100 years (or more) in some countries, car replacements typically happen every 10 years. For aircrafts, it may be 25 years, and 30 to 40 years for trains and ships.[7] As a result, fleet replacements or upgrades can have significant

---

[7] While fleet rotations may be quicker in industrialized countries, actual lifetimes may be extended in emerging countries, thus negating worldwide decarbonization efforts. It is thus desirable that road vehicles and aircraft be recycled, whenever possible, after their first useful lifetime, which, in turn, raises technical and economic challenges.

emissions impacts by 2035, i.e., more readily so than in the housing sector. Renovations, which are important for housing, are quite different; they primarily involve maintenance, modernization, and occasionally, the extension of infrastructures such as airports, train stations or harbors.

Emission reduction efforts must primarily focus on cars since, combined, they have the highest fossil fuel consumption and thus cause the highest emissions. The consumption and emissions of other road vehicles like buses, trucks and delivery vehicles, are also important. Having discussed road transport first, rail, maritime and air transport are examined next but in lesser detail.

"Soft" transportation means, such as bicycles and scooters — which are increasingly electric — and gasoline mopeds, are not considered here since their overall consumptions are quite low.[8]

Road transport — cars

As of 2020, cars around the world overwhelmingly used internal combustion engines — gasoline or diesel. Two technologies are presently available to enable significant CO_2 emission reductions in most countries, along with significant pollution and noise reductions; ref. "Transport and Travel" chapter:

- Full electric vehicles (FEVs), which are still primarily used for daily urban and/or restricted area applications. While their ranges are increasing, routinely reaching 200 km as of 2019, their charging times, even under fast recharge, remain above 20 minutes.
- Rechargeable hybrid vehicles, for which several configurations exist. Range and charging time issues are largely circumvented by these vehicles.

A successful deployment of these two technologies requires that holistic public policies be in place to ensure that sufficient and reliable electricity supply is available, both in terms of energy and power, along with suitable recharging infrastructures. Such policies have societal, economic, industrial and regulatory implications which are beyond the scope of this book.

---

[8]This does not imply that related public policies should overlook them. The noise and pollution they cause, if using internal combustion engines, should be curtailed.

Energy considerations

Electrifying the entire car fleet in an industrialized country, including the existing FEVs and rechargeable hybrid vehicles along with the internal combustion engine cars, would require, depending on the particular country, an increase of the country's electricity consumption by 20 to 30% as further discussed in the next insert. Spread over some 20 years, such an increase would not represent a major challenge. Instead, it represents an opportunity to valorize low-carbon electricity such as produced using renewable energies.

Additional electricity consumption to electrify a car fleet

In industrialized countries, the transport end-user sector typically represents one-third of the overall final energy consumption. Furthermore, half of the transport sector's consumption, i.e., 1/6 or 16.7% of the total final consumption, is consumed by cars, while trucks, buses, trains, ships, aircraft, etc., consume the other half.

As seen in the "Transport and Travel" chapter's insert dealing with propulsion system comparisons, an FEV consumes slightly less than 30% of the energy consumed by a similar internal combustion engine car. As a result, electrifying the entire car fleet would require some $30\% \cdot 16.7\% = 5\%$ of the total final energy consumed by a particular country.

If, in turn, the country's electrification rate was 20%, for example, this would mean a 25% electricity consumption increase compared to a situation without electric vehicles. Spread over the 20 years after 2015, this represents an annual 1.1% increase.

Power considerations

While the additional energy consumption required for a complete electrification of an industrialized country's car fleet is well within other yearly increases, the distribution between day and night of the additional power required to charge the FEVs and rechargeable hybrid vehicles needs careful attention to avoid potentially excessive demand peaks. Vehicle charging concentrated between 18:00 and 20:00, for example, could induce network operation difficulties as illustrated by the next insert. Suitable communications between electric vehicles and the electric system must thus be further developed along with well-adapted business models and regulations.

Additional power requirements induced by FEVs and hybrid electric vehicle recharging

For the sake of simplicity, the situation described in the previous insert is revisited. When the entire car fleet is electrified, the final electricity consumption would have increased by 25%; all other consumptions remaining unchanged. As a result, the final electricity consumption of the car fleet would represent 20% of the overall electricity consumption. If the charging remains constant the entire day, it would thus represent P/5, with P being the average overall electricity demand (in GW).

In most industrialized countries, the ratio between the installed electricity production capacity and the average demand typically is 2.5, to reliably cover high demand periods. If the FEV and hybrid electric vehicles were all charged within two hours, rather than evenly over 24 hours, the required power would be: $(24/2) \cdot (P/5) = 2.5\ P$. This would mean that during the two hours, the country's entire installed power production capacity would have to be allocated to vehicle recharging, leaving none for the other users!

This extreme situation illustrates the need to spread the charging over the entire day while avoiding concentrated charging between 18:00 and 20:00 when people return from work.

Network and infrastructure considerations

Ensuring a broad development of FEVs and rechargeable hybrid vehicles implies that suitable charging stations be installed in residential houses and building garages as well as a broad deployment of public charging stations in cities, shopping centers, at work locations and along main thoroughfares. Their numbers must be sufficient to limit related access delays. The development of charging system standards is progressing to provide for inter-city and international travel; this will also reduce manufacturing and maintenance costs.

The quest for fast charging, which remains a major obstacle for longer distance trips, is also making progress. Fast charging has been normalized at around 40 kW for alternating currents (AC) and 50 kW for direct currents (DC). Recently, the Combined Charging System (CCS) norm provides for combined AC and DC charging stations with an overall capacity of 350 kW; over 8 000 were available across Europe early 2020.

CO_2 emissions reduction considerations

Higher efficiencies of electric propulsion systems, compared to internal combustion engines (ref. "Transport and Travel" chapter), and low-carbon content electricity mix both contribute to the emissions reductions brought by FEVs and rechargeable hybrid vehicles. However, the manufacturing of the required batteries consumes more energy than for traditional lead-acid batteries.[9] The initial differences depend not only on the type of battery but also on the emissions levels in the country where the battery was manufactured and where it is used.

The potential impact of autonomous vehicles

Along with the IT industry, car manufacturers worldwide are allocating significant resources, engineering-wise and financially, to develop autonomous vehicles. The level of autonomy increases steadily with the development of hands-off automatic parking and radar-enabled adaptive cruise control. As of 2020, a number of pilot and demonstration projects were underway in North America, Europe and Asia. While the strictly technical challenges are likely to be resolved well before 2035, the legal aspects (i.e., resolution of the responsibilities between "driver-operators", manufacturers, infrastructure owners and operators, in case of accidents) are likely to be far more complex.

One concern is that a broad penetration of autonomous vehicles could actually increase traffic and thus induce a need for further road infrastructures. Indeed, why use public transport to commute to work when the same trip could be done, door to door, while reading a book, emptying your email inbox, or even chatting with your friends and colleagues?

---

[9] Considerations as to the quantities, availabilities and qualities of the raw materials required to manufacture the batteries are beyond the scope of this book.

Road transport — passenger and freight

The three main avenues to rapidly decarbonize both passenger and freight road transport are:

- *Full-electric* — using an embarked battery. The vehicle itself is zero-emission; its noise pollution is also considerably reduced. This option is particularly attractive if the local electricity production mix is low-carbon. If any, only incremental upgrades of the local electricity distribution infrastructures are required. The disadvantages include reduced duty cycles due to battery recharging, which may prevent broad deployment for some applications, and the weight of the battery, which reduces the vehicle's useful load capabilities.
- *Hydrogen* — using an embarked fuel cell "burning" hydrogen stored onboard. As for the full electric option, the vehicle itself is zero-emission and considerably reduces noise pollution. The hydrogen option is particularly attractive if the hydrogen is produced using electrolysis supplied from a low-carbon electricity mix. The main advantage over the full-electric option is the short refueling times, of the same order as for internal combustion engine vehicles. The main challenge toward broad deployment remains a sufficient densely geographical coverage of the hydrogen-fueling infrastructure.
- *Natural gas*, if the battery recharging duty cycle reduction is problematic or electrolysis-produced hydrogen is not locally available.

Passenger road transport

Full-electric solutions are to be favored for urban bus applications as it is already the case in a growing number of cities and airports for which the range requirements are less constraining. The range of full-electric buses remains problematic for long distance routes. Hydrogen long distance bus propulsion systems should be favored over diesel ones; if not technically feasible, the natural gas option should be favored.

Full-electric buses with reduced battery capabilities are demonstrated in several cities; they rely on partial battery recharging along the route using a pantograph which connects to a supply catenary implemented at some or all stops. The need for a full catenary along the entire route is thus avoided.

Freight road transport

Whenever possible, full-electric solutions should be favored for urban delivery applications such as mail and online shopping deliveries as well as for intra airport and harbor transportation for which the full-electric solution is already the preferred option. Given the range requirements of typically well over 500 kilometers, full-electric long-distance solutions were not available as of 2020. Hydrogen alternatives are thus increasingly explored by way of a number of pilot projects in several countries.

Despite performance enhancements of modern trucks (i.e., high-efficiency diesel engines, lightweight materials, aerodynamics), long-distance freight road transport needs to be further reduced to reduce the associated fuel consumption and emissions. Full replacement by rail transport, while desirable, especially where electrified, often remains problematic in terms of "first and last mile" requirements.

Passenger rail transport

Passenger rail transport is encouraged and developing worldwide — metros and trams in cities and high-speed intercity travel. High traffic lines are increasingly being electrified; where difficult to cost-justify, hydrogen trains with embarked fuel cell electric locomotives are increasingly being implemented as alternatives to diesel locomotives.

The development of city rail transport hubs integrating high-speed and commuter trains with local metro and tramway infrastructures, while complex and requiring heavy investments, are required to further solidify multimodality and thus smoothening the transition from individual to public transport.

The expanding deployment of high-speed train infrastructures will gradually displace point-to-point air transport for distances up to 800 km. The development of high-speed train systems on all continents would further reduce overall transport emissions.

Telecommuting

As mentioned above, while autonomous vehicles could lead to a rebound in the use of personal cars, the hope is that information and communication technologies will provide for convenient and user-friendly telecommuting practices, thus reducing the need for some in-person meetings without hampering the conviviality of such gatherings.

Energia's 2015–2035 Energy Mutation in the Transport and Travel Sector

The primary objective of Energia's successive governments is to promote low-carbon passenger and freight transport while facilitating a mutation from fossil fuels toward electricity, biofuels, and where feasible, hydrogen, at an overall cost affordable to the country and thus its citizens.

Rail transport evolution

To reduce emissions, rail transport quality improvements remain a priority, so too is the promotion of its usage, not only within cities but also for passenger and freight intercity transport. In that spirit, the rail network will be extended and electrified where needed. Some abandoned tracks will also be put back into service. As a result, rail transport electricity consumption will increase by 20% between 2015 and 2035.

Car fleet evolution

The 2015 car fleet consisted of 24 million vehicles driven an average of 15 000 km/yr and consuming 8 l/100 km of a fuel that was composed of 95% gasoline and 5% ethanol, i.e., a 5% renewable and zero-emission content.

By 2035, an increasing number of urban families will have renounced owning a car; the overall car fleet will thus have decreased to reach 20 million in spite of the 2.5 million population increase. Sustained emissions reduction policies combined with local pollution and noise reduction regulations, including fees levied for internal combustion engine cars in several towns, also contribute to the decrease.

The car fleet will have the following composition by 2035:

- 6.5 million internal combustion engine cars that are driven an average of 15 000 km/yr, consuming 4 l/100 km,[10] of which 3.2 liters is gasoline and 0.8 liters, ethanol, i.e., a 20% ethanol content.
- 6.5 million FEVs that are driven an average of 17 000 km/yr and consuming an average of 16 kWh/100 km. Some of these cars are shared, hence the higher yearly driving distance.
- 7 million rechargeable hybrid vehicles that are driven an average of 12 000 km/yr:
  ○ Electric at 16 kWh/100 km for 9 000 km/yr.
  ○ Internal combustion engine consuming 4 l/100 km of the same fuel for 3 000 km/yr.

Taking the 111 gCO_2/kWh 2035 emissions of the electricity mix into account, the yearly emissions of Energia's entire car fleet will be reduced from 61.6 $MtCO_2$ in 2015 to 8.5 $MtCO_2$ in 2035!

Bus fleet evolution

In 2015, the fleet consisted of 100 000 buses covering an average of 40 000 km/yr and consuming 28 l/100 km of diesel fuel.

By 2035, the overall fleet size will remain the same, but each bus will be driven an average of 50 000 km/yr as a result of public transport promotion efforts. The share among the buses will be:

- Half will use diesel fuel with an average consumption of 24 l/100 km.
- The other half will be hydrogen consuming 8 $kgH_2/100$ km.

---

[10] As an exception, for the 2035 fuel consumption of internal combustion engine cars, we have elected to use the likely situation by then as opposed to present-day data. Some readers may consider this value optimistic; others may consider it pessimistic!

Heavy truck fleet evolution

In 2015, the fleet consisted of 400 000 heavy trucks covering an average of 65 000 km/yr and also consuming 28 l/100 km of diesel fuel.

By 2035, thanks to the increased use of both rail and river transport, the overall fleet size will have been reduced to 300 000 units, each covering an average of 75 000 km/yr and also consuming an average of 24 l/100 km of diesel fuel.

Delivery vehicles fleet evolution

In 2015, the fleet consisted of three million vehicles covering an average of 40 000 km/yr and consuming 14 l/100 km of the same fuel as cars, i.e., 95% gasoline and 5% ethanol.

In 2035, the fleet will still consist of three million vehicles consisting of:

- Two million full-electric vehicles, each covering an average of 36 000 km/yr and consuming 23 kWh/100 km.
- One million vehicles each covering an average of 40 000 km/yr and using natural gas with a consumption equivalent to 14 l/100 km and 9 kWh/l of gasoline. The related emissions are counted at 200 gCO_2/kWh.

Energia's overall 2015–2035 transport and travel energy mutation

TWh	Gasoline/Diesel	Natural Gas	Ethanol	Electricity	Total	*MtCO$_2$*
2015	465.4	0	16.0	30.0	**511.4**	*129.1*
2035	88.1	50.4	6.7	90.4[11]	**235.6**	*42.1*

Industry and Agriculture

Emissions from the industry and agriculture sector can be reduced by:

- Redesigning some processes and/or by converting them to electricity, such as the use of three-dimensional printing manufacturing or electrolysis to produce hydrogen.

---

[11] The 90 TWh of electricity includes the 10 TWh required to produce, using electrolysis, the hydrogen for the buses.

- For heat requirements, which are significant in industry:
  - ○ Replacing heating systems that burn fossil fuels with electric systems or implementing biomass and/or waste combustion.
  - ○ Recuperating waste heat from various processes, using heat pumps, for example.
  - ○ Implementing more holistic approaches within entire factories or industrial sites to leverage synergies between processes operating at different temperatures and/or at different times.
- Improving the use of mechanical energy by replacing, for example, constant-speed motors with variable-speed systems or replacing hydraulic actuators with electric systems.
- Improving monitoring and control systems either by implementing additional sensors or indirect measurements. An in-depth analysis of processes and their key parameters is also beneficial.
- Implementing recycling processes, as well as circular economy approaches.
- Leveraging heating and cooling synergies, for example, between industrial processes and office buildings on an industrial site.

Within agriculture, potential measures include: electrification of greenhouses, barns and building heating.

Energia's 2015–2035 Energy Mutation in the Industry and Agriculture Sector

The primary objective of Energia's successive governments remains to promote electro-technologies and their deployment within a number of industrial applications.

Evolutions within industry

1. As the best domestic iron ore mines are being depleted, several blast iron plants will close. Coal imports will decrease as a result; the related natural gas consumption will also decrease slightly. To compensate for the related job losses and in line with the circular economy development goal, cast iron and steel recycling will be developed inducing additional electricity consumption.

2. International competition along with circular economy developments will lead to a reduction of the chemical and cement industries. In turn, this will lead to a decrease in fossil fuel consumption.
3. Electricity consumption will increase slightly as recycling of a growing number of materials, especially non-metallic ones, improves and becomes better organized at the national level.
4. The development of three-dimensional printing manufacturing will be ongoing, which will lead to increased electricity consumption.
5. By 2035, the aggressive target of using 100 000 tons of hydrogen outside the transport sector and manufactured using electrolysis will have been reached.

Process evolutions

6. Electric ovens, arc furnaces and induction heating will replace natural gas heating processes in the chemical, glass and metal industries.
7. A systematic promotion of industrial heat pumps will be deployed for heating, drying and cooling applications to replace fossil fuels but also in an effort to facilitate waste heat leveraging.
8. The replacement of pneumatic hydraulic systems with electric ones and of internal combustion engineers with electric ones will be pursued.
9. Upgrading existing refrigeration systems and heat pumps for better performances will decrease both natural gas and electricity consumptions. This will be important for heat and more so for cold requirements in the food industry and for medical applications.

District heating network developments

10. Connecting several industrial sites to district heating networks, supplied by a combination of waste heat recuperation and waste incineration or using cooling systems of new and nearby nuclear plants, as planned for under the 2035 electricity mix, will accelerate the elimination of fossil fuels.

Agriculture

11. Replacing crop greenhouse and poultry brooder fossil fuel heating systems with electric ones and using electric motors, such as for a number of pumping applications, will provide for the displacement of oil and gas consumptions by biofuels and/or by renewable electricity.

Note: The ongoing road transport electrification will induce a significant reduction of petroleum consumption, which, in turn, will reduce the refining industry's energy consumption of roughly 1 liter of oil for 7 liters of gasoline produced; ref. "Industry and Agriculture" chapter. This reduction is not included under the industry and agriculture's final energy consumption since it is not final. Rather, it is included under transformation energy, ref. "Energy and Emissions — Where We Are" chapter.

Energia's overall 2015–2035 industry and agriculture energy mutation

TWh	Coal	Oil	Natural Gas	Biomass and Waste	Electricity	District Heating	Total	*MtCO$_2$*
2015	40	115	120	35	100	20	**430**	*118*
2035	15	38	85	38	137	26	**339**	*50*

Tertiary and Services

Aside from the scales involved and aside from specific applications — such as medical apparatus, operating rooms, computer centers and shopping center cooling systems — the end-uses in the tertiary and services sector are the same as those in the housing sector. Larger application scales make it possible to use more advanced equipment with better efficiencies and controls, which thus reduces emissions and pollutions. The smaller number of decision-makers often also makes it easier to reach and implement decisions, especially regarding renovations; valorizing possible synergies is also easier, including for mixed-use facilities.

The smaller number of decision-makers also facilitates the implementation of a systemic vision, leading to an integrated management of a building or an ensemble of buildings from an energy point of view. At the level of an entire building that features, for example, offices, a computer center and a canteen, an integrated heating and cooling system can be implemented, including PV panels and energy storage, air-air heat pumps, geothermal contribution, and hot or cold water storage reservoirs.

Energia's 2015–2035 Energy Mutation in the Tertiary and Services Sector

The sector will grow due to the development of the economy and population increase. The overall floor space occupied will increase by 20%, primarily in new developments. During this same period, 20% of the space existing in 2015 will have been replaced by 2035. As a result, two-third of all tertiary and services buildings will have been built before 2015 and one-third afterward.

For the sake of simplicity, it is assumed that the distribution among activities will remain the same in 2035 and that the replacement percentages will also be the same for all activities. In reality, office and shopping center floor spaces grow faster and are replaced more often than public buildings.

Existing buildings constructed before 2015 and which will still exist in 2035

The primary focus of Energia's successive governments remains on heating, air-conditioning, hot water production and end-uses specific to electricity, which among them represent 80% of the sector's consumption in 2015.

The following measures are to be implemented:

1. To set an example for the population, especially young persons, schools, universities, hospitals as well as public buildings that use

coal or oil heating will be renovated to improve their insulations by 20%. In addition, the furnaces are to be replaced with heat pumps with an average COP of 3 over a year's operation. To enhance the comfort of students and the public, the heat pumps will be reversible to also provide for air-conditioning.

PV panels will be installed on the roofs, coupled with suitably-sized batteries, to facilitate self-consumption. To contribute to the public's awareness and education of students, energy dashboards will be installed in the entry halls of public buildings.

2. The insulation of some buildings that use natural gas for heating will also be improved by 20% while their furnaces will be replaced by heat pumps, mostly reversible, with an average COP of 3 over a year's operation.

3. For the buildings mentioned under the first two measures, over half of the hot water heating that relied on natural gas will be replaced by solar thermal systems with supplemental electric heating.

4. The development of district heating networks will also make it possible to provide a few TWh for office buildings. Heat will also be recovered from computer centers, if any, in the building; this will allow for a decrease in their overall consumption.

5. The promotion of LED lighting, motion detectors along with energy monitoring and control systems, BEMS, will limit the increase of electricity consumption.

New buildings

Mixed activities and energy synergies will be taken advantage of to reduce consumptions, for example, by way of seasonal heat/cold underground storages. PV panels and shorter-term energy storage systems will also be installed; their contributions are included in the related productions, as further discussed.

The energy consumption of the new floor space — replacements and surface increases — will be reduced to 60% of the consumption for the same space in 2015. This consumption will be 67% electric, 13% solar thermal and 10% each for biomass and district heating.

Energia's overall 2015–2035 tertiary and services energy mutation

TWh	Oil	Natural Gas	Solar Thermal	Biomass and Waste	Electricity	Distric Heating	Total	$MtCO_2$
2015	35	55	10	10	100	10	**220**	*66.6*
2035	0	21	17	13	132	17	**200**	*20.7*

Energia's Final Energy Consumptions in 2015 and 2035

Once the final energy consumptions have been determined for each end-use sector and vector, the overall consumptions of each vector can be summed up over all four end-use sectors; the results are summarized in the following table. In the table, the first two rows provide the consumptions in TWh; the third row provides the ratio, in percent, of the 2035 values with respect to the 2015 values, while the last two rows provide the relative contribution, also in percentages, of each vector in 2015 and 2035, respectively.

	Unit	Coal	Oil	Natural Gas	Solar Thermal	Biofuels and Waste	Electricity	Heat	Total
2015	TWh	50	685	310	25	66	380	45	1 561
2035	TWh	15	126	219	39	62	534	59	1 054
2035 over 2015	%	30	18	71	156	94	141	131	68
2015	%	3.2	43.9	19.9	1.6	4.2	24.3	2.9	100
2035	%	1.4	12.0	20.8	3.7	5.9	50.7	5.5	100

The table calls for the following comments:

- The overall final energy consumption will decrease by 32%, primarily due to the decrease in oil consumption, driven in turn by:
  - The significant reduction of the number of fossil fuel furnaces replaced by heat pumps.
  - The electrification of the car fleet.

- Electricity consumption will increase by 41%. Its share will double from 24.3% to 50.7%. It is thus important to also focus on the decarbonization of the electricity production mix.

Emissions

At this stage, Energia's CO_2 emissions, due to the direct consumptions of coal, oil and natural gas are known but not those due to the production of electricity and heat; neither those due to the transformation industry, which will be known once the electricity and heat production mixes have been determined.

Electricity Production Mix Planning — A Structured Process

The focus of this second part is on the electricity production mix.[12] The heat production mix for district heating systems is also important in the context of an overall low-carbon energy production mix. Most of the observations related to the electricity production mix also apply to the heat production mix, with one important distinction: while electrical systems are interconnected over vast territories, such as Europe, district heating systems are typically local, covering a city neighborhood or an entire city where the load densities are high.

The elaboration of an electricity production mix requires that the final consumption and its probable evolution be known as precisely as possible. System losses and interconnections with neighboring systems must also be included. Typically, one does not start from a clean sheet of paper but an existing situation and previously elaborated consumption, production and network expansion planning trajectories. The planning process thus consists of updates of previously elaborated trajectories based on new information and/or to extend them further out into the future. Several trajectories can be devised, depending on which assumptions are made for future developments, such as related to consumptions and penetration of renewable energies, and arbitrages between several realistic outcomes.

---

[12] The electricity production mix generally describes the actual contribution — in GWh or TWh — of each type of production: hydro, fossil, nuclear, solar, wind, etc. The electricity mix provides the same information but as percentages of the total production.

Time is all-important in the planning process. Indeed, the construction of power plants and transmission lines can last a decade or more, depending on the licensing process. As a result, the uncertainties inherently attached to several input information categories can be quite significant. Proper planning seeks to reduce the sensitivities of the planning trajectories toward the highest uncertainties. Planning results are thus compromises between several considerations leading to "what is possible".

The two primary intents of this presentation are to highlight: (a) the necessity of a systemic approach to electricity production planning as opposed to an electricity source by source approach, and (b) the complexities inherent with the massive and necessary integration of renewable energies within future electricity mixes. While still somewhat complex, for the sake of simplicity, the following presentation intentionally overlooks multiple intricacies of full-fledged electricity production planning as it is routinely carried out within electric utilities worldwide.

Key Operational Requirements to Be Fulfilled by Future Electricity Mixes

While maintaining the electricity cost at an affordable level, as per the United Nations Sustainable Development Goals, the key operational requirements of future electricity mixes are:

Low-carbon

An overall emissions target, in gCO_2/kWh, should be set for the electricity mix based on strategic objectives in line with global environmental commitments, such as under the 2015 Paris Agreement.

Complete supply of the overall yearly electricity consumption, typically in TWh[13]

Final energy consumptions of each end-use sector vary at different time scales:

---

[13] An analogy can be made with an automobile trip — one needs to ensure that enough gasoline or diesel, i.e., energy, is available in the reservoir.

- *Seasonally*: For example, in the northern hemisphere, residential consumption is higher in the winter in northern countries to cover heating requirements. However, it is higher in the summer in the south to cover cooling requirements. Similar observations can be made for all end-use sectors.
- *Weekly*: The consumptions of several sectors are not the same during weekdays or weekend days or holidays.
- *Daily*: The consumptions change between day and night and also with the time of day. In turn, the daily variations also change with the seasons and days of the week.

Full coverage of the overall electricity demand,[14] *typically in GW*

The total end-user power, i.e., the demand as referred to further on, must be covered with acceptable qualities of service in terms of availability,[15] voltage and frequency.

Prioritizations of Electricity Production Technologies — Guiding Principles

In line with the low-carbon overarching priority:

- Avoiding coal power plants should be the highest priority. Depending on the situation, existing coal plants could be converted into gas plants[16] while still using existing electric system infrastructures such as transmission lines. Another possibility is to implement regulations

---

[14] Returning to the automobile analogy, one also needs to ensure that enough power is available at all times, irrespective of the on-board load, to drive up hills or to accelerate to merge onto a freeway, etc.

[15] As seen in the "Electric Power Systems" chapter, in many countries, residential, commercial or industrial customers have come to expect availabilities exceeding 99.96%, which corresponds to total outages not exceeding three hours per year.

[16] Such conversions have been and continue to be implemented in a growing number of countries. In the United States (US), the increased use of shale and tight gas in power plants has contributed to the recent reductions of CO_2 emissions.

to gradually reduce the number of hours a coal plant is authorized to operate annually.

- Once all coal power plants have been taken out of service, oil power plants should be eliminated as the next priority. Except on rare occasions, as mentioned in the "Combustion Thermal Power Plants" chapter, the use of oil for electricity production has already been eliminated.

- If the reliance on fossil fuels cannot be eliminated, natural gas should be favored. As seen in the "Combustion Thermal Power Plants" chapter, combined cycle gas power plants can reach efficiencies above 60%, thus causing far less than half of the emissions caused by coal plants. Whenever possible, combined heat and power (CHP) plants should be implemented to reduce overall emissions. If biogas is available, gas power plant emissions can be further curtailed.

- Hydro power plants, be they storage or run-of-the-river, should be prioritized and remain one of the main renewable electricity sources, if not the major source. The role of pump-storage hydro plants should be increased to accommodate the variabilities of both solar and wind energy.

- While their overall contributions will remain small in most regions of the world, geothermal and marine energies need to be explicitly included during electricity mix planning studies. Both energy sources are renewable, even though geothermal energy is not always emissions-free, and their productions can be accurately forecasted long in advance.

- As the reliance on fossil fuels decreases, the need for base-loaded power plants will remain. As long as both the short and longer-term storage infrastructures are not fully developed such that solar and wind power plants can be scheduled as base-loaded units, the need for nuclear energy will remain as a zero-emissions and base load electricity source.

Reducing existing nuclear energy contributions can be done by way of: (a) increasing renewable energy contributions, which will generally require additional energy storage and may also lead to network stability difficulties and/or (b) increasing the contributions of gas power plans, which, in turn, leads to increased emissions.

- Once the productions from fossil fuels, hydro, geothermal, marine and nuclear power plants have been determined, the role to be played by solar and wind energy can be determined.

Iterations are often needed during the planning process between, on one hand, realistic renewable energies (especially solar and/or wind) and storage possibilities, and, on the other, the need for the other sources mentioned above. In essence, planning future electricity mixes inherently implies political arbitrations between natural resource availabilities and strategic choices, on one hand, and societal, public acceptance and economic, industrial, environmental and technical considerations, on the other.

Electricity Production Mix Elaboration Process

While the process is quite complex, it can be simplified and structured in three major stages that are illustrated in the Energia context in the following part:

1. Determination of the yearly production (in TWh) of each type of power plant required to cover the overall yearly consumption and the system losses while also satisfying a set emissions objective.[17] The outcomes of this first stage include the annual productions of the:

   - Combustion thermal power plant fleet.
   - Hydro, geothermal and marine energy power plants.
   - Nuclear power plant fleet.
   - Solar and wind power plants.

   If the geothermal power plants do not cause any CO_2 emissions, then the electricity production emissions are entirely due to the combustion thermal power plant fleet, which, in turn, sets its maximum contribution given the emissions objective.

---

[17] Including the economical dimensions of the decisions, while essential, are beyond the scope of the book. Many factors come into play during the planning analysis for a new power plant: not only raw investment, operational, fuel and deconstruction costs but also explicit and implicit subsidies, the rate of return and lifetime assumptions, to name but a few. The planning assumptions used affect the relative importance of the various costs.

Next, the operational possibilities of the hydro, geothermal, marine, solar and wind resources in the country being considered, their installed capacities can be deduced.[18] Since nuclear power plants are usually operated as base loaded, their overall fleet installed capacity can also be deduced. Given the possibilities to adjust their productions, the installed capacities of the thermal combustion power plant fleet are determined in the following stages.

At this stage of the process, the contributions of each type of source have been determined provided the following steps can also be satisfied.

2. Determination of the seasonal electricity productions, in TWh, of each electricity source required to cover the seasonal electricity consumptions of all end-use sectors. Seasonal variations are particularly important, not only for hydro resources but also for solar and wind resources.

The overall installed capacity of the intermediate gas power plants can also be determined during this stage.

3. The third stage verifies that the system demand, including system losses, in GW, are met at all times — referred to as the demand-power equilibrium. Potential excess productions due to solar and/or wind productions must also be considered. The best compromises are sought between peaking units and storage while also considering consumption flexibilities[19] such as:

- Reducing the consumption of selected residential, commercial or industrial loads during a few minutes or hours.
- Considering short demand increases by way of energy storage in batteries, pump-storage plants and/or heat in hot water heaters, for example.

One must also consider:

- The need for sufficient reserve margins to ensure that the impacts of unexpected events, such as power plant and/or line outages,

---

[18] Such as statistical information related to river flows, wind speeds and solar radiation.

[19] The levels of flexibility depend on the technologies available to the consumers as well as the specific contracts offered to those willing to participate in such actions, which leads to holistic consumption-production strategies.

remain at acceptable levels for all end-users in terms of the number of yearly occurrences and their durations.

- That the overall power system stability requirements are met, i.e., that sufficient mechanical system inertia be available; ref. "Electric Power Systems" chapter. Based on present technology levels, this could limit the amount of solar, and to a lesser extent, wind energy that can safely be included in the production mix.

Energia's 2015–2035 Electricity Production Mix Mutation

In this third part, of the four outlined in the chapter's onset, the three-stage electricity production mix elaboration process is applied to Energia. One possible outcome, provided as an illustration and not indicative of any authors' preferences, is provided in the table below; other solutions can be proposed depending on the assumptions used. Energia's 2015–2035 overall CO_2 emissions reductions are highlighted in the fourth and last part.

In the table below, the 2015 and 2035 electricity productions, in TWh and percent of the total production, i.e., the electricity mix, are provided in the second and third, respectively fourth and fifth, columns, followed by the installed capacities, in GW, in the last two columns.

Power Plant	2015 TWh — %	2035 TWh — %	2015 GW	2035 GW
Coal	130.2 — *31%*		21.2	
Oil	8.4 — *2%*		1.4	
Gas	105.0 — *25%*	190.0[20] — *32%*	31.1	54.2
Total Fossil	**243.6 — *58%***	**190.0 — *32%***	**53.7**	**54.2**

(Continued)

---

[20] Not including peaking units, kept as reserves and not included in the generation mix.

(Continued)

Power Plant	2015 TWh — %	2035 TWh — %	2015 GW	2035 GW
Nuclear	75.6 — *18.0%*	106.8 — *18.0%*	10.2	14.3
Hydro — Storage and Run-of-the river	54.6 — *3.0%*	61.5 — *10.4%*	15.2	16.0
Hydro Pump-storage			1.0	5.0
Geothermal[21]	6.5 — *1.6%*	19.6 — *3.3%*	1.0	3.0
Marine	1.0 — *0.2%*	2.0 — *0.3%*	0.45	0.9
Biomass and Waste	<u>14.7</u> — <u>*3.5%*</u>	<u>16.2</u> — <u>*2.7%*</u>	<u>2.1</u>	<u>2.3</u>
Total Zero Emission excluding Solar and Wind	**152.4 — *36.3%***	**206.1 — *34.7%***	**30.0[22]**	**41.5**
Solar	13.9 — *3.3%*	97.2 — *16.4%*	10.8	71
Wind	<u>10.1</u> — <u>*2.4%*</u>	<u>100</u> — <u>*16.9%*</u>	<u>5.5</u>	<u>42</u>
Total Solar and Wind	<u>**24.0** — **5.7%**</u>	<u>**197.2** — **33.3%**</u>	<u>**16.3**</u>	<u>**113.0**</u>
Total Production	**420.0 — *100%***	**593.3 — *100%***	**100.0**	**208.7**

The following key 2015–2035 evolutions are noteworthy:

- The electricity production will increase by 41%; the overall installed capacity will more than double.
- The combined solar and wind energy production will increase by a factor of 8.2. It will represent 33.3% of the overall production in 2035 versus only 5.7% in 2015.
- The combined installed capacity of the solar and wind power plants will increase by a factor of 6.9[23] and will represent 54.1% of the entire

---

[21] Energia's geothermal power plants' emissions are considered to be negligible, hence their inclusion under zero-emission.

[22] Rounded up from 29.95.

[23] The increase of the combined production of solar and wind will be higher than the increase of the combined installed capacity. The reasons are: (a) the hypothesis on the increasing performance of PV panels, and (b) the increasing role of off-shore wind production.

power plant fleet installed capacity but will produce 33.3% of the entire production.

- In addition to the 190 TWh produced by intermediate gas power plants, 7.8 TWh will be produced by peaking units. The 197.8 TWh electricity production is the only one causing emissions:
  - ○ The resulting emissions will be 65.9 MtCO$_2$, a 63% reduction from the 179.9 MtCO$_2$ in 2015.
  - ○ The overall carbon content of the electricity production will decrease from 426 gCO$_2$/kWh in 2015 to 111 gCO$_2$/kWh, a 74% reduction, i.e., higher than the overall emissions reduction since the share of fossil productions will have decreased from 58% in 2015 to 32% in 2035.

Stage 1: Annual Consumption and Electricity Production Mix Determination

As seen in this chapter's first part, Energia's total electricity consumption will increase from 380 TWh in 2015 to 534 TWh in 2035.

Assuming that the power system losses remain at 10%, **Energia's overall electricity production needs to be 593.3 TWh[24] in 2035**.

The overall production fleet can then be dimensioned[25]:

Coal and oil steam power plants

As a reminder, in 2015 Energia had 32 coal and 4 oil power plants with a combined production of 138.6 TWh and a combined installed capacity of 22.6 GW.

They will all have been taken out of service by 2035.

---

[24] Indeed: 534 TWh / 90% = 593.3 TWh.

[25] Coal, oil, gas, nuclear, solar and wind industries will be affected by these decisions. While beyond the scope of this book, properly addressing the conversion of the related workforces will be a key — yet sometimes overlooked — dimension to ensure the successful implementation of energy mutations.

Gas power plants

In 2015 Energia had:

- 67 intermediate plants with a combined production of 102 TWh and a combined installed capacity of 29.1 GW, such that their average capacity factor[26] was 40%.
- 40 peaking units with a combined production of 2.9 TWh and a combined installed capacity of 2 GW, such that their average capacity factor was 16.7%, i.e., 4 hours per day.

By 2035, all intermediate gas power plants that were in operation by 2015 and had sufficient remaining lifetime and on-site space to justify the required investment will have been upgraded to become combined cycle plants. All intermediate gas power plants constructed between 2015 and 2035 will be combined cycle plants with efficiencies of 60% or better.

As a result, by 2035 the average efficiency of all *intermediate* gas power plants will be 60%.[27]

The combustion of natural gas results in emissions of 200 gCO_2/kWh; ref. "Units and Reference Values" annex. At 60% efficiency, the resulting average gas power plant emissions will be: 200 gCO_2/kWh / 60% = 333 gCO_2/kWh.

Since the CO_2 content of the overall generation mix is to be limited to 111 gCO_2/kWh, and since the only remaining electricity production emissions will be due to the gas power plant fleet, its annual electricity production is to be limited to one-third (111 / 333) of the entire electricity production, i.e., 593.3 / 3 = 197.8 TWh, which is the value retained for the dimensioning of the rest of the generation mix.

---

[26] Capacity factor; ref. "Introductory Remarks" chapter under "How Electricity Is Produced". In this case, 102 000 GWh/(29.1 GW · 8 760 h) = 40%.

[27] To simplify the presentation, the average efficiency of all intermediate plants and peak units is assumed to be 60%. In reality, peaking units' efficiencies are generally lower since they are mostly not combined cycles. As a result, their emissions are higher; however, on average, they only operate a few hours daily.

The combined production of the gas power plant fleet is to be 197.8 TWh. The total emissions of the gas power plant fleet will be 65.9 tCO$_2$.[28]

The 2035 197.8 TWh gas power plant production will be segmented into[29]:

- 190 TWh for intermediate plants, i.e., an increase of 86% from 102 TWh in 2015. The winter–summer segmentation of the 190 TWh will be determined following the seasonal segmentations for all other power plant productions. Contrary to the production segmentations for other plants, such as hydro, solar and wind power plants, which are decided by nature, the winter–summer production segmentation for gas power plants can be decided upon to suit overall planning requirements.
- 7.8 TWh to be operated as peaking units, i.e., an increase from 2.9 TWh in 2015. As their designation suggests, peaking units are only operated during peak demand or emergencies. The 7.8 TWh value is a strategic decision; it should be high enough to provide for peak demand and emergencies. However, limiting the peaking units' production also limits their emissions. Setting both the production and the installed capacity of the peaking unit fleet requires adequate experience with the peak demands to be faced while also taking other measures available to face them into account, as will be discussed later.

The yearly energy production of all peaking units can only be summed up at the end of each operating period, typically yearly, when all actual peak load and emergency situations have been accounted for.

Hydro power plants

In 2015, Energia had:

- A total of ten storage hydro power plants, with a combined production of 25.2 TWh and a combined installed capacity of 9.6 GW, such that their average capacity factor was 30%.

---

[28] 197.8 TWh at 333 gCO$_2$/kWh.

[29] The segmentation between intermediate power plants and peaking units is tied to the planning objectives.

- A total of 35 run-of-the river power plants with a combined production of 29.4 TWh and a combined installed capacity of 5.6 GW, such that their average capacity factor was 60%.
- Two pump-storage power plants with installed capacities of 400 and 600 MW. Their natural inflows cover the pump-storage losses, such that they did not contribute to the electricity production.

All hydro power plants that were operational in 2015 are expected to be still fully operational by 2035.[30]

The dams at three of the existing storage hydro power plants will be elevated and their water collection infrastructures will be updated to collect more water over enlarged collection basins.

The overall electricity production of the hydro storage power plant fleet will increase by 10% to reach 27.7 TWh annually. Since the power plants themselves will not be affected, the overall installed capacity will remain unchanged at 9.6 GW.

A number of new run-of-the-river hydro power plants will be constructed along smaller rivers while also developing other possibilities such as water networks.

The overall production and installed capacity of the run-of-the-river plants will increase by 15% to reach 33.8 TWh and 6.4 GW, respectively, still assuming a 60% capacity factor.

Geothermal, tidal, biomass and waste incineration plants

In 2015, Energia had:

- One geothermal power plant with geothermal wells distributed over several square kilometers. The installed capacity was 1 GW for a production of 6.5 TWh at an average capacity factor of 74%.

---

[30] Some of these plants will have been renovated by 2035 to prolong their expected lifetimes. Increasingly strict environmental regulations, applicable as renovations are implemented, are such that any increase in electricity production will generally be negligible.

- Two tidal power plants with a combined installed capacity of 450 MW for a production of 1 TWh at an average capacity factor of 25%.
- A number of waste incineration and biomass power plants with a combined installed capacity of 2.1 GW for a production total of 14.7 TWh at an average capacity factor of 80%.

By 2035:

- **The geothermal production will have reached 19.6 TWh**, i.e., slightly more than triple its 2015 production. **The installed capacity will reach 3 GW** such that the average capacity factor will reach 74.6% rather than 74.2% in 2015. Their CO_2 emissions will be negligible based on the characteristics of the geothermal sites.
- New in-stream tidal power plants will have been constructed in Energia's archipelago. **The installed capacity will reach 900 MW,** with an average capacity factor of 24%. Including the existing plants, the **total tidal and in-stream tidal power plant fleet production is to reach 2 TWh**.
- By 2015, Energia had already implemented a broad waste incineration plan to eliminate the need for landfills. As a result, only limited expansion possibilities were still available such that by 2035, only a 10% increase is expected. The installed capacity will reach 2.3 GW; the average capacity factor will remain at 80%. **The waste incineration production will reach 16.2 TWh.**

Nuclear power plants

In 2015, Energia had a total of ten reactors located at three sites with a combined production of 75.6 TWh and a combined installed capacity of 10.2 GW, such that their average capacity factor was 84.6%. This represented 18% of 2015's overall electricity production of 420 TWh.

Among several possible strategic options, the political decision of Energia's government is to keep the nuclear power plant contribution to the overall electricity production mix at 18% as it was in 2015, i.e., as base-loaded.

By 2035, the nuclear power plant production is to be 106.8 TWh.[31]

---

[31] 593.3 TWh · 18%.

Assuming a capacity factor at 85%, the 2035 **total installed capacity of the nuclear power plant fleet will be 14.3 GW**, i.e., an increase of 4.1 GW from 2015. At 1 400 MW per unit, this implies the construction of three generation III PWR reactors, on existing sites if possible.

Overall solar and wind electricity production requirements

The yearly electricity productions determined above are summarized here (in TWh):

- Gas   190.0
- Nuclear   106.8
- Hydro   61.5
- Geothermal   19.6
- Tidal and in-stream tidal   2.0
- Waste incineration and biomass   16.2
  Subtotal   396.1 TWh

To reach the total production of 593.3 TWh, the solar and wind production must thus be 197.2 TWh.

Based on detailed solar and wind natural resource and simulation models and given the surface of suitably oriented roofs for PV installations, the following segmentation has been decided upon for 2035:

- **Solar PV**   **87.2 TWh**
- **Solar CSP**   **10.0 TWh**
- **Wind — on-shore**   **70.0 TWh**
- **Wind — off-shore** **30.0 TWh**

Energia's 2035 overall electricity production will thus be almost evenly distributed among:

- **Gas**, which have the only emissions   **190.0 TWh**
- **Hydro, other renewables and nuclear** **206.1 TWh**
- **Solar and wind**   **197.2 TWh**

Once the overall winter and summer solar and wind productions have been determined, the specifications of the respective fleets can be completed.

PV power plant fleet

Mainland Energia is located between the 35th and 45th latitude north. For the following computations, the average countrywide sun radiation is taken to be: 4.5 kWh/m²·day, i.e., 1 642 kWh/m²·yr.

Assuming the average efficiency of photovoltaic panels to have reached 25% by 2035, as per the computations detailed under the *"PV Energia 2035"* Companion Document, 283 million m², will be required to produce the 87.2 TWh given above.

The total PV installed capacity will be 69 GW.

Energia's public policy requires that all PV panels be installed on rooftops or building façades and not on the ground. A detailed survey of Energia's rooftops has revealed that some 700 million m² are suitable for PV installations under good conditions. At the 283 million m² required by 2035, this leaves ample space not only for solar thermal and HVAC installations but also for future PV expansions.

CSP power plant fleet

A combination of ten solar towers and solar cylindrical-parabolic technologies will be installed. For industrial manufacturing reasons, each of the ten CSP plants will have an installed capacity of 200 MW such that the

The total CSP installed capacity will be 2 GW.

Each plant will have enough integrated energy storage such that it can operate at full output during an average of 13 hours per day, achieving an overall capacity factor of 57%. As a result, each one will produce 1 TWh for a total CSP power plant fleet production of 10 TWh as planned. At a typical surface requirement of 1 ha/GWh of annual production, on average, each plant will occupy 1 000 ha.

Wind power plant fleet

Detailed measurements and climate models have shown that at favorable wind energy locations, the average capacity factors are 25% for on-shore installations and 34% for off-shore installations.

- Since 70 TWh are to be produced using on-shore installations with a capacity factor of 25%, the **total installed capacity for on-shore wind turbines needs to be 32 GW.**[32] With an average on-shore wind turbine rating of 8 MW, a total of 4 000 machines will be required.
- Since 30 TWh are to be produced using off-shore installations with a capacity factor of 34%, the **total installed capacity for off-shore wind turbines needs to be 10 GW.**[33] With an average off-shore wind turbine rating of 10 MW, a total of 1 000 machines will be required.

Stage 2: Winter–summer Consumption and Production Segmentation and Gas Power Plant Dimensioning

To simplify the following discussion, the year will be segmented into:

- Winter, from October to March, i.e., 130 weekdays and 52 weekend days, i.e., 4 368 hours.
- Summer, from April to September, i.e., 131 weekdays and 52 weekend days, i.e., 4 392 hours.

Using this segmentation, each end-use sector consumption, as determined in the first part, can be further subdivided as follows:
- Housing:

Consumption — TWh	Winter	Summer	Total
Heating	51.4	—	51.4
Air-conditioning	—	12.0	12.0
Hot Water	17.2	17.2	34.4

(Continued)

---

[32] 32 GW = 70 10^3 GWh / (25% · 8 760 h).
[33] 10 GW = 30 10^3 GWh / (34% · 8 760 h).

(Continued)

Consumption — TWh	Winter	Summer	Total
Others, including Lighting	39.4	37.8	77.2
Total	**108**	**67**	**175**
	62%	**38%**	

- Similarly, the other three end-use sectors can be segmented as summarized in the following table:

Consumption — TWh	Winter	Summer	Total
Housing	108	67	**175**
Transport and Travel	49	41	**90**
Industry and Agriculture	75	62	**137**
Tertiary and Services	79	53	**132**
Totals	**311**	**223**	**534**

As a result, the 593.3 TWh of production can, in turn, be segmented into 345.5 TWh during winter and 247.8 TWh during summer.[34]

The production of each power plant category can also be seasonally segmented:

- **Hydro**: Set by the natural water cycle — winter, 40%, and summer, 60%.
- **Geothermal, tidal, in-stream tidal and biomass**: No significant seasonal variation — winter, 50%, and summer, 50%.
- **Nuclear**: In countries such as Energia where winter consumptions are higher than during summers, the maintenance and refueling operations are primarily done during summers — winter, 55%, and summer, 45%.
- **Wind**: Given the sites retained and based on the available resources — winter, 60%, and summer, 40%.
- **Solar**: Similarly, winter, 35%, and summer, 65%.

---

[34] 345.5 TWh = 311 TWh/90% – 247.8 TWh = 223 TWh/90%. Taking 10% losses into account.

The required winter and summer productions have thus been determined for 2035:

- Overall production required resulting from the overall electricity consumption.
- Productions from the nuclear, hydro, marine, geothermal and waste incineration power plant fleet.
- Solar and wind productions.

The winter–summer production required from the intermediate gas power plant park can thus be determined as the difference between the overall production, on one hand, and the sum of the productions from all other power plant fleets, on the other.

Type of Plant	Winter — TWh	Summer — TWh	Year — TWh
Nuclear	58.7	48.1	106.8
Hydro — Storage	11.1	16.6	27.7
Hydro — Run-of-the-river	13.5	20.3	33.8
Geothermal	9.8	9.8	19.6
Tidal and In-stream Tidal	1.0	1.0	2.0
Waste Incineration and Biomass	<u>8.1</u>	<u>8.1</u>	<u>16.2</u>
Zero Emissions excluding Solar and Wind	**102.2**	**103.9**	**206.1**
Solar — PV	30.5	56.7	87.2
Solar — CSP	3.5	6.5	10.0
Wind — On-shore	42.0	28.0	70.0
Wind — Off-shore	<u>18.0</u>	<u>12.0</u>	<u>30.0</u>
Solar and Wind	**<u>94.0</u>**	**<u>103.2</u>**	**<u>197.2</u>**
Intermediate — Non-gas	**196.2**	**207.1**	**403.3**
Gas required	**<u>149.3</u>**	**<u>40.7</u>**	**<u>190.0</u>**
Total	**345.5**	**247.8**	**593.3**

As can be seen from the table above, along with the wind and nuclear energy fleets, the winter electricity consumption is primarily covered by the gas power plant fleet. Indeed, 78.6% of the yearly gas power plant production will need to happen during the winter.

Overall installed capacity of the intermediate gas power plant fleet

Once the overall gas power plant fleet's summer and winter productions have been determined, its installed capacity can be determined. The summer production being small, the focus needs to be on the winter production. If all plants were operating at full capacity during the 4 368 winter hours, the resulting required installed capacity would be 34.2 GW[35]; this is the lowest possible value to ensure full coverage. In reality, a higher installed capacity will be required to accommodate load variations and other system incidences. A compromise needs to be reached:

- On one hand, increasing the installed capacity (and thus reducing the capacity factor)[36] to accommodate the load variations as well as the variability of the solar and wind resources while also limiting the required energy storage capacities. While improving the overall electricity supply reliability, increasing the installed capacity of the intermediate gas power plant fleet reduces its overall capacity factor and thus the financial attractiveness of such investments.
- On the other hand, limiting the installed capacity and thus increasing the capacity factor. While potentially affecting the overall reliability of the electricity supply, decreasing the installed capacity of the intermediate gas power plant fleet increases the financial attractiveness of such investments.

The 2015 overall capacity factor of the intermediate gas power plant fleet was 40%. **The political decision is to maintain the said capacity factor at 40% for 2035.**

---

[35] 149 300 GWh / 4 368 h = 34.2 GW.

[36] For any given production level, increasing the installed capacity decreases the capacity factor and vice versa.

The 2035 installed capacity of the intermediate gas power plant fleet will be 54.2 GW.[37]

As a reminder, it was 29.1 GW in 2015. The increase is primarily required to ensure suitable overall supply reliability despite the elimination of all coal and oil-based production, on one hand, and to accommodate the substantial increase of the solar and wind production, on the other.

Stage 3: Demand–Power Equilibrium — GW

In addition to the winter–summer consumption segmentation, one must also consider the following demand segmentations:

- At the weekly level between weekdays and weekends.
- At the daily level. To simplify matters, the high, low and average demand level segmentations are assumed to be identical during winter and summer and across all end-use sectors[38]:
  - High demand: 10:00–16:00, i.e., 6 h/d.
  - Low demand: 21:00–03:00, i.e., 6 h/d.
  - Average demand otherwise, i.e., 12 h/d.

Further demand segmentation assumptions are detailed in the following insert, leading to the demand values in the next table.

Demand distribution assumptions

Housing

- The consumption during a weekday is 90% of the consumption during a weekend day when occupants spend more time at home.
- The high demand is 120% of the average demand and the low demand is 80% of the average demand. These values account for peak shaving

(Continued)

---

[37] 54.5 GW = 190 TWh / (8 760 h · 40%).

[38] The daily demand variations are generally quite different from one end-use sector to the next. For example, residential demands typically show two high levels in the morning and early evening hours with a lull in-between. For offices, the demand is generally high during office hours only. In addition, the variations can significantly differ from one country to the next.

(Continued)

measures such as the use of electric water heaters with integrated energy storage. The same is true for the tertiary and services sector.

Transport and travel

- The consumption during a weekday is double that during a weekend day when commutes to work are significantly reduced.
- The high demand is 130% of the average demand and the low demand is 80% of the average demand.

Industry and agriculture

- The consumption during a weekday is 2.5 times higher than a weekend day when most factories do not operate.[39]
- During weekdays, the high demand is 160% of the average demand and the low demand is 80% of the average demand; during weekend days the ratios are 120% and 80%, respectively.

Tertiary and services

- The consumption, which includes shops and shopping centers, is 2.5 times higher during a weekday than for a weekend day when most businesses, administrations and educational institutions do not operate.
- The high demand is 140% of the average demand and the low demand is 80% of the average demand.

Incorporating the information above and based on the computations further detailed under *"Demand Level Computations — Energia 2015"* in the Companion Document, the following demand data is computed for the four end-use sectors taken together for 2035:

	High — GW	Average — GW	Low — GW
Winter weekday	104.6	76.3	61.1
Winter weekend day	61.1	49.7	39.7
Summer weekday	75.7	54.7	43.8
Summer weekend day	42.6	34.1	27.3

---

[39] Exceptions include industrial processes which cannot be interrupted without incurring significant restart costs such as glass and cement factories or aluminum smelters.

Remains the verification that the installed production capacities, determined above and summarized in the initial table at the onset of this third part, are compatible with the demand levels during winter and summer as well as for week and weekend days as provided in the table above. The two main issues to be addressed are:

- Coverage of the high demand during winter weekdays.
- Absorption of the peak production during summer weekend days.

High demand during winter weekdays

As can be seen from the table above, the high production requirement is 116.2 GW during six hours daily.[40]

To assess the adequacy of the production situation during the winter months, sound precaution is to assume that during an entire week no solar energy production and only 15% of the wind resources are available.[41]

As a result, the available production capacities would be:

- Gas: 54.2 GW
- Nuclear: 14.3 GW
- Hydro: 16.0 GW — storage, 9.6, and run-of-the-river, 6.4
- Geothermal: 3.0 GW
- Marine: 0.9 GW
- Other: 2.3 GW
- Solar: 0
- Wind: 6.3 GW — 15% of 42 GW
  Total   **97.0 GW**

Therefore, assuming that the two pump-storage plants with a combined capacity of 1 GW are available, a production shortfall of $116.2 - 97 - 1 = 18.2$ GW can be expected during the six weekday high demand hours.

---

[40] 116.2 GW = 104.6 GW/90%.

[41] The likelihood of falling below 15% of the wind production is quite small, given the 5 000 turbines dispersed over the entire country and off-shore.

To cover this production shortfall without relying on electricity importations from neighboring countries,[42] the following measures will be implemented:

- 6.4 GW of load shaving will be possible through different contracts:
  - Stopping electrolyzers used to produce hydrogen for transportation, at a level of 1.1 GW, which assumes that proper storage capacities are available.[43]
  - Reducing the consumption of customers under contracts providing for such reductions (generally implying lower electricity rates) at an overall level of 3.9 GW.
  - Reducing the charging of electric vehicles during the six high demand hours, again by way of adapted tariffs, for an overall reduction of 1.4 GW.
- By valorizing suitable geographical sites, 4 GW will be covered by way of new pump-storage facilities:
  - 2 GW will be added by utilizing the dams of some of the ten existing storage facilities as upper or lower reservoirs for new pump-storage power plants and by constructing new lower or upper dams, or better, using existing natural lakes with sufficient capacity.
  - 2 GW will come from new pump-storage power plants, including using the sea or ocean as lower reservoirs.

---

[42] This is a strategic decision of Energia's successive governments. Indeed, weather conditions within neighboring countries may also curtail their available solar and wind productions, thus curtailing their export capabilities at the same time as shortages occur within Energia. In addition, unforeseen events must also be faced using domestic production infrastructures.

[43] The yearly electricity required to produce the hydrogen for transportation purposes is 10 TWh, which corresponds to a level power of 10 000 GWh/8 760 h = 1.14 GW.

Compensation of the pump-storage plant consumptions during the six hours of high demand

The overall installed capacity of the pump-storage facilities will be 5 GW — 1 GW from the existing units and 4 GW from the new ones. The electricity that can be produced by the pump-storage plants during each of the six hours of high demand periods is: $5 \cdot 6 = 30$ GWh.

Assuming that the efficiency of the pump-storage plants is 80%, the electricity production required to restore the energy will be 37.5 GWh, which corresponds to 6.25 GW during the six hours daily. This is a small amount compared to the available production margins during the average and low demand periods, i.e., 10.0 GW and 30.4 GW.[44] The electricity consumed by the pump-storage plants during the six hours of high demand can thus be readily restored daily during the remaining 18 hours of average and low demand, without needing to wait for the even lower demands during weekend days.

- Gas peaking units can also be called upon. To fully cover the winter weekday peak production deficit, their overall installed capacity will have to be 7.8 GW.[44] Under stage 1 of Energia's electricity production mix mutation, the strategic decision to limit the production of the peaking unit fleet to 7.8 TWh was explained, primarily to limit their emissions. As a result, their yearly operations will be limited to only 1 000 hours — which remains a reasonable target from an economic viewpoint.

By implementing peak shaving, adding new pump-storage capacities and calling on peaking units when required, the winter week day peak consumption can be covered. It is thus a systemic approach.

High production during the summer

As determined above and summarized in the initial table of the third part, the combined installed capacity of solar (71 GW) and wind (42 GW) is 113 GW. As a result, especially during weekend days, when the required

---

[44] 18.2 GW – 6.4 GW (load shaving) – 4.0 GW (new pump-storage facilities) = 7.8 GW.

production is 30.3 GW,[45] the solar and wind production alone can occasionally exceed — even by wide margins — Energia's overall demand. This is even possible under partially cloudy and low wind situations.

To avoid wasting available solar and/or wind production, three avenues can be pursued:

- The outputs of hydro storage, intermediate gas and biomass, and waste incineration power plants can be readily reduced or brought to zero with only short time delays. This is less so for nuclear power plants and run-of-the-river plants to avoid wasting available water resources.
- Excess solar and/or wind production can be valorized by pumping water up into the upper dams (if not already full) of pump-storage power plants.
- Any excess can also be valorized by way of other energy storage systems such as batteries and/or hot water heaters.

Even when fully deployed, depending on the situations, these measures may not be sufficient to avoid wasting potential electricity production. It is not reasonable, from several viewpoints, to assume that any excess electricity can simply be exported to neighboring countries; indeed, they may face similar renewable energy integration challenges as they implement lower carbon electricity mixes.

As seen in the "Multi-Energy Systems" chapter, a range of technologies, such as power-to-gas, are available to valorize excess electricity productions to supply power plants, vehicles and the industry, or hydrogen production for transportation applications. While these technologies are expected to contribute to the zero-emissions by the 2050 target, they do consume energy that needs to be produced with zero emissions, hence requiring higher renewable energy capacities or the prolonged need for nuclear energy, if required.

---

[45] 30.3 GW = 27.3 GW / 90%.

Concluding Remarks Concerning Energia's 2015–2035 Electricity Production Mix Mutation

While the overall electricity production will have increased by 41% over the 2015 level, the overall emissions will decrease by 63% to reach 65.9 $MtCO_2$. This drastic emissions reduction will be possible by deploying technologies already available[46] while not waiting for future technologies, which, when available, will amplify decarbonization. Implementing the "triple one" policy leading to the 111 gCO_2/kWh emissions level by 2035 is realistic.[47]

The contribution of nuclear energy to the electricity production mix is an important decision. In the context of Energia's 2035 energy mutation described above, the decision was reached to keep its contribution at 18% as base-loaded, which, in turn, leads to an increased production in line with the overall electricity consumption increase. While leaving the overall CO_2 emissions unaffected, two options can be considered:

- Further increasing the nuclear contribution would not induce any technical problems since the solar PV and wind energy contributions could be decreased.
- Decreasing the nuclear energy contribution would require a commensurate increase in the contributions of other sources, in particular, solar PV and wind, which, in turn, would require additional energy storage. In addition, the reduction of the mechanical system inertia brought by the turbine–generator assemblies of nuclear power plants could induce network stability difficulties.

---

[46] The only exception relates to the average PV panel efficiency, which we have assumed will have reached 25% by 2035. Should this turn out not to be the case, more rooftop surfaces will need to be allocated. As the related computations show, ample space would be available.

[47] During the summer, the gas intermediate power plants contribute 40.7 TWh to the overall 247.8 TWh electricity production. As a result, the emissions are: $(40.7/247.8) \cdot 333 \ gCO_2/kWh = 55 \ gCO_2/kWh$, i.e., lower than the 111 gCO_2/kWh yearlong average.

During the winter, the gas intermediate and peaking power plants contribute 149.3 + 7.8 TWh respectively to the overall 345.5 TWh electricity production. As a result, the emissions are: $((149.3 + 7.8)/345.5) \cdot 333 \ gCO_2/kWh = 151 \ gCO_2/kWh$, i.e., higher than the 111 gCO_2/kWh yearlong average but still below the 200 gCO_2/kWh when using gas for heating.

Energia's 2015–2035 Overall Emissions Reductions

The following emissions have already been determined:

- Due to the direct consumptions of coal (50 TWh in 2015 and 15 in 2035), oil (685 TWh in 2015 and 126 in 2035) and gas (310 MWh in 2015 and 219 in 2035): 251 $MtCO_2$ in 2015[48] and 81 $MtCO_2$ in 2035.
- Due to the production of electricity, 178.9 $MtCO_2$ in 2015[49] and 65.9 $MtCO_2$ in 2035.

The district heating networks provided 45 TWh in 2015 and will provide 59 TWh in 2035. Neglecting the related losses and assuming that the heat production emissions content is the same as the electricity production for both 2015 and 2035, the resulting emissions can be computed to be 19 $MtCO_2$ in 2015 and 6.6 $MtCO_2$ in 2035.

As a result, the emissions due to the production of electricity and heat, taken together, will reach 72 $MtCO_2$ rather than 198 $MtCO_2$ in 2015.

The emissions due to the transformation energies, such as oil refining and coke production for the steel industry, must also be taken into account. Assuming them to be 10% of Energia's overall emissions as per the hypothesis outlined in the "Energy and Emissions — Where We Are" chapter, i.e., 1/9 of the direct emissions and those due to the production of electricity and heat, one can compute the emissions due to the transformation energies.

The results for 2015 and 2035 are summarized in the following table. The first two rows provide values in $MtCO_2$ while the last two provide data in percent.

---

[48] Using TWh and $MtCO_2$/TWh: $(50 \cdot 0.360) + (685 \cdot 0.250) + (310 \cdot 0.200) = 251.25$ $MtCO_2$

[49] 593.3 TWh $\cdot$ 0.111 $MtCO_2$/TWh = 65.9 $MtCO_2$ and 420 TWh $\cdot$ 0.426 $MtCO_2$/TWh = 178.9 $MtCO_2$

Emissions	Fossil, Excluding Electricity and Heat (A)	Electricity and Heat Production (B)	Transformation C = (A + B)/9	Total Emissions A + B + C
2015 (MtCO$_2$)	251	198	51[50]	**500**
2035 (MtCO$_2$)	81	72	17	**170**
2015 (%)	50	40	10	100
2035 (%)	51	39	10	100

This table calls for the following comments:

- The emissions will be divided by almost three, both resulting from the direct use of fossil fuels and those due to the production of electricity and heat. The same is also true for the emissions due to the transformation energies; this is largely due to the decrease in the refining volumes.
- Thanks to the "triple one society" policy, the emissions in 2035 due to the direct use of fossil fuels, i.e., 81 MtCO$_2$, will remain above those due to the production of electricity, i.e., 72 MtCO$_2$.
- This latter outcome is primarily due to the reduction of the electricity production emissions, from 426 gCO$_2$/kWh to 111 gCO$_2$/kWh, which largely compensates for the significant increase in the electricity consumption observed above.

[50] 51 rather than 50, such that the total comes out to 500 for greater reader impact!

To Conclude

Four Families in 2035

To illustrate the impact the energy technologies presented throughout the book could have by 2035, four households with identical compositions, living at the same locations and having identical activities as those summarized in the four cartoons at the beginning of the book, are revisited as far as their 2035 energy consumptions at home and overall emissions are concerned. A comparison between their 2015 and 2035 energy consumptions and emissions is then provided.

Overall Assumptions for 2035

In the following presentation, only the key 2015–2035 changes related to their energy consumptions are mentioned for each family.

To simplify the comparisons, while historically different in each of the four countries, it is assumed that their 2035 electricity production mixes will have the same emissions at 111 gCO_2/kWh, i.e., the same as for Energia;[1] the 2035 district heating network emissions are also assumed to be at 111 gCO_2/kWh.

To confirm their potential impact, the technologies used by the 2035 families are essentially those already available in 2015; i.e., presented throughout the book.

---

[1] The emissions of the French electricity mix varies between 50 and 90 gCO_2/kWh from one year to the next; in 2017, it was 74 gCO_2/kWh when the nuclear power plant contributions were at 75%. Also, in 2017, it was 453 gCO_2/kWh for the United States and 640 gCO_2/kWh in China where it is decreasing rapidly. In Cameroon, the emissions vary between 150 and 200 gCO_2/kWh due to the significant variations of hydro energy around an average of 60%.

The Duponts

In 2035, the Duponts, a retired couple, will still be living in Lille in a house similar in size and features to that of their 2015 elders, aside from two evolutions: (a) by 2035, the house will have been renovated such that its G coefficient will have been reduced from 1.2 to 1.0, and (b) the natural gas furnace will have been replaced by a heat pump.

The electric hot water heaters will also have been replaced by heat pumps with a yearly average COP of 3 which reduces the related electricity consumption by a factor of three.

Since the house's lighting already used LEDs in 2015, further electricity consumption reductions cannot be expected to counterbalance an increase due to the broader use of communicating appliances and a second TV.

The Duponts will have a fully electric car that they will drive the same distances as their elders.

The emissions due to the electric public transportation system in France will increase based on the overall emissions assumption retained.

The Jones

In 2035, the Jones — two parents and their three children — will still be living in Peachtree City in a house similar in size and features to that of their 2015 elders. However, the house will have been recently constructed primarily using wood with improved insulation leading to a G factor of 0.9 rather than 1.4; it will be heated and air-conditioned using reversible geothermal heat pumps, rather than a natural gas furnace for heating.

The natural gas hot water heater will have been replaced by a solar unit with electrical add-on heating covering one-third of the energy required yearly.

The enhanced use of LED lighting will compensate for the increased use of leisure technologies and connected appliances, such that their combined electricity consumption will not be affected.

By 2035 all three cars in the Jones household will be fully electric.

The Lis

In 2035, the Lis — two parents and their two children — will still be living in Beijing in a high-rise building apartment similar in size and features to that of their 2015 elders. The building's maintenance will have improved, especially its heating infrastructure, providing for improved comfort during winters. However, the Lis will still rely on the same amount of supplemental electric heating as their elders — the rebound effect at work! In addition, they will rely on electric add-on heating for the living room as well as the three bedrooms; a heating lamp will have been installed in the bathroom.

Air-conditioning in the entire apartment, not only in the children's bedrooms, will further improve the quality of life. Since the building's gas distribution network will have been suppressed, the hot water heating will rely on a heat pump. The kitchen will have an induction cooktop and an electric oven.

Even though a washer–dryer and a dish washer will have been installed and their leisure appliances will be more heavily used (second computer, video games with screens for the children), the related electricity consumption will remain stable thanks to the installation of LEDs.

The Lis will use shared electric vehicles for 4 000 km per year.

The Menyes

In 2035, the Menye household — two parents, their five children, a live-in aunt and two live-in nieces — will still be living in Djoum in a house recently built using "permanent materials" and having running drinking water and proper sanitation. Heating will still not be required.

By 2035, the local electricity distribution network will reliably supply the house and the grocery store. Some cooling will be provided using ceiling fans in the main room and bedrooms.[2] The house will have a solar hot water heater without add-on electric heating.

---

[2] One 60 W fan and three 30 W fans running a total of 2 000 hours consume 300 kWh/yr.

The household will have a refrigerator. An electric cooktop will be used to prepare the food; this will eliminate the wood fireplace, which significantly reduces the related consumption, as listed in the following tables, and, in turn, the emissions.

All rooms in the house will feature LEDs. The television will be on regularly. The two parents and their two eldest children will have a computer as will the aunt; all computers will be connected to the Internet. Aside from the youngest ones, all members of the household will have mobile phones while the youngest ones will have game stations.

By 2035, all moto-taxis, tricycles and bush taxis will be electric. In addition to the trips taken by their elders, the oldest daughter will attend the University of Yaoundé; she will return home four times a year on a diesel-powered bus, a 550-km round trip.[3]

Four times a year, the family will travel to the country's capital, Yaoundé, or national parks, traveling a total of 5 000 km using electric cars shared by the neighborhood.

Energy and Emissions Comparisons

Based on the overall assumptions and evolutions for each family mentioned above, the 2015 and 2035 energy consumptions are summarized in the following table. Except where noted, all consumptions are electric given in MWh. The data for 2015 is summarized in the cartoons for each family at the beginning of the book. The results for 2035 have been derived using the 2015 approach presented under *"Families 2015 Worksheets"* in the Companion Document.

---

[3] Moto-taxis and tricycles are assumed to consume 8 kWh/100 km and bush taxis 16 kWh; diesel-powered buses are assumed to consume 24 l/100 km. The emissions are "shared" between the 10 passengers in the bush taxi or 30 passengers on the diesel-bus.

Dwelling Consumption	Dupont		Jones		Li		Menye	
MWh	**2015**	**2035**	**2015**	**2035**	**2015**	**2035**	**2015**	**2035**
Heating	22.2	5.6	45.4	8.7	5.4[1]	6.0[1]	0	0
Cooling	0	0	25.7	5.5	0.2	0.65	0	0.3[2]
Hot Water	2.8	0.9	6.4	2.1	4.45	1.5	0	1.0[3]
Cooking	1.1	1.1	3.7	3.7	1.2	0.95	6.3[4]	2.1
Other	2.2	2.4	10.4	10.4	2.75	2.7	0.8	1.4
Total MWh	**28.3**	**10.0**	**91.6**	**30.4**	**14.0**	**11.8**	**7.1**	**4.8**
2035/2015	**35.3%**		**33.2%**		**84.3%**		**67.6%**	

Notes: [1]Including 5.0 TWh from district heating. [2]For ventilation. [3]Solar hot water heater. [4]Wood.

As evident by the table above:

- The dramatic reductions in the heating energy requirements for the Duponts and Jones are due to the enhanced insulation coupled with the conversion from gas furnaces to heat pumps.
- The slight increase of the heating requirements for the Lis is due to the rebound effect in conjunction with the increased add-on electric heating while the district heating use remains at 5 MWh. However, the related emissions will decrease as the CO_2 content of the district heating and electricity mixes decrease.
- The air-conditioning requirements for the Jones will also significantly decrease due to the installation of a heat pump.
- The 32.4% decrease on the Menye household consumption is somewhat counter-intuitive. However, aside from the cooking requirement, which decreases from 6.3 to 2.1 MWh, taken together, all other consumptions increase from 0.8 MWh in 2015 to 2.7 MWh in 2035, confirming the enhanced living conditions.

Similarly, the 2015 and 2035 CO_2 emissions are summarized in the following table — first in total emissions followed by the per capita data.

CO_2 Emissions	Dupont		Jones		Li		Menye	
tCO_2/year	2015	2035	2015	2035	2015	2035	2015	2035
Dwelling	4.74	1.11	26.6	3.7	4.28	1.3	2.64	0.42
Transport and Travel	1.74	1.44	22.6	5.2	2.33	1.59	0.61	0.17
Total	**6.48**	**1.55**	**49.2**	**8.9**	**6.61**	**2.89**	**3.25**	**0.59**
tCO_2/yr·cap	**3.24**	**0.78**	**9.84**	**1.78**	**1.65**	**0.72**	**0.325**	**0.059**
2035/2015	**24.1%**		**18.1%**		**43.6%**		**18.2%**	

As evidenced by the second table, the emissions reduction percentages are higher than for the consumptions. This confirms that to reach the overall emissions reduction objective, it is not only important to deploy low-emission end-use technologies but also to decarbonize the electricity production mix.

Conclusion

*We Have the Technologies
Let's All Get Going!*

Our priority must be to safeguard our ecological niche, broadly speaking, including the fauna and flora that surround us. Energy-wise, the priority is to decarbonize both the production and consumption of energy while also ensuring improved access to reliable and affordable energy for those less favored. Using energy always has an impact on the environment. It therefore behooves us to use it as rationally as possible. To reduce the negative impacts of the production of energy, it is important to decouple them, whenever possible and as much as possible, from the energy production itself.

We know that our remaining "carbon budget" decreases year by year; we also know that we will have spent it well before the end of this century if we simply move on along the trajectory we have followed this far.

We have hopefully demonstrated that there are no "excuses" for not acting without further delay to reduce our CO_2 emissions by deploying technologies readily available, especially those that can massively decarbonize electricity, thereby enabling its use for an ever-growing range of applications we rely upon, hence the book's title — *Electricity: Humanity's Low-Carbon Future.*

The 2050 objective of a zero-carbon world, as set by the 2015 Paris accord, seems to us to be both ambitious but also feasible. However, the 2050 horizon is too remote, given the urgency to implement the energy mutation. We thus believe that it is necessary to set an intermediate horizon at 2035 for our case studies around the Dupont, Jones, Li and Menye families, and Energia. Based on these case studies, we wish to put forward a strong message, supported by data-based estimations we believe to be realistic, that a significant portion of the path toward a zero-carbon world can already be accomplished by 2035.

Any energy mutation in line with a carbon neutral outcome by 2050 requires that CO_2-emitting energy consumption and production equipment and systems be replaced by available equivalent low-carbon ones, or be suitably adapted, as they reach their useful life times or well before. Decarbonation opportunities are available for a number of industrial processes. The integration of low-carbon energy sources, particularly of intermittent electricity sources, also require additional production and grid investments. Key international entities (World Bank, IEA) estimate[1] these incremental investments at 1–2% of the average worldwide GDP, yearly, during several decades to come. How they are to be distributed among industrialized and emerging countries remains to be settled within international instances. Within each country, the distribution of the necessary financial efforts among stakeholders is subject to debates which lead to public policies intended to orient future energy-related decisions reached by consumers and industry. The range of subsidies, incentives, taxes, penalties and economic development initiatives embedded in them often obscure the true cost of any apparatus, energy supply or service.

Research and development activities, broadly supported over the long term by both the public and private sectors, are and will remain indispensable. They will lead to new action paradigms toward a zero-carbon world more respectful of the environment. These new technologies and services will be carried forward by present and future leaders and stakeholders.

We are well aware of the complexities of the required energy mutations; challenges will be addressed and overcome while striving to reach balanced compromises. Transparent and durable public policies are required and will remain a necessity such that everybody can participate and act: citizen-consumers, industry stakeholders, teachers, scientists and public servants.

While underway, it behooves us all not to forget the decarbonization urgency, even as our attention will occasionally be called upon by other urgencies requiring shorter-term efforts.

---

[1] Such estimates include a number of assumptions related to population evolution, fuel and production costs, etc.

Annex

Companion Document List

Alternative Nuclear Reactors
Carbon Tax and Carbon Cap-and-Trade
Coal Power Plant Layout
Demand Level Computations — Energia 2035
Desalination
Earth's Structure and Geothermal Sources
Elements of Classical Mechanics
Energia Worksheet
Energy and Water Nexus
Families 2015 Worksheets
Feed-In Tariffs
Fossil Fuel Supply Chain Accidents
HDD
Hydro Power Plant Accidents
Integrated CCS Projects
Key Energy Indicators
Major Nuclear Power Plant Accidents
Multimodal Shipment
Nuclear Power Plant Dismantling
Phase Shifts
Power Plant Indicators
Power Plant Scheduling — Swiss Case Study
Power System Blackouts
PV Energia 2035
Radioactivity and Chain Reactions
Thermodynamic Cycles
Types of Gases and Their Processing
Vehicle Refueling Versus Recharging Comparisons
Wind Turbine Engineering

Units and Reference Values

Topics Covered in this Annex

The annex is structured in four parts. The first one supplements the comments in the book's preamble regarding the numerical information provided. The second part presents the international system of units, which is consistently used. The third part provides some physical and chemical reference values; the fourth and final one provides information as to the average heat, humidity and emissions from each of us!

Numerical Information and Its Quality

The physical and chemical characteristics of pure materials are precisely known as are the constants in physical laws. They are provided in a number of documents from scientific organizations in different countries; they are accessible on the web.

The physical and chemical characteristics of various ores, minerals or metals, as well as climate information may vary somewhat depending on the geographical location. The information is available in data bases of scientific and technical national entities; they are also publicly available. When useful, we have elected to use average values as for the data given in the third part of this annex.

For worldwide or national statistics, we have primarily relied on websites from international organizations — International Energy Agency [*iea. org*], United Nations [*un.org*], World Health Organization [*who.int*], World Meteorological Organization [*wmo.int*] — as well as from some national websites. The definitions of the information and units used are also given

447

on the sites as well as the methodology used to obtain aggregated data. The inherent reason for any data uncertainties is also provided; for example, the uncertainties attached to data from 1850 are higher than regarding data from 2018. This can help explain the differences between the same information provided by two different organizations which should, at first sight, be identical. Aggregated consumption information — for example: how many consumers in Beijing used air-conditioning during a summer Sunday in 2016? — is often the result of data gathering campaigns of varying precisions. The advent of the Internet of Things should gradually provide for the compilation of more precise and more recent information. Energy production information — for example: what was the electricity production using coal power plants in Europe in 2016? — is generally more precise and up-to-date since they are provided by a smaller number of industries, which are often required to declare their individual information. The increasing deployment of distributed energies, such as solar, increases the number of actors but in a context where the data collection is easier than related to the number of housing insulations, for example.

Significant differences may appear regarding equipment performance information such as for appliances, refrigerators, furnaces or electrolyzers. The differences may be intrinsic; indeed, the performances of practically all technologies are progressing. The efficiency of solar photovoltaic panels during the past 40 years is but one example. One must also differentiate between technologies still at the laboratory stage from the best ones already on the market. The efficiency of systems already in operation, sometimes since several decades, is not the same as for similar systems but put into service in 2019. It thus important to date the information whenever possible to avoid citing as present-day performance a 1970 record performance or that hoped for in ten years when market penetration could be initiated upon the completion of on-going laboratory tests!

Similar comments also apply to the differences which may appear related to the actual performances of systems already in operation such as power plants or heat pumps, for example. In addition, yearlong performances generally depend on exogenous factors such as the weather during the year, operating conditions and energy requirements. Should the best ever performance be cited or an average over several years? If information is provided publicly, as required by a regulator, such as open data related to power plant productions, or for market transparency reasons, it is generally only global whereas more detailed information would be of greater interest such as seasonal or daily variations. We have tried to use illustrative examples of actual installations for which transparent information is publicly available.

Differences between data from several sources may be due to the implicit use of different definitions for the same information. For example, is the rated power for a turbine that of the turbine alone or does it also include the consumption of its auxiliaries? Some gaps can be due to actual errors; others can be tied to commercial interests. Advocates of a particular technology, in their enthusiasm, may occasionally refer to laboratory results as already deployable on a large scale. When referring to apparently coherent data from several manufacturer sites, or from a number of national and international industry association sites and provided by certified laboratories, we have systematically cross-referenced them. These sites are generally not mentioned in the text; however, they are easy to find. Public websites, for example from national US laboratories, mention performance ranges for almost all systems specific to a type of operation while excluding extreme values. These ranges are generally obtained by compiling refereed publications supplemented or not by explicit data. The ranges we provide for temperatures, for performances, etc., correspond, except when stated otherwise, on one hand, to the lower end of the range for systems already in service and, on the other, to the value corresponding to the best systems in 2019–2020. Finally, some sensitive information in a competitive context may be difficult to find; we then provide a range or a value designated as "roughly" or "some".

International System of Units

The international system of units, referred to as the SI (Système International), is built on seven base units:

- Second for time. Usual abbreviation: s
- Meter for length. Usual abbreviation: m
- Kilogram for mass. Usual abbreviation: kg

- Ampere[1] for electric current intensity. Usual abbreviation: A
- Kelvin for temperature. Usual abbreviation: K
- Candela for luminous intensity. Usual abbreviation: cd
- Mole amount of substance. Usual abbreviation: N

As of May 2019, a new set of definitions of the base units was implemented, all based on physical or chemical phenomena and no longer only on samples.

All other units can be derived from these seven base units, such as:

- Velocity: m/s
- Force and weight: $1 \text{ kg·m/s}^2 = 1$ Newton, designated as N
- Energy, work, heat: $1 \text{ kg·m}^2/\text{s}^2 = 1$ Joule, J = 1 N·m
- Power: $1 \text{ kg·m}^2/\text{s}^3 = 1$ Watt,[2] W = 1 Joule/s

Using electrical units:

- Volt[3] for electrical potential across the terminals of a load or a source: Usual abbreviation: E = V
- Power: P = V·A, measured in Watt, W
- Energy: E = V·A·s, generally measured in Wh

Pressure — Definition and Units

Pressure is the ratio of a force and an area; it is therefore measured in N/m^2.
The scientific unit of pressure is the Pascal, Pa.[4]

$$1 \text{ Pa} = 1 \text{ N/m}^2$$

The most commonly used barometric unit, the bar, is equal to 10^5 Pa.
By definition, the air pressure at sea level is: 1 atm = 1.013 bar.
Often, bar and atm are treated as equal.

---

[1] Named after **André-Marie Ampère** (1775–1836), French physicist, chemist and mathematician.
[2] Named after **James Watt** (1736–1819), Scottish mechanical engineer.
[3] Named after **Alessandro Volta** (1775–1827), Italian physicist and chemist.
[4] Named after **Blaise Pascal** (1623–1662), a French mathematician and philosopher.

Weather reports often refer to millibars. The air pressure at sea level is 1 013 hPa = 1 013 millibars.

Pressure can also be measured in PSI (Pound per Square Inch). Tire pressures are also measured in PSI or bar. For example: 36 PSI = 2.5 bar.

Absolute and Relative Temperature Units

The most commonly used unit is the Celsius degree, designated as °C[5]. Using this scale, at the sea level and at atmospheric pressure, water freezes at 0°C and boils at 100°C.

Absolute temperatures are measured in kelvins,[6] K, which is the temperature unit used in the international system of units. The scale is defined such that all thermal motions cease at 0 K. By definition, 1°C = 1 K.

$$T[°C] = T[K] - 273.15$$

Proposed in 1724, the Fahrenheit[7] scale is the oldest one. At this time, it is practically only used in the United States. The degree Fahrenheit is noted °F. The conversions between the Celsius and Fahrenheit scales are:

$$T[°C] = (T[°F] - 32) \cdot 5/9 \quad \text{and} \quad T[°F] = (T[°C] \cdot 9/5) + 32$$

Reference Values

Physical Characteristics of Water — At Atmospheric Pressure

- Specific heat of ice[8]:   0.6 Wh/kg·°C
- Specific heat of water: 1.16 Wh/kg·°C
- Latent melting heat of ice[9]:   92 Wh/kg
- Latent vaporization heat of water: 627 Wh/kg.

The specific heat values vary by a few percent with temperatures.

---

[5] Named after the Swedish astronomer and physicist, **Anders Celsius** (1701–1744).

[6] Named after the Irish engineer and physicist, **William Lord Kelvin** (1824–1907).

[7] Named after the German physicist **Gabriel Fahrenheit** (1686–1736).

[8] The specific heat of a body is the amount of heat energy required to increase its temperature by one degree (°C) — rigorously, it also depends on the pressure and the temperature.

[9] The latent melting (vaporization) heat of a solid (liquid) body is the heat required to melt (vaporize) it.

Physical, Chemical and Energy Characteristics of Selected Fuels

When a numerical value may vary, either due to natural characteristics or due to different interpretations of the definitions depending on the context, we have opted to provide an average value to familiarize the reader with orders of magnitude.

The emission values given below correspond to the location where the combustion actually takes place. If one also includes the extraction, the processing and the transport of the particular fuel, the corresponding emissions will typically increase by a proportion which depends on the fuel and on its supply chain.

Selected molecular masses: Hydrogen: 1; Carbon: 12; Oxygen: 16.

Combustion of natural gas

The combustion of methane, CH_4, at atmospheric pressure and at 15°C, produces 13.9 kWh/kg and 2.75 kg of CO_2 as well as 2.25 kg of H_2O. The resulting emissions are 198 gCO_2/kWh.

Natural gas contains not only over 80% of methane but also other constituents, which may affect the values above.

When relevant, one may include the recuperation of the heat in the vapor produced by the combustion. This increases the energy obtained by some 10% and thus reduces the emissions by 10%.

Combustion of natural gas — the following values are used:
- **Energy produced: 14 kWh/kg.**
- **Emissions: 200 gCO_2/kWh.**

Combustion of gasoline, diesel and kerosene

The composition of petroleum depends on the location of extraction. In turn, the compositions of gasoline, diesel and kerosene depend on the refining processes used and on the norms enforced locally for various applications.

Gasoline contains several hydrocarbons, for example octane, C_8H_{18}, and heptane, C_7H_{16}. Its specific mass slightly depends on their proportions; 0.75 kg/liter is typically used. The energy produced by its combustion also depends on these proportions, as do the resulting emissions. Gasoline combustion produces between 8.5 and 10 kWh/liter and results in emissions of roughly 2.4 kgCO$_2$/liter.

Diesel contains heavier hydrocarbons than in gasoline with between 15 and 20 carbon atoms per molecule — between, $C_{15}H_{32}$ and $C_{20}H_{42}$. Its specific density, around 0.84 kg/liter, also depends on these proportions. It produces roughly 10% more energy per liter and 10% more emissions per unit energy.

Kerosene contains intermediary hydrocarbons, between those of gasoline and diesel, between $C_{10}H_{22}$ and $C_{14}H_{30}$. Its main advantage for aviation applications is that it remains liquid down to temperatures around –40 °C.

Combustion of gasoline, diesel and kerosene[10] **— the following values are used:**

- **Energy produced: 9 kWh/liter, i.e., 12 kWh/kg.**
- **Emissions: 2.4 kgCO$_2$/liter.**

Combustion of coal

The combustion of pure carbon produces 9.7 kWh/kg. For pure carbon, the proportion between the weight of CO$_2$ emitted and the weight of the carbon burned is equal to the proportion of the corresponding molecular masses, i.e., 44 / 12 = 3.67. There are many qualities of coal, from anthracite, which contains more than 90% of carbon, to lignite, i.e., brown coal, which contains roughly 50%.

Combustion of coal — the following values are used:

- **Energy produced: 9 kWh/kg.**
- **Emissions: 360 gCO$_2$/kWh.**

---

[10]To simplify, we have taken the same values, the aim of the book being not to study the differences between these three families of fuel. These values could be perceived as being slightly too low. It does not change the order of magnitudes from the different results with which we want to familiarize our reader, nor the conclusion concerning the overarching necessity to reduce the CO$_2$ emissions.

Combustion of wood

The composition and texture of wood depend on the particular kind and on the local climate. Chemically, wood is an aggregate of organic molecules containing some 50% of carbon, 45% of oxygen and 5% of hydrogen. The rest, mineral elements, is usually less than 1%. The humidity degree, specially "free water", i.e., which is not part of organic molecules, can vary between 10% and 60%; it is crucial for the inflammation and combustion capabilities. Fine particles are produced during the combustion. The energy produced during the combustion depends on precise combustion circumstances. The average values below derive from the carbon content of wood.

Combustion of wood — the following average values are used:
- **Energy produced: 4.5 kWh/kg.**
- **Emissions: 400 gCO_2/kWh.**

If one includes the entire life cycle of the wood, the resulting emissions are zero since during its life time the wood has naturally absorbed the CO_2 emitted during its combustion.

Selected Values Per Person

Typically, a person who is awake and at rest, not performing any particular physical exercise:

- Produces 100 Wh of heat per hour.
- Causes 50 ml of humidity per hour.
- Emits 1 kg of CO_2 per day due to her or his respiration.

Website Acronyms

alphaliner.com	Data service for the shipping industry
atag.org	Air Transport Action Group
bfe.admin.ch	Bundesamt für Energie — Swiss Federal Energy Secretariat
bpie.eu	Building Performance Institute Europe
calepa.ca.gov	California Environment Protection Agency
carbontax.org	CTC — Carbon Tax Center
clcouncil.org	Climate Leadership Council
csx.com	Chessie-Seaboard Merger — Railroad corporation, United States
ctg.com.cnn	China Three Gorges Corporation
degreedays.net	Degree days calculation for locations worldwide
dlsc.ca	Drake Landing Solar Community
ec.europa.eu/eurostat	European Commission statistics data base
eea.europa..eu	European Environment Agency
eosweb.larc.nasa.gov	Atmospheric Science Data Center, NASA
eia.gov	Energy Information Agency, United States
electricitymap.org	List of emissions for selected countries — in CO_{2e}/kWh
entsoe.eu	European Network Transmission System Operators — Electricity
epa.gov	Environmental Protection Agency, United States

fao.org	Food and Agriculture Organization of the United Nations
ferc.gov	Federal Energy Regulatory Commission, United States
geothermal-energy.org	International geothermal associations
gov.uk/government/ statistics	UK Government Statistics
greyhound.com	Greyhound bus transport corporation, United States
GWEC.net	Global wind energy council
hydropower.org	International Hydropower Association
iaea.org	International Atomic Energy Agency
icao.int	United Nations International Civil Aviation Organization
icglass.org	International Commission on Glass
icold-cigb.org	International Commission on Large Dams — Commission Internationale des Grands Barrages
idadesal.org	International Desalination Association
iea.org	International Energy Agency
iea-shc.org	Solar heating and cooling programme, IEA
iifiir.org	International Institute of Refrigeration
imo.org	International Maritime Organization
ipcc.ch	Intergovernmental Panel on Climate Change
irena.org	International Renewable Energy Agency
iru.org	International Road Transport Union
itaipu.gov.br	Itaipu Binacional
itu.org	International Telecommunication Union
nasa.gov	National Aeronautics and Space Administration
nrc.gov	United States Nuclear Regulatory Commission
nrel.gov	National Renewable Energy Laboratory

oica.net	International Organization of Motor Vehicle Manufacturers
ornl.gov	Oak Ridge National Laboratory
population.un.org/wpp	United Nations; Department of Economic and Social Affairs
posiva.fi/en	Posiva Oy; a nuclear waste disposal expertise entity
pvwatts.nrel.gov	PVWatts calculator; NREL
ren21.net	REN21 — renewables now
riotinto.com	Rio Tinto
sapp.co.zw	Southern African Power Pool
stuk.fi/web/en	Radiation and Nuclear Safety Authority in Finland
sdgs.un.org	United Nations sustainable development goals — knowledge platform
tedb.ornl.gov	Transportation Energy Data bank, United States
top500.org	Top500 list of Supercomputers
uic.org	International Union of Railways
un.org	United Nations
unep.org	UN Environment Programme
unscear.org	UN Scientific Committee on the Effects of Atomic Radiation
wano.inf	World Association on Nuclear Operators
who.int	World Health Organization
wmo.int	World Meteorological Organization
world-aluminium.org	International Aluminum Institute
worldbank.org	World Bank
worldsteel.org	World Steel Association
wri.org	World Resources Institute

Selected Electric Energy Installations Mentioned

Alta, *US*	Wind farm	254
Bath County, *US*	Pump-storage plant	180
Belchatow, *Poland*	Coal power plant	192
Bonneville, *US*	Run-of-the-river hydro power plant	169
Bruce, *Canada*	Nuclear power plant	159
Cortes La Muela, *Spain*	Pump-storage plant	180
Drake Landing, *Canada*	Interseasonal heat storage system	285
Enerbois, *Switzerland*	Multi-energy system	336
Fortuna, *Germany*	Combined Cycle Gas Turbine power plant	197
Geysers, *US*	Geothermal power plant	260
Grande Dixence, *Switzerland*	Gravity dam	170
Göteborg, *Sweden*	District heating network	349
Hellisheidi, *Iceland*	Combined heat and power geothermal power plant	262
Hoover, *US*	Arch dam	171
Jinping-I, *China*	Arch dam	171
Kashiwazaki-Kariwa, *Japan*	Nuclear power plant	159
La Rance, *France*	Tidal power plant	183
Lünen, *Germany*	Coal power plant	193

Index

Endorsement

Life without electricity is unimaginable. Yet, if our society is to survive, we cannot go on producing and consuming it in the way we do today. In this book, Hans B. Püttgen and Yves Bamberger look at where we are now and what needs to change — through technology, investment and government policy, as well as through shifts which each one of us can make as a consumer and a citizen.

They bring the problem to life both by looking at the experience of real families around the world and by creating a new world — Energia — a modern, industrialized country which has put electricity at the top of its priority list. Thoroughly researched, the book shows what is at stake and how important it is to consider the longer-term impact of the choices we make today.

Over the past four decades, global energy consumption has more than tripled — reflecting rapid population growth and even more rapid electrification of our lives. In that time, the contribution of electricity and heat production to global CO_2 emissions has risen to nearly half of all fuel combustion, from a third back in 1980, according to World Bank data.

That is simply not sustainable. But change is not easy — for many reasons, including the confusing plethora of possible technological solutions, the skepticism of some governments and consumers and the high levels of expenditure and investment needed now to secure the benefits over the long term.

Investment in clean energy is a key theme for us at Pictet Asset Management, and we have particularly valued the insights of Hans B. Püttgen, who is a member of our Clean Energy Thematic Advisory Board. Those insights deserve to be shared more widely.

Today, more than ever, there is a need for books, such as this one, which tackle complex and strategic societal problems — in this case the need for reliable and sustainable electricity, without which the modern economy as we know it would cease to function.

Dr. Philippe Rohner
Sr. Investment Manager, Thematic Equities
Pictet Asset Management

Made in the USA
Middletown, DE
15 October 2022

12686928R00270